Signals and Communication Technology

For further volumes:
http://www.springer.com/series/4748

Chiara Buratti · Marco Martalò
Roberto Verdone · Gianluigi Ferrari

# Sensor Networks with IEEE 802.15.4 Systems

## Distributed Processing, MAC, and Connectivity

Springer

Dr. Chiara Buratti
Dipto. Elettronica, Informatica e Sistemistica (DEIS)
Università di Bologna
Viale Risorgimento 2
40136 Bologna
Italy
e-mail: c.buratti@unibo.it

Prof. Roberto Verdone
Dipto. Elettronica, Informatica e Sistemistica (DEIS)
Università di Bologna
Viale Risorgimento 2
40136 Bologna
Italy
e-mail: roberto.verdone@unibo.it

Dr. Marco Martalò
Dipto. Ingegneria dell'Informazione
Università di Parma
Viale Usberti 181/A
43124 Parma
Italy
e-mail: Marco.Martalo@unipr.it

Prof. Gianluigi Ferrari
Dipto. Ingegneria dell'Informazione
Università di Parma
Viale Usberti 181/A
43124 Parma
Italy
e-mail: gianluigi.ferrari@unipr.it



ISSN 1860-4862

ISBN 978-3-642-17489-6 e-ISBN 978-3-642-17490-2

DOI 10.1007/978-3-642-17490-2

Springer Heidelberg Dordrecht London New York



*Cover design:* eStudio Calamar, Berlin/Figueres

Printed on acid-free paper

Springer is part of Springer Science+Business Media (www.springer.com)

*To my parents and my love.*

Chiara Buratti

*To my parents, the lights that lead my search.*

Marco Martalò

*To my parents and whom loves me.*

Roberto Verdone

*To my new twinkling little star.*

Gianluigi Ferrari

# Foreword

This book provides an in-depth theoretical look at the data processing and communication aspects of wireless sensor networks, as well as a practical look at how data processing can be done in real ZigBee/IEEE 802.15.4 sensor networks. The book provides enough background for the newcomer to the field to understand the fundamentals of wireless sensor networks, while also providing novel analytical frameworks for distributed detection, joint detection and communication, and cross-layer design, making this an excellent choice for those looking for a deeper understanding of the interaction between data processing and communication.

Rochester, September 2010

Wendi Heinzelman
Associate Professor
University of Rochester, NY, USA

When Roberto told me that he was co-authoring a book on Wireless Sensor Networks and asked me if I would be willing to write a foreword for it, my first thought was: "Anything to come back to this beautiful city of Bologna!". So I said "Sure!", before realizing that it is perfectly possible to write a foreword without visiting the authors in their "natural habitat" ever again. My bad. I guess Roberto and I will have to organize together another Joint Workshop and Summer School on Sensor Networks very soon to have an excuse to visit Bologna again.

The book you have in your hands contains first-class information and crisp explanations about the state-of-the-art in Wireless Sensor Networks. Having done research myself in this field for several years and being the coordinator of CONET, the Cooperating Objects Network of Excellence, that has managed to edit and publish two research roadmaps on sensor networks and related fields, I can only applaud the choice of topics the authors have done. Indeed three of the most fundamental challenges in this area are a proper definition and characterization of distributed processing, MAC protocols and connectivity and the authors have done a tremendous job at them in this book.

Additionally, the IEEE 802.15.4 MAC protocol is used as the main wireless communication mechanism for the models and analysis performed in each chapter. This protocol is becoming the de-facto standard for Sensor Networks and is gaining importance as we speak (or better said, as you read). Another highlight is the thorough analysis of multi-sink networks that have been traditionally neglected in the literature in favor of simpler problems to tackle. The authors manage to consider multi-sink networks in almost each chapter of the book, which is a tremendous achievement.

Of course, this in-depth analysis of topics also has its price. If you were hoping for a light bedtime read or despise greek letters and formulas, this book is not for you. It is a book written for professionals by professionals and it shows in the quality of each page. It combines not only a theoretical perspective and foundation with numerical analysis, but also with experimental evaluations using simulation tools. Therefore, each chapter contains a performance evaluation of the algorithms that helps in understanding the operating characteristics of each approach. This is extremely important since the field of Sensor Networks is all about practical implementations and solving real-world problems that, sometimes, cannot be approximated by theoretical models.

I can very much appreciate the effort put by the authors in writing this book and the clarity they have achieved in explaining these three fundamental aspects of the Sensor Network field: Distributed processing, MAC protocols and Connectivity. Young researchers willing to understand the intricacies of Wireless Sensor Networks will be thrilled by the insights in this book, and more experienced scientists will surely recommend its read. I definitely do.

Duisburg, September 2010

Pedro José Marròn
Full Professor
University of Duisburg-Essen in Germany
Coordinator of CONET, the Cooperating
Objects Network of Excellence

This book provides a comprehensive overview of the major theoretical and practical issues of IEEE 802.15.4 and ZigBee, which nowadays represent a key technology for wireless sensor networks. The contents are organized into a roadmap which starts with a tutorial-oriented description of the basic foundations, and continues with some chapters offering an in-depth analysis of performance-related problems. This structure makes the book an ideal reference for everyone wants to approach the study of sensor networks and their applications according to a cross-layer design perspective which takes both data processing and communication aspects into the due account: on one hand, Ph.D. students and researchers in the field of wireless sensor networks are provided with an extensive coverage of major theoretical issues relevant to IEEE 802.15.4 technology, while professionals and networking system developers will also find it an invaluable primer guide.

Catania, September 2010

Sergio Palazzo
Full Professor
University of Catania, Italy

# Preface

Wireless Sensor Networks (WSNs) have become an increasingly active field of research in recent years. The very idea of making many small objects with limited capabilities (the sensors) collaborate to create a very versatile and powerful system (a WSN) has stimulated the intellectual and scientific "fantasies" of many researchers. In fact, WSNs can be studied from several perspectives. Moreover, besides their scientific appeal, they hold the promise of playing a key role in future communication and networking systems, such as machine-to-machine communication systems and the Internet of Things.

IEEE 802.15.4 is a Physical (PHY)/Medium Access Control (MAC) air interface commonly considered as a de facto standard for WSNs. While IEEE released the current version of the standard years ago and many books cover issues related to the use of IEEE 802.15.4 for WSNs, there is still a lack in the understanding of the true performance achievable in large WSNs using IEEE 802.15.4 when distributed processing techniques are applied to estimate the values taken by physical instances.

During spring 2009, after years of research performed on the topic of WSNs, Chiara Buratti and Marco Martalò received their Ph.D. degrees from the University of Bologna (under the supervision of Prof. Roberto Verdone) and the University of Parma (under the supervision of Prof. Gianluigi Ferrari), respectively. While research was carried out separately and the two Ph.D. theses were prepared independently, it was immediately clear that the two works were addressing the two faces of the same coin. Therefore, it was decided to put together the contents of the two theses to give birth to a coherent text providing the deepest possible knowledge of how the IEEE 802.15.4 MAC layer can be accurately modeled and of the limits of distributed processing, inherently connected to the MAC behavior, in WSNs. This book is the outcome of that will. However, it is even more.

This book is indeed composed of four parts. After a short introductory chapter on the fundamentals of WSNs, the second and third parts represent the core of Marco's and Chiara's theses, respectively, after proper alignment was found. More precisely, the second part (Chaps. 2 and 3) is dedicated to distributed processing in

WSNs, whereas the third part is dedicated to MAC and connectivity of WSNs. The fourth part is completely new, as it reports on the outcomes of the effort to integrate the models proposed and validated in the two Theses, to create a novel cross-layer framework for IEEE 802.15.4 system design.

It is our view that the book, in its final form, can assist a WSN designer, whose aim is the estimation of physical instances, in understanding all mechanisms lying behind IEEE 802.15.4 MAC and the application of distributed processing techniques. The book provides many performance charts and all means to generate a complete evaluation tool which is able to compute the performance, in terms of reliability, latency, energy efficiency, of an IEEE 802.15.4-based WSNs for environmental monitoring applications.

We would like to thank Dr. Cristoph Bauman, our Springer Engineering Editor, for immediately supporting our idea and allowing us to finalize this project. We are also indepted to several collaborators, whose help was instrumental. In particular: C. Buratti and R. Verdone would like to thank those scientists whose activities have had an impact on the content of this book: John Orriss, from the University of Manchester, and Flavio Fabbri from the University of Bologna; M. Martalò and G. Ferrari would like to thank Dr. Roberto Pagliari, Dr. Paolo Medagliani, and Marco Sarti for their contributions (while at the University of Parma) to Part II of the book. Authors would also like to thank Flavia Martelli for reading the draft of this book.

Bologna and Parma, September 2010

Chiara Buratti
Marco Martalò
Roberto Verdone
Gianluigi Ferrari

# Contents

# Acronyms

| | |
|---|---|
| ACK | ACKnowledgement |
| AODV | Ad-hoc On-demand Distance Vector |
| AP | Access Point |
| aU | arbitrary Unit |
| AWGN | Additive White Gaussian Noise |
| BER | Bit Error Rate |
| BIBO | Binary Input Binary Output |
| BPSK | Binary Phase Shift Keying |
| BSC | Binary Symmetric Channel |
| CAP | Contention Access Period |
| CDF | Cumulative Distribution Function |
| CFP | Contention Free Period |
| CH | Cluster Head |
| CRC | Cyclic Redundancy Check |
| CSMA | Carrier-Sense Multiple Access |
| CSMA/CA | Carrier-Sense Multiple Access with Collision Avoidance |
| CSS | Chirp Spread Spectrum |
| dB | deciBel |
| DML | De-Moivre Laplace |
| DS-SS | Direct Sequence Spread Spectrum |
| EC | European Commission |
| ED | Event Detection |
| FC | Fusion Center |
| FEC | Forward Error Correction |
| FFD | Full Function Device |
| fm | firing mote |
| GFSK | Gaussian Frequency Shift Keying |
| GTS | Guaranteed Time Slot |
| IF | Intermediated Frequency |
| IR-UWB | Impulse Radio UWB |
| ISM | Industrial Scientific Medical |

| | |
|---|---|
| LDPC | Low-Density Parity-Check |
| LLR | Logarithmic Likelihood Ratio |
| LQI | Link Quality Indication |
| MAC | Medium Access Control |
| MAP | Maximum A posteriori Probability |
| MFR | MAC Footer |
| MHR | MAC Header |
| MMSE | Minimum Mean Square Error |
| MPDU | MAC Payload Data Unit |
| MSDU | MAC Service Data Unit |
| NIST | National Institute of Standards and Technologies |
| O-QPSK | Offset Quadrature Phase Shift Keying |
| p.d.f. | probability density function |
| PAN | Personal Area Network |
| PBPO | Person-By-Person Optimization |
| PEG | Progressive Edge Growth |
| PHR | Physical Header |
| PHY | Physical |
| p.m.f. | probability mass function |
| POI | Phenomenon Of Interest |
| PPDU | Physical Protocol Data Unit |
| PPM | Pulse Position Modulation |
| PPP | Poisson Point Process |
| PSDU | Physical Service Data Unit |
| QoS | Quality of Service |
| r.v. | random variable |
| RF | Radio Frequency |
| RFD | Reduced Function Device |
| RREP | Route Reply |
| RREQ | Route Request |
| RSSI | Received Signal Strength Indication |
| SHR | Synchronization Header |
| SNR | Signal-to-Noise Ratio |
| SP | Sum–Product |
| SPE | Spatial Process Estimation |
| UMTS | Universal Mobile Telecommunications System |
| UWB | UltraWide Bandwidth |
| WPAN | Wireless Personal Area Network |
| WSN | Wireless Sensor Network |
| ZC | Zigbee Coordinator |
| ZED | Zigbee End Device |
| ZR | Zigbee Router |
| 6LowPAN | IPv6 over Low-power PAN |

# Part I
# Fundamentals

# Chapter 1
# Wireless Sensor Networks

This chapter introduces the topic of Wireless Sensor Networks (WSNs), providing a definition and describes the main characteristics and features of this kind of networks. A brief overview of the state of the art of the research in this field, with particular attention toward the main European projects, is also provided. The IEEE 802.15.4 standard, reference technology of this book for the realisation of such networks, is described. This chapter is structured as follows. In Sect. 1.1, fundamental features and issues of WSNs are discussed. In Sect. 1.2., instead, the main application fields are briefly presented. Sections 1.3 and 1.4 introduce the main characteristics of the IEEE 802.15.4 and Zigbee protocols, respectively. In Sect. 1.5, we discuss on research trends in WSNs, whereas in Sect. 1.6 further readings on the topic are recalled.

## 1.1 Wireless Sensor Networks Fundamentals

A WSN [1–8] can be defined as a network of devices, denoted as *nodes*, which can sense the environment and communicate the information gathered from the monitored field (e.g., an area or volume) through wireless links [9]. The data is forwarded, possibly via multiple hops, to a *sink* (sometimes denoted as *controller* or *monitor*) that can use it locally or is connected to other networks (e.g., the Internet) through a *gateway*. The nodes can be stationary or moving. They can be aware of their locations or not. They can be homogeneous or not. Monitoring and communication are performed cooperatively by the nodes.

In Fig. 1.1 (left part) a traditional single-sink WSN is shown. This single-sink scenario suffers from the lack of scalability: by increasing the number of nodes, the amount of data gathered by the sink increases and once its capacity is reached, the network size can not be augmented. Moreover, for reasons related to MAC and routing aspects, the network performance cannot be considered independent from the network size.

C. Buratti et al., *Sensor Networks with IEEE 802.15.4 Systems*,
Signals and Communication Technology, DOI: 10.1007/978-3-642-17490-2_1,

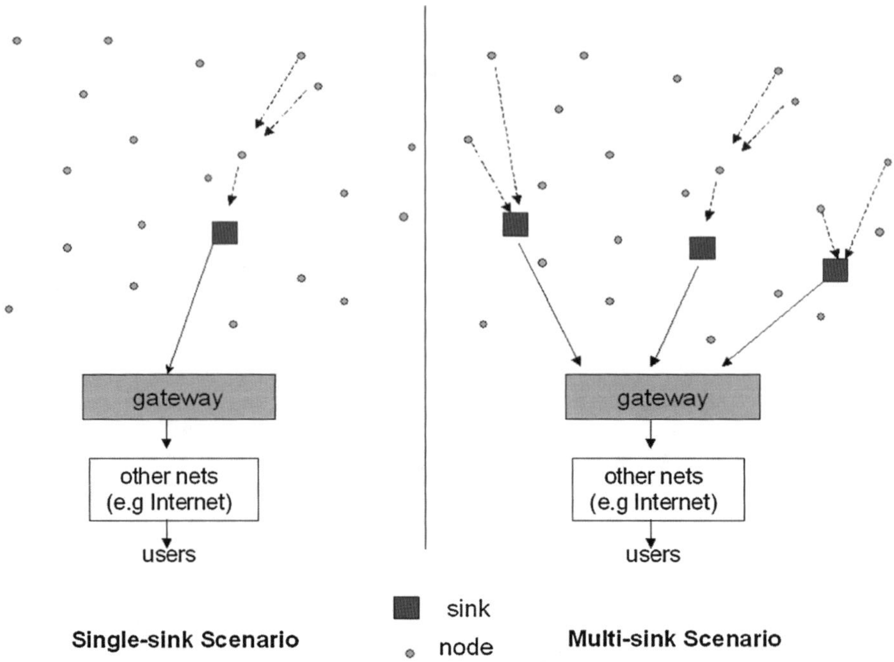


**Fig. 1.1** *Left part*: single-sink WSN. *Right part*: multi-sink scenario

A more general scenario includes multiple sinks in the network (see Fig. 1.1 right part) [10]. Given a level of node density, a larger number of sinks will decrease the probability of isolated clusters of nodes that cannot deliver their data owing to unfortunate signal propagation conditions. In principle, a multiple-sink WSN can be scalable (i.e., the same performance can be achieved even by increasing the number of nodes), while this is clearly not true for a single-sink network. However, a multi-sink WSN does not represent a trivial extension of a single-sink case for the network engineer. In many cases, nodes send the data collected to one of the sinks, selected among many, which forwards the data to the gateway, toward the final user (see Fig. 1.1). From the protocol viewpoint, this means that a selection can be carried out on the basis of a suitable criterion, such as, for example, minimum delay, maximum throughput, minimum number of hops, etc.. Therefore, the presence of multiple sinks ensures better network performance with respect to the single-sink case (assuming the same number of nodes is deployed over the same area), but the communication protocols must be more complex and should be designed according to suitable criteria.

## *1.1.1 Main Features of WSNs*

The main features of WSNs are: scalability with respect to the number of nodes in the network, self-organisation, self-healing, energy efficiency, a sufficient degree of connectivity among nodes, low complexity, low cost, and small size of nodes. Protocol architectures and technical solutions providing such features can be considered as a potential framework for the creation of these networks. Unfortunately, the definition of such a protocol architecture and a corresponding technical solution is not simple, and research is still needed [5].

The massive research on WSNs started after the year 2000. However, it took advantage of the outcome of the research on wireless networks performed since the middle of the previous century. In particular, the study of ad hoc networks attracted a lot of attention for several decades, and some researchers tried to apply their skills, acquired in the field of ad hoc networking, to the study of WSNs.

According to some general definitions, wireless ad hoc networks are formed dynamically by an autonomous system of nodes connected via wireless links without using an existing network infrastructure or centralised administration. Nodes are connected through "ad hoc" topologies, set up and cleared according to user needs and temporary conditions [11]. Apparently, this definition can include WSNs. However, this is not true. In fact, wireless ad hoc networks are characterised by features which are very different from those of WSNs: they are unplanned and highly dynamical; nodes are normally "smart" terminals (laptops, etc.); typical applications include real-time or non real-time data, multimedia, voice; every node can be either source or destination of information; every node can be a router toward other nodes; energy is not the most relevant matter; capacity is the most relevant matter.

But for the very first item, which is common to WSNs, in all other cases there is a clear distinction between WSNs and wireless ad hoc networks. In WSNs, nodes are simple and low-complexity devices; the typical applications require few bytes sent periodically or upon request or according to some external event; every node can be either source or destination of information, not both; not all nodes play the role of routers; energy efficiency is a very relevant matter, while capacity is not for most applications. Therefore, WSNs are not a special case of wireless ad hoc networks. Thus, a lot of care must be exercised when taking protocols and algorithms, which are effective for ad hoc networks, and using them in the context of WSNs.

### 1.1.2 Issues Related to Energy Management

As stated above, energy efficiency is a key issue for many WSNs applications and many works in the literature are devoted to this issue [12–16]. Network lifetime must be kept as long as possible. Clearly, it depends on the potential length of the period of time starting with network deployment and ending when the battery of sensor nodes is no longer able to provide the energy needed for communication, sensing, or processing. Energy consumption issues are discussed throughout the entire book. However, a brief discussion about some important aspects of energy management is introduced in this subsection.

A node is basically composed of a battery, a microprocessor, a memory, the sensor(s), and the transceiver. Normally, when in transmit mode, the transceiver drains much more current from the battery than the microprocessor in active state, or the sensors and the memory chip. As a conclusion, the transceiver is the part responsible for the consumption of most of the energy. This justifies the energy consumption model adopted in the remainder of the book.

Intuitively, the state that requires more current drain from the battery is the transmit state, as both the baseband and Radio Frequency (RF) parts of the transceiver are active. However, the same is true for the receive state, during which a node can consume as much energy as in the transmit one. Owing to the hardware design principles, sometimes in the receive state the transceiver can consume even more energy than in the transmit state. Therefore, receive and transmit states are both highly energy consuming, and the transceiver must be kept in those two states for the shortest possible percentage of time. For this reason, also the energy consumed for receiving packets and for doing carrier sensing is accounted for in this book.

Clearly, permanence in the transmit state is needed only when a data burst needs to be transmitted. The smaller is the number of data bursts to be transmitted, the longer is node life. This suggests to avoid using protocols based on complex handshakes. As an example, in some cases it could be better to avoid ACKnowledgement (ACK) mechanisms. However, a transceiver might need to stay in receive mode for longer periods of time, if proper scheduling of transmit times is not performed. Protocols should avoid a phenomenon, called *overhearing*, such that nodes need to stay in receive time for long periods waiting for a packet while listening to many data bursts sent to other nodes. However, this is not enough. In fact, many Medium Access Control (MAC) protocols consider channel sensing mechanisms: the transceiver senses the wireless channel for some periods of time in order to determine whether it is busy or free. Depending on the specific hardware platform, channel sensing can be very energy consuming, almost as the transmit and receive states. Thus, protocols must not abuse of the channel sensing mechanism and when using a Carrier-Sense Multiple Access (CSMA) protocol, long (in the order of 95–99% of time) intervals of time with the transceiver in sleep state are required. During such periods, a data burst sent to the node cannot be detected. Therefore, the management of sleep mode is a very relevant issue for WSNs.

A final consideration is related to the use of power control. This technique, aiming to set the transmit power at the minimum level needed to allow correct signal detection at the receiver, is often used in wireless networks to reduce the impact of interfering transmissions and the useless emission of radiowaves with high power. However, setting a proper power level requires information on the channel gain, which might be difficult to achieve in applications with very bursty data transmissions. Therefore, it is worthwhile wondering whether power control is a useful technique for WSNs. Moreover, looking at the data reported on the datasheet of a sample transceiver used in many commercial platforms, such as CHIPCON CC2420 [17], one can draw an interesting conclusion. When transmitting at the highest power level (0 dBm), about 17 mA are drained from the battery. At minimum transmit power (−25 dBm), the drained current is 8.5 mA.

Therefore, there is no relevant energy saving, when decreasing the transmit power level by 25 dB. Even if this example is given with reference to a specific chip, there are reasons to state that the conclusion is general. The energy consumed in transmission state is not proportional to the transmit power level used and, therefore, power control is not an efficient technique to reduce energy consumption. For this reason, power control is not considered in this book.

## 1.2 Applications

The variety of possible applications of WSNs to the real word is practically unlimited: environmental monitoring [18], health care [19], positioning and tracking [20], logistics, localization, etc. One of the possible classifications distinguishes applications according to the type of data that must be gathered in the network. Almost any application, in fact, could be classified into one out of two categories: Event Detection (ED) and Spatial Process Estimation (SPE).

In the first case, sensors are deployed to detect an event, for example a fire in a forest, a quake, etc. [21–23]. Signal processing within devices is very simple, owing to the fact that each device has to compare the measured quantity with a given threshold and to send the binary information to the sink(s). The density of nodes must ensure that the event is detected and forwarded to the sink(s) with a suitable probability of success while maintaining a low probability of false alarm. The detection of the Phenomenon Of Interest (POI) could be performed in a decentralized (or distributed) way, meaning that sensors, together with the sink, cooperatively undertake the task of identifying the POI. However, unlike in classical decentralized detection problems, greater challenges exist in a WSN setting. There are stringent power constraints for each node, communication channels between nodes and the fusion center are severely bandwidth-constrained and lossy (e.g., fading, noise, and, possibly, external sources of interference are present), and the observation at each sensor node is spatially varying. In the context of decentralized detection, cooperation allows the exchange of information among sensor nodes to continuously update their local decisions until consensus is reached across the nodes.

In SPE, the WSN aims at estimating a given physical phenomenon (e.g., the atmospheric pressure in a wide area, or the ground temperature variations in a small volcanic site), which can be modelled as a bi-dimensional random process (generally non stationary). In this case, the main issue is to obtain the estimation of the entire behavior of the spatial process based on the samples taken by sensors which are typically placed in unknown positions [24–28]. The measurements will then be subject to proper processing which might be performed either in a distributed manner by the nodes or centrally at the sink. The estimation error is strictly related to nodes density as well as on the spatial variability of the process. Higher nodes' density leads to a more accurate scalar field reconstruction at the expense of a larger network throughput and cost.

There are also applications that belong to both categories. As an example, environmental monitoring applications could be ED- or SPE-based. For instance, the location of a fire in a forest or the detection of a quake belong to the first category. The estimation of the temperature in a given area, instead, belongs to the second category. In general, these applications aim at monitoring indoor or outdoor environments, where the supervised area may be few hundreds of square meters or thousands of square kilometers, and the duration of the supervision may last years. Natural disasters such as floods, forest fires, earthquakes may be perceived earlier by installing networked embedded systems closer to places where these phenomena may occur. Such systems cannot rely on a fixed infrastructure and have to be very robust, because of the inevitable impairments encountered in open environments. The system should respond to environmental changes as quickly as possible. The environment to be observed will mostly be inaccessible by humans all the time. Hence, robustness plays an important role. Also security and surveillance applications have some demanding and challenging requirements such as real-time monitoring and high security.

Another application that could belong to both the above defined categories is devoted to the realisation of energy efficient buildings. In this application, in fact, sensor nodes could aim at estimating a process (SPE), but also events (ED). In this case, the WSN is distributed in buildings (residential or not) to manage efficiently the energy consumption of all the electrical appliances. To this aim, nodes have to continuously monitor the energy consumed by all appliances connected to the electrical grid. Therefore, sensors have to estimate a process (i.e., the energy consumption) which varies with time. However, in some cases they could be used to detect some events. As an example, sensors could detect the arrival of a person in a room to switch on some electrical appliances.

## 1.3 IEEE 802.15.4 Technology

The IEEE 802.15.4 wireless technology is a short-range communication system intended to provide applications with relaxed throughput and latency requirements in Wireless Personal Area Networks (WPANs). The key features of the IEEE 802.15.4 wireless technology are low complexity, low cost, low power consumption, low data rate transmissions, to be supported by cheap (either fixed or moving) devices. The main field of application of this technology is the implementation of WSNs: IEEE 802.15.4, in fact, can be considered the de facto standard for WSNs.

The IEEE 802.15.4 Working Group focuses on the standardization of the bottom two layers of ISO/OSI protocol stack: Physical (PHY) and MAC. There are mainly two options for the upper layers definition: Zigbee protocol stack, specified by the industrial consortia ZigBee Alliance, and IPv6 over Low-power PAN (6LowPAN) [29]. In this book, only the Zigbee solution will be accounted for.

In the following, some technical details on the PHY and MAC sublayers defined in the standard are reported.

### *1.3.1 IEEE 802.15.4 Physical Layer*

The IEEE 802.15.4 technology offers low rate services that enable the connection of possibly mobile low-complexity devices based on a Carrier-Sense Multiple Access with Collision Avoidance (CSMA/CA) channel access technique.

The IEEE 802.15.4 physical layer operates in three different unlicensed bands (and with different modalities) according to the geographical area where the system is deployed. However, Direct Sequence Spread Spectrum (DS-SS) is mandatory to reduce the interference level in shared unlicensed bands.

The PHY provides the interface with the physical medium. It is in charge of radio transceiver activation and deactivation, energy detection, link quality, clear channel assessment, channel selection, and transmission and reception of the message packets. Moreover, it is responsible for establishment of the RF link between two devices, bit modulation and demodulation, synchronization between the transmitter and the receiver, and, finally, for packet level synchronization.

The IEEE 802.15.4 standard specifies a total of 27 half-duplex channels across the three frequency bands, whose channelization is depicted in Fig. 1.2 and is organized as follows.

- The 868 MHz band, ranging from 868.0 to 868.6 MHz and used in the European area, implements a raised-cosine-shaped Binary Phase Shift Keying (BPSK) modulation format, with DS-SS at chip-rate 300 kchip/s (a pseudo-random sequence of 15 chips transmitted in a 25 $\mu$s symbol period). Only a single channel with data rate 20 kbit/s is available and, with a required minimum

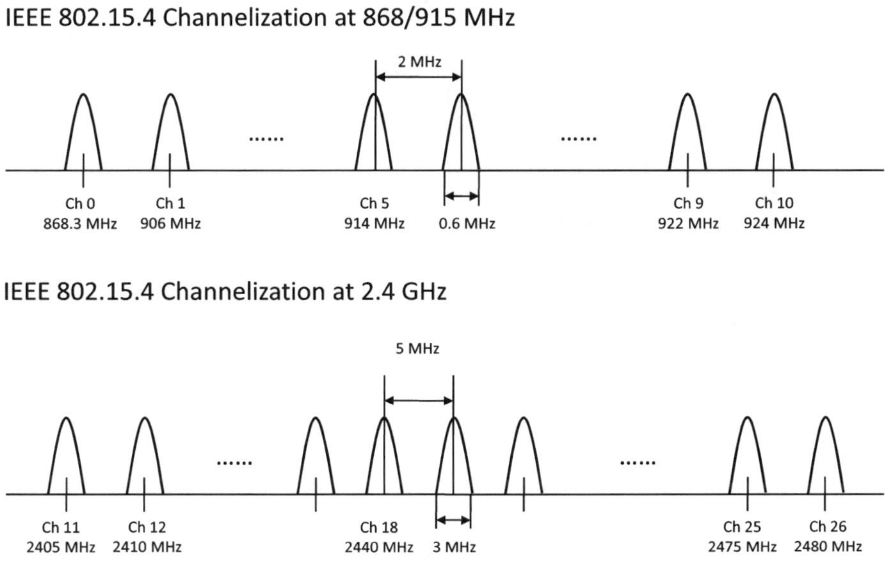


**Fig. 1.2** Channelization at the 868/915 MHz bands and at the 2.4 GHz band

−92 dBm RF sensitivity, the ideal transmission range (i.e., without considering wave reflection, diffraction, and scattering) is approximately 1 km.
- The 915 MHz band, ranging between 902 and 928 MHz and used in the North American and Pacific area, implements a raised-cosine-shaped BPSK modulation format, with DS-SS at chip-rate 600 kchip/s (a pseudo-random sequence of 15 chips is transmitted in a 50 μs symbol period). Ten channels with rate 40 kbit/s are available and, with a required minimum −92 dBm RF sensitivity, the ideal transmission range is approximatively 1 km.
- The 2.4 GHz Industrial Scientific Medical (ISM) band, which extends from 2400 to 2483.5 MHz and is used worldwide, implements a half-sine-shaped Offset Quadrature Phase Shift Keying (O-QPSK) modulation format, with DS-SS at 2 Mchip/s (a pseudo-random sequence of 32 chips is transmitted in a 16 μs symbol period). Sixteen channels with data rate 250 kbit/s are available and, with minimum −85 dBm RF sensitivity required, the ideal transmission range is approximatively 200 m.

The ideal transmission range is computed considering that, although any legally acceptable power is allowed, IEEE 802.15.4-compliant devices should be capable of transmitting at −3 dBm. The true transmission ranges are often much lower than the ideal ones, owing to real-world propagation impairments.

Since power consumption is a primary concern, in order to guarantee a long battery lifetime the energy must be taken continuously at an extremely low rate, or in small amounts at a low power duty cycle: this means that IEEE 802.15.4-compliant devices are active only during a short time. The standard allows some devices to operate with both the transmitter and the receiver inactive for over 99% of time. Therefore, the instantaneous link data rates supported (i.e., 20, 40, and 250 kbit/s) are high with respect to the data throughput in order to minimize the device duty cycle.

According to the IEEE 802.15.4 standard, transmission is organized in frames, which can differ according to the relevant purpose. In particular, there are four frame structures, each designated as a Physical Protocol Data Unit (PPDU): a beacon frame, a data frame, an ACK frame and a MAC command frame. They are all structured with a Synchronization Header (SHR), a Physical Header (PHR), and a Physical Service Data Unit (PSDU), which is composed of a MAC Payload Data Unit (MPDU), which, in turn, is constructed with a MAC Header (MHR), a MAC Footer (MFR), and a MAC Service Data Unit (MSDU), excepting the ACK frame, which does not contain an MSDU. The structure of each possible frame is depicted in Figs. 1.3, 1.4, 1.5 and 1.6. To detect that a message has been received correctly, a Cyclic Redundancy Check (CRC) is used. The meaning of the four possible frame structures will be clear in the following, after introducing the possible network topologies and the possible MAC channel access strategies.

### *1.3.2 IEEE 802.15.4 Network Topologies and Operational Modes*

To overcome the limited transmission range, multihop self-organizing network topologies are required. These can be realized taking into account that IEEE

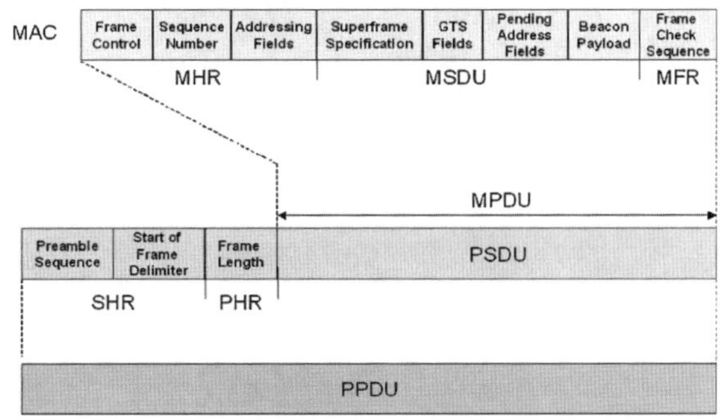


**Fig. 1.3** Beacon frame structure

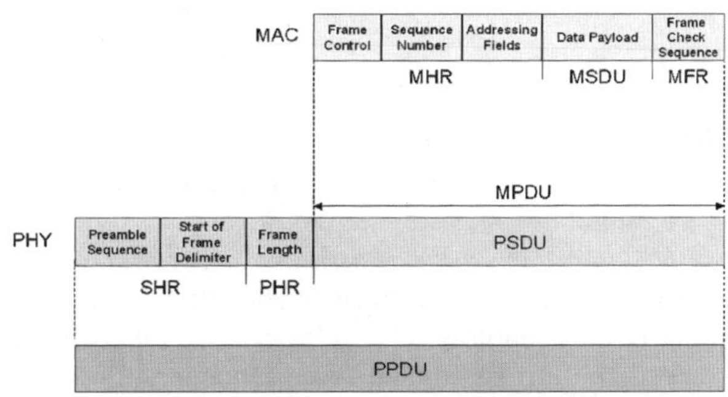


**Fig. 1.4** Data frame structure

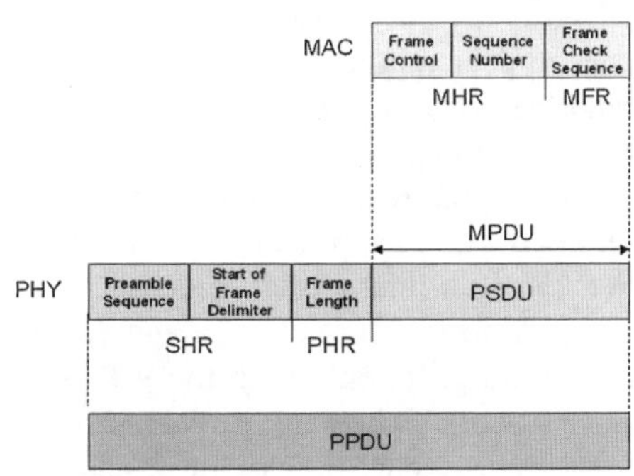


**Fig. 1.5** ACK frame structure

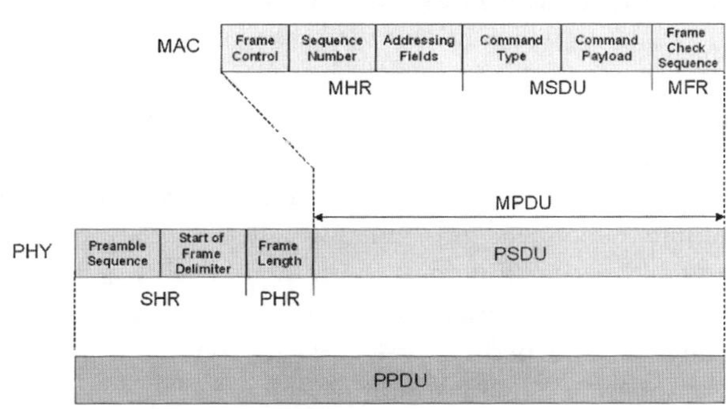


**Fig. 1.6** MAC command frame structure

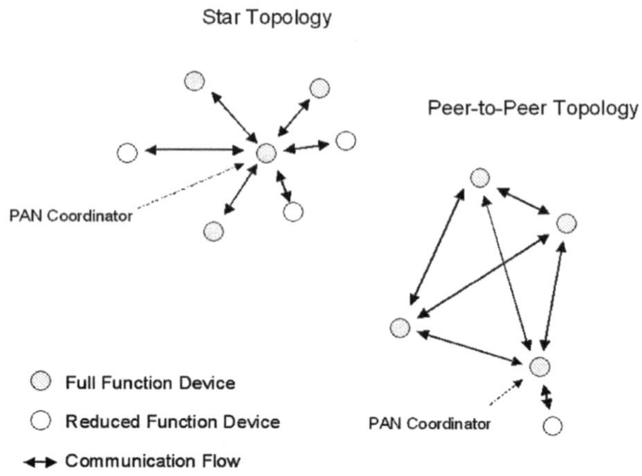


**Fig. 1.7** The two IEEE 802.15.4-compliant network topologies: star and peer-to-peer topologies

802.15.4 defines two types of devices: the Full Function Device (FFD) and the Reduced Function Device (RFD) The FFD contains the complete set of MAC services and can operate as either a network coordinator (hereafter also denoted as WPAN *coordinator*) or as a simple network device. The RFD contains a reduced set of MAC services and can operate only as a network device.

Two basic topologies are allowed, but not completely described by the standard since definition of higher layers functionalities are out of the scope of IEEE 802.15.4: the star topology, formed around an FFD acting as a WPAN coordinator, which is the only node allowed to form links with more than one device, and the peer-to-peer topology, where each device is able to form multiple direct links to other devices so that redundant paths are available. An example of both the IEEE 802.15.4-compliant network topologies is shown in Fig. 1.7.

Star topology is preferable when the coverage area is small and a low latency is required by the application. In this topology, communication is controlled by the WPAN coordinator that acts as network master, sending packets, denoted as *beacons*, for synchronization and device association. Network devices are allowed to communicate only with the WPAN coordinator and any FFD may establish its own network by becoming a WPAN coordinator according to a predefined policy. A network device wishing to join a star network listens for a beacon message and, after receiving it, the network device can send an association request back to the WPAN coordinator, which allows the association or not. Star networks support also a non-beacon-enabled mode. In this case, beacons are used only for the purpose of association, whereas synchronization is achieved by polling the WPAN coordinator for data on a periodic basis. Star networks operate independently from their neighboring networks. As will be clarified in Sect. 1.3.5, in fact, each WPAN will operate in different channels and, generally, they will not interfere among themselves.

Peer-to-peer topology is preferable when a large area has to be covered and latency is not a critical issue. This topology allows the formation of more complex networks and permits any FFD to communicate with any other FFD within its transmission range via multiple hops. Each device in a peer-to-peer structure needs to proactively search for other network devices. Once a device is found, the two devices can exchange parameters to recognize the type of services and features each supports. However, the introduction of multihop requires additional device memory for routing tables.

All devices, regardless of the type of topology, belonging to a particular network use their unique IEEE 64-bit addresses and a short 16-bit address is allocated by the WPAN coordinator to uniquely identify the network.

Finally, the WPAN coordinator election can be performed in different ways according to the application. In particular, for the applications in which only one device can be the coordinator (e.g., a gateway) it is preferable to have a dedicated WPAN coordinator. In other applications, it could be better to have several eligible FFD and an event-determined WPAN coordinator. There can also be applications where it is not relevant which particular device is the WPAN coordinator, which can be self-determined. Moreover, the WPAN coordinator may be selected because it has special computation capabilities, a bridging capability to other network protocols, or simply because it was among the first participants in the formation of the network.

### *1.3.3 IEEE 802.15.4 MAC Layer*

The MAC layer provides access control to a shared channel and reliable data delivery. IEEE 802.15.4 uses a protocol based on a CSMA/CA algorithm, which requires listening to the channel before transmitting to reduce the probability of collisions with other ongoing transmissions. The main functions performed by the MAC sublayer are: association and disassociation, security control, optional star network topology functions, such as beacon generation and Guaranteed Time Slots (GTSs) management, generation of ACK frames (if used), and, finally, application support for the two possible network topologies described in the standard.

IEEE 802.15.4 defines two different operational modes, namely *beacon-enabled* and *non beacon-enabled*, which correspond to two different channel access mechanisms.

In the non beacon-enabled mode, nodes use an unslotted CSMA/CA protocol to access the channel and transmit their packets [30]. The algorithm is implemented using units of time called backoff periods. Each node maintains two variables for each transmission attempt: *NB* and *BE*. *NB* is the number of times the CSMA/CA algorithm was required to backoff while attempting the current transmission; this value will be initialized to 0 before each new transmission attempt and cannot assume values larger than $NB_{\max}$. *BE* is the backoff exponent related to the maximum number of backoff periods a node will wait before attempting to assess the channel. *BE* is initialized to the value of $BE_{\min}$ and cannot assume values larger than $BE_{\max}$. Figure 1.8 illustrates the steps of the CSMA/CA algorithm, starting from when the node has data to be transmitted. First, *NB* and *BE* are initialized and then the MAC layer will delay all activities for a random number of backoff

Fig. 1.8 The IEEE 802.15.4 CSMA/CA algorithm in the non beacon-enabled mode

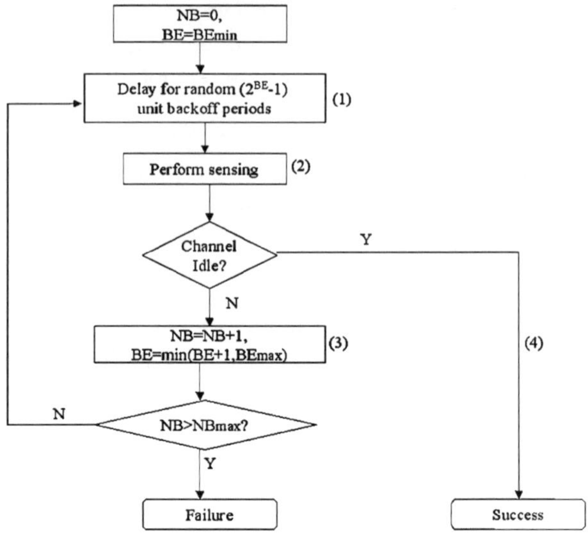


Fig. 1.9 The IEEE 802.15.4 CSMA/CA algorithm in the beacon-enabled mode

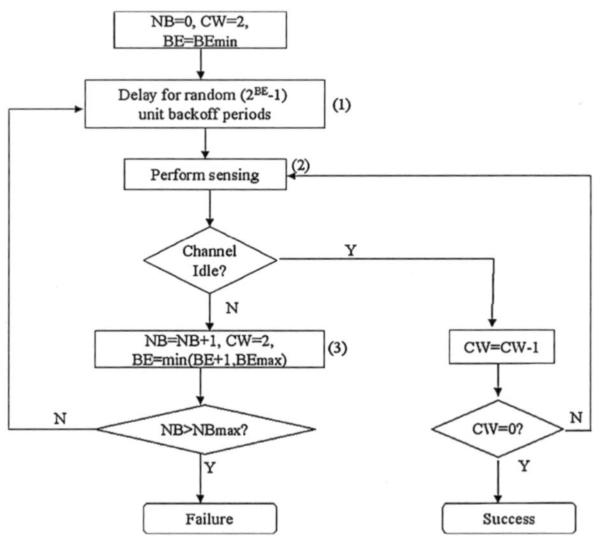


periods in the range $\{0, \ldots, 2^{BE} - 1\}$ [step (1)]. After this delay, channel sensing is performed for one unit of time [step (2)]. If the channel is assessed to be busy [step (3)], the MAC sublayer will increase both *NB* and *BE* by one, ensuring that *BE* is not larger than $BE_{\max}$. If the value of *NB* is less than or equal to $NB_{\max}$, the CSMA/CA algorithm will return to step (1). If the value of *NB* is larger than $NB_{\max}$, the CSMA/CA algorithm will terminate with a "Failure," meaning that the node does not succeed in accessing the channel. If the channel is assessed to be idle [step (4)], the MAC layer will begin transmission of data immediately ("Success" in accessing the channel) (Fig. 1.9).

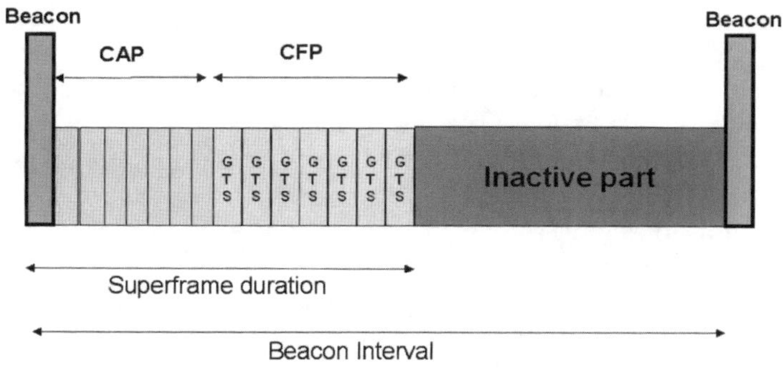


**Fig. 1.10** The IEEE 802.15.4 superframe defined in the beacon-enabled mode

In the beacon-enabled mode [30], instead, the access to the channel is managed through a superframe, starting with a packet, called *beacon*, transmitted by the WPAN coordinator. The superframe may contain an inactive part, allowing nodes to go in sleeping mode, whereas the active part is divided into two parts: the Contention Access Period (CAP) and the Contention Free Period (CFP), composed of GTSs, that can be allocated by the WPAN coordinator to specific nodes (see Fig. 1.10). The use of GTSs is optional.

The duration of the active part of the superframe, containing CAP and CFP, is called *superframe duration*, will be denoted, in the following, as *SD*, and can be expressed as

$$SD = 960 \cdot 2^{SO} \cdot T_s, \tag{1.1}$$

where *SO* is the *superframe order*, an integer parameter ranging from 0 to 14, and $T_s$ is the symbol time. In this book, the 2.4 GHz band is considered, so that the symbol rate is equal to 62.5 ksymbol/s, which brings to $T_s = 16\,\mu s$.

The duration of the whole superframe, i.e., the interval of time between two successive beacons, is called *beacon interval*, will be denoted, in the following, as *BI*, and can be expressed as

$$BI = 16 \cdot 60 \cdot 2^{BO} \cdot T_s, \tag{1.2}$$

where *BO* is the *beacon order*, an integer parameter ranging from 0 to 14. Note that *BO* must be not smaller than *SO*.

According to the standard, each GTS must have a duration multiple of $60 \cdot 2^{SO} \cdot T_s$ and must contain the packet transmitted by the node to which the GTS is allocated to and an inter-frame space, that is the minimum interval between two subsequent packets received [30]. The WPAN coordinator may allocate up to seven GTSs, but a sufficient portion of the CAP must remain for contention-based access. The minimum CAP duration is equal to $440\,T_s$.

For what concerns the CSMA/CA algorithm used in the CAP portion of the superframe, the only difference with the non beacon-enabled mode is that nodes have to find the channel free for two subsequent backoff periods before transmitting the packet (see Fig. 1.9). To this aim, each node maintains another variable, denoted as *CW*, indicating the number of backoff periods that need to be clear of channel activity before the transmission can start. First, *CW* is initialized to 2. Then, once the node senses the channel [step (2)] (see Fig. 1.9), if the channel is found free, *CW* is decremented by 1 and compared with 0: if $CW > 0$, the algorithm returns to step (2) and another sensing phase is implemented; otherwise a transmission may start.

The other difference with the non-beacon-enabled case is that backoff period boundaries of every node in the WPAN must be aligned with the superframe slot boundaries of the coordinator. Therefore, the beginning of the first backoff period of each node is aligned with the beginning of the beacon transmission. Moreover, all transmissions will start on the boundary of a backoff period.

### 1.3.4 Data Transfer Protocol and MAC Frames

As a consequence of the different type of topologies and the possibility of implementing beacon- and non beacon-enable modes, three different MAC data transfer protocols are defined by the IEEE 802.15.4 standard.

- In the case of beacon-enabled star topology, a network device wishing to send data to the WPAN coordinator needs to listen for a beacon. If it does not have a GTS assigned, the device transmits its data frame in the CAP through CSMA/CA. If the device has a GTS assigned, it waits for the appropriate one to transmit its data frame. Afterwards, the WPAN coordinator sends back an ACK to the network device, as shown in Fig. 1.11. When the WPAN coordinator has data for a network device, it sets a special flag in its beacon. Once the appropriate network device detects that the PAN coordinator has pending data for it, it sends back a "Data Request" message. The WPAN coordinator responds with an ACK followed by

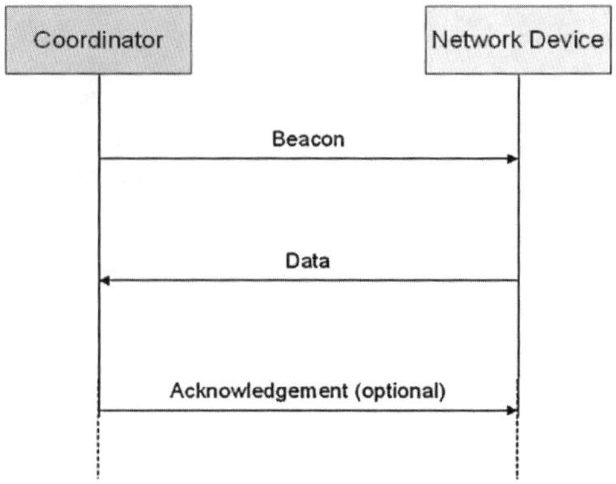


**Fig. 1.11** Communication from a network device to the WPAN coordinator in a beacon-enabled network

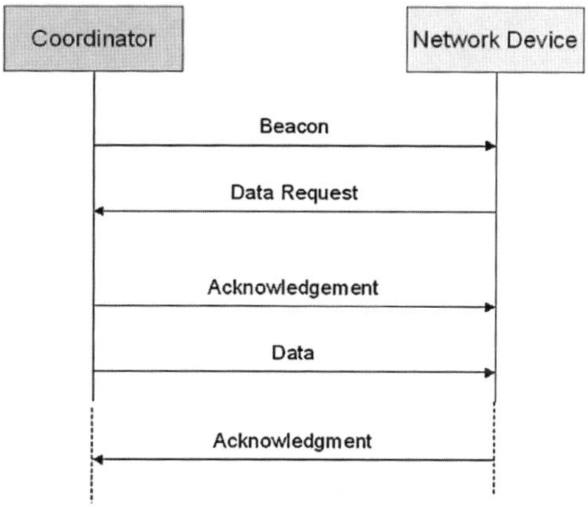


**Fig. 1.12** Communication from the WPAN coordinator to a network device in a beacon-enabled network

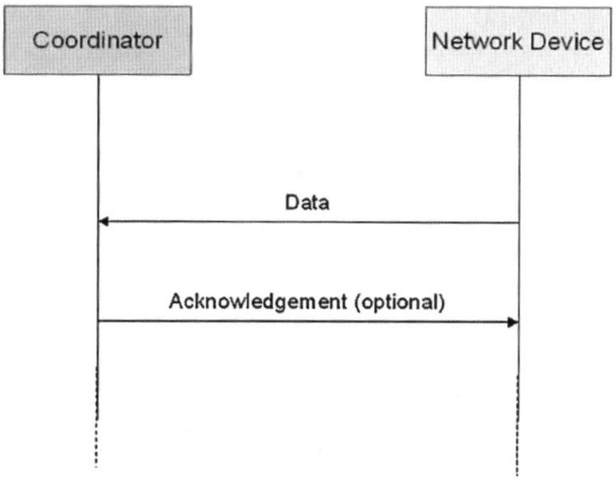


**Fig. 1.13** Communication from a network device to the WPAN coordinator in a non beacon-enabled network

the data frame and, finally, an ACK is sent by the network device, as shown in Fig. 1.12.

- In the case of non beacon-enabled star topology, a network device wishing to transfer a data frame to the WPAN coordinator uses CSMA/CA. The WPAN coordinator responds to the network device, sending an ACK message, as shown in Fig. 1.13. When a WPAN coordinator requires making a data transfer to a network device, it shall keep the data until the network device sends a data request message. If there are data pending, the ACK message from the WPAN coordinator will contain information indicating the corresponding network device, which will send the data immediately after receiving the ACK. Finally, the network device acknowledges reception of the data frame, as shown in Fig. 1.14.
- In the case of peer-to-peer topology, the strategy is governed by the specific network layer managing the wireless network. A given network device may stay in reception mode scanning the radio channel for on-going communications or

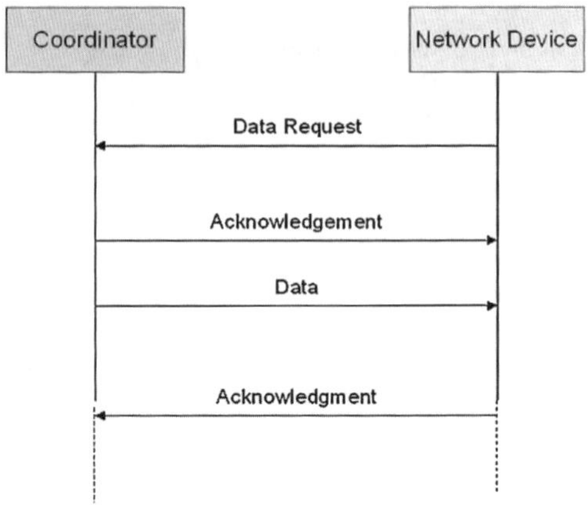


**Fig. 1.14** Communication from the WPAN coordinator to a network device in a non-beacon-enabled network

can send periodic "hello" messages to achieve synchronization with other potential listening devices.

Finally, as far as the MAC frame structure is concerned, a MAC frame consists of three parts: header, variable length payload, and footer. The MAC header contains a frame control field and an addressing field. The MAC payload contains information specific to the type of transaction being handled by the MAC. The MAC footer consists of a 16-bit CRC algorithm. The maximum MAC payload size is 127 bytes. Four types of MAC frames are defined: beacon, data, ACK, and MAC command.

### *1.3.5 The IEEE 802.15.4 Topology Formation Procedure*

The IEEE 802.15.4 Working Group defined a mechanism to support a WPAN coordinator in channel selection when starting a new WPAN, and a procedure, called *association procedure*, which allows other devices to join the WPAN. A WPAN coordinator wishing to establish a new WPAN needs to find a channel which is free from interference that would render the channel unsuitable (e.g., in a multi-sink network, a channel may be already occupied by other WPANs). The channel selection is performed by the WPAN coordinator through the energy detection scan, which returns the measurement of the peak energy in each channel. It must be noticed that the standard only provides the energy detection mechanism but it does not specify the channel-selection logic. The operations accomplished by a device to discover an existing WPAN and to join it can be summarised as follows: (i) search for available WPANs; (ii) select the WPAN to join; (iii) start the association procedure with the WPAN coordinator or with another FFD device, which has already joined the WPAN. The discovery of available WPANs is performed by scanning beacon frames broadcasted by the coordinators. Two different types of scan, that can be used in the association phase, are proposed:

1. passive scan: in beacon-enabled networks, the associated devices periodically transmit beacon frames, so that the information on the available WPAN can be derived by eavesdropping the wireless channels;
2. active scan: in non beacon-enabled networks, the beacon frames are not periodically transmitted but shall be explicitly requested by the device by means of a beacon request command frame.

After the scan of the channels, a list of available WPANs is used by the device to choose the network to try to connect to. In the standard, no specific procedure to select a WPAN is provided and, therefore, this selection among potential parents is open for different implementations. Hence, the device sends an association request frame to the coordinator device by means of which the selected network was discovered. The association phase ends with a successful association response command frame to the requesting device. This procedure basically results in a set of MAC association relationships between devices named, in the following, parent–child relationship.

## 1.4 Zigbee Upper Layers

As stated at the beginning of Sect. 1.3, one possible protocol stack solution to be used on top of the IEEE 802.15.4 has been defined by the Zigbee Alliance [31]. The purpose of the ZigBee Alliance is to univocally describe the ZigBee protocol standard in such a way that interoperability is guaranteed also among devices produced by different companies, provided that each device implements the ZigBee protocol stack.

The ZigBee stack architecture, shown in Fig. 1.15, is composed of a set of blocks called layers. Each layer performs a specific set of services for the layer above. Given the IEEE 802.15.4 specifications on PHY and MAC layers, the ZigBee Alliance provides the network layer and the framework for the application layer.

The responsibilities of the ZigBee network layer include: mechanisms to join and leave a network, frame security, routing, path discovery, one-hop neighbours discovery, neighbour information storage.

The ZigBee application layer consists of the application support sublayer, the application framework, the ZigBee device objects and the manufacturer-defined application objects. The responsibilities of the application support sublayer include: maintaining tables for binding (defined as the ability to match two devices together based on their services and their needs), and forwarding messages between bound devices. The responsibilities of the ZigBee device objects include: defining the role of the device within the network (e.g., WPAN coordinator or end device), initiating and/or responding to binding requests, establishing secure relationships between network devices, discovering devices in the network, and determining which application services they provide.

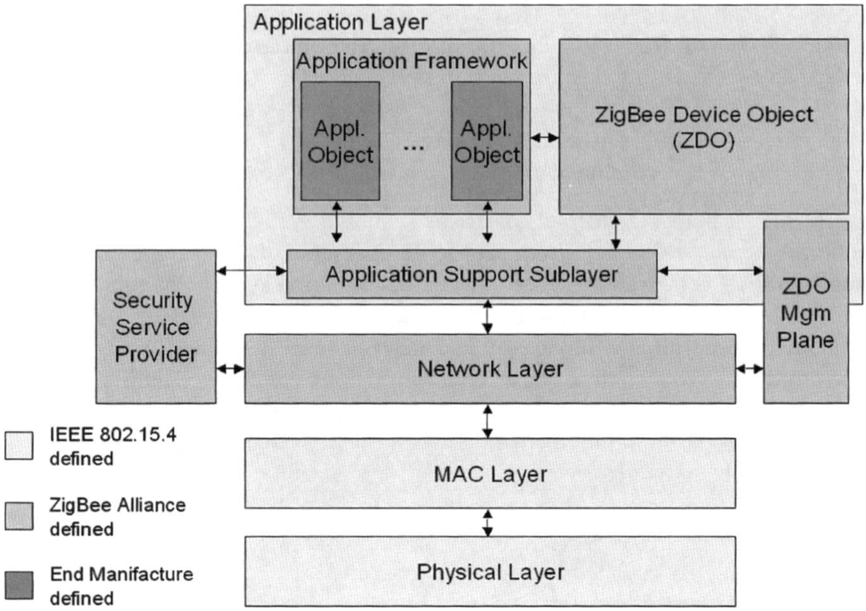


**Fig. 1.15** A detailed overview of ZigBee stack architecture

## *1.4.1 Zigbee Topologies*

Three different types of topologies are supported by Zigbee: star, mesh, and tree-based. Star topologies are the simplest, but they are not scalable with respect to the number of nodes in the network and the area to be covered. Mesh and tree-based topologies, instead, are suitable also for large networks, distributed over large areas. For WSNs, where the set of destination nodes, that are the sinks, is separated by those of sources, namely sensor nodes, tree-based topologies seem to be more efficient than the others: in fact, routing is much simpler, and also distributed data aggregation mechanisms are more efficient. In tree-based topologies, in fact, only one path between each couple of nodes in the network is established, resulting in a very simple routing (each node transmits packets to its parent in the tree). On the other hand, the absence of redundant paths implies that a tree topology is not robust to link failures: if a parent dies all the packets coming from its children are lost. In mesh topologies, instead, multiple *virtual* paths between couples of nodes may exist, requiring the use of efficient routing protocols, but also bringing to a more flexible and reconfigurable network, more robust to link failures.

ZigBee supports three types of devices: ZigBee Routers (ZRs), able to perform all the duties described in IEEE 802.15.4, including routing; ZigBee Coordinators (ZCs), particular ZRs that manage the WPAN, and ZigBee End Devices (ZEDs), that do not have routing capabilities. The ZC corresponds to the 802.15.4 WPAN coordinator, ZRs are the FFDs, whereas the ZEDs are the RFDs.

The rest of the subsection is devoted to the Zigbee tree-based and mesh topologies.

### *1.4.2 The Zigbee Tree-Based Topology*

The Zigbee specifications [31] define a beacon-enabled tree-based topology as a particular case of the IEEE 802.15.4 peer-to-peer networks (shown in Fig. 1.16). A tree, rooted at the WPAN coordinator (the ZC), is formed, and nodes at a given level transmit data to nodes at a lower level, to reach the WPAN coordinator, which is at level zero, in the example shown in the figure. Two different types of nodes are present in the tree: the ZRs, which receive data from their children and forward them to their parent, toward the final ZC; and the ZEDs, the leaves of the tree, that only transmit packets to parents.

The topology formation procedure is started by the WPAN coordinator, which broadcasts beacon packets to neighbour nodes. A candidate node receiving the beacon may request to join the network at the WPAN coordinator. If the WPAN coordinator allows the node to join, it will start transmitting periodic beacons so that other candidate nodes may join the network.

As stated above, nodes must be in beacon-enabled mode: each child node tracks the beacon of its parent (see Fig. 1.17, where the tracking period is outlined as a dashed rectangle). A core concept of this tree topology is that the child node may transmit its own beacon at a predefined offset with respect to the beginning of its parent beacon: the offset must always be longer than the parent superframe duration and shorter than the beacon interval (see Fig. 1.17). This implies that the beacon and the active part of child superframe reside in the inactive period of

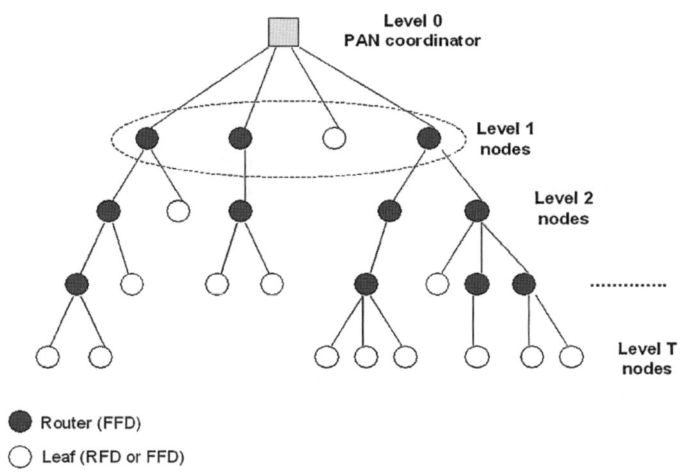


**Fig. 1.16** Tree-based network topology

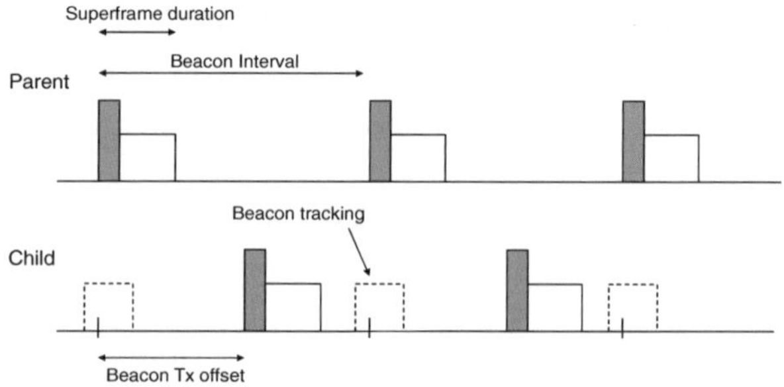


**Fig. 1.17** The tracking of the beacon's parent, performed by a generic child

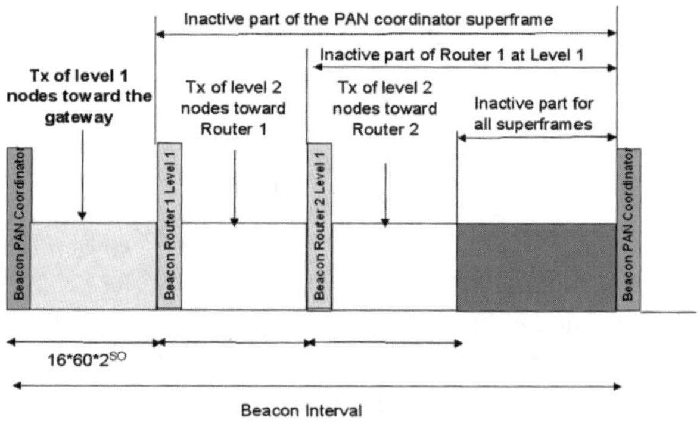


**Fig. 1.18** The superframe structure used in the tree-based topology

the parent superframe; therefore, there is no overlap at all between the active portions of the superframes of child and parent. This concept can be expanded to cover more than two nodes: the selected offset must not result in beacon collisions with neighbouring nodes. This implies that the node must record the time stamp of all neighbouring nodes and select a free time slot for its own beacon. Obviously, a child will transmit a beacon packet only if it is a ZR in the tree; if the child is a ZED, it has only to transmit the packet to its parent. Each child will transmit its packet to the parent in the active part (CAP or CFP) of the parent superframe.

Therefore, each router in the tree, after the reception of the beacon coming from the parent, will select the time interval to transmit its beacon (see Fig. 1.18). Beacon scheduling is necessary to prevent the beacon frames of one device from colliding with either the beacon frames or data transmissions of its neighboring devices. This topology will be analysed in Chap. 5.

### *1.4.3 The Zigbee Mesh Topology*

In the Zigbee mesh topology, devices do not have hierarchical relationships, but they are allowed to be connected with any other device in the network directly or via ZRs. In general, more than one path connects each couple of devices in the network, such that in the case of link failures or if the environment changes, the source device can find an alternative path on demand. The route discovery is similar to that defined in the Ad-hoc On-demand Distance Vector (AODV) algorithm [32]. The ZC is responsible for starting the network and with the ZRs for discovering and maintaining the routes. In particular, to find a route toward a destination node, a source node broadcasts a Route Request (RREQ) packet and then intermediate nodes, having routing capabilities (i.e., ZRs), rebroadcast the RREQ. Once the RREQ is received by the destination, it will reply with a unicast Route Reply (RREP) packet sent to the source node passing through the selected path. Paths are selected as follows. Every path has a length defined as the number of devices in the path and a cost, which is the sum of the costs of all links composing the path. In the ZigBee standard, the link cost is a function of the probability of successful packet delivery through the equation:

$$C\{\ell\} = \min\{7, \mathrm{round}(P_\ell^{-4})\}, \tag{1.3}$$

where $P_\ell$ is the probability of packet delivery on link $\ell$, that could be estimated through the measure of the Link Quality Indication (LQI) provided by the IEEE 802.15.4 MAC and PHY. Links with a cost larger than three are assumed not to exist, meaning that the two corresponding nodes are not connected. To find the optimal route to a destination, each path is associated with a path cost which is the sum of the costs of the single links that belong to the path: $C\{P\} = \sum_i C\{\ell_i\}$. The route which minimizes the path cost is the optimal route. Path costs are recorded in the route discovery table which contains also the address of the device that requests a route and the address of the device that relayed the request to the current device. This address is used for relaying the result of the route discovery back to the source device. To relay a message along the path, the ZC and each router create and record tables, denoted as routing tables, that contain the next-hop address for all possible destinations.

## 1.5 Current and Future Research on WSNs

Basically, the research in the field of WSNs started very recently with respect to other areas of the wireless communication society, like broadcasting or cellular networks. The first IEEE papers on WSNs were published after the turn of the Millennium.

The first European projects on WSNs were financed after year 2001: during the sixth and seventh Framework Programmes, some projects were financed by the European Commission (EC), with explicit activities dedicated to communication protocols, architectural and technological solutions for embedded systems: among them, the first to be launched were WISENTS [33], e-SENSE [34], CRUISE [35], and CONET [36]. In the US, the research on WSNs was boosted a few years before.

Standardization is a key issue for the success of WSN markets. For low data rate applications (250 kbit/s on the air), IEEE 802.15.4 seems to be the most flexible technology currently available, while also Bluetooth Low Energy can be attractive for applications demanding higher data rates. However, IEEE is also currently developing a new standard specifically oriented to WSN for Body Area Networks, through the Task Group 802.15.6. While this shows the perceived relevance of standards in the research arena, it also sets the basis for the possible creation of a heterogeneous WSN environment and opens the field to new technical solutions. In fact, many technical topics of WSNs are still considered by research, as the current solutions are known to be non optimized, or too much constrained.

Then, it is also relevant to mention that there exist two European Technology Platforms, gathering all stakeholders in the field, related to the area of WSNs: e-Mobility and ARTEMIS. They have drawn research agendas that will drive the selection of large cooperative projects in the next years in Europe.

For a proper discussion on future WSN research directions, it is useful to split this issue in two: some research efforts are in fact application-agnostic, while others are mainly driven by market trends and future application needs.

### 1.5.1 Application-Agnostic Research Trends

From the physical layer viewpoint, clearly the need to have low-complexity and low-cost devices does not push short term research in the direction of advanced transmission techniques, while in the medium-long term the application of some novel concepts, such as, for example, *cognitive radio networking*, might be applied to WSNs. The wide use of unlicensed bands for WSNs leads to the situation that they often need to be deployed in environments were many other wireless devices operate, e.g., WiFi systems, etc. A proper use of the radio resource in such unplanned and dynamic environment requires the ability to adapt transmission techniques to the current use of the spectrum. Therefore, some adaptability to the spectrum usage needs to be implemented in future platforms for WSNs.

MAC and network layers have attracted a significant attention in the past years and still deserve investigation. In particular, combined approaches that jointly consider MAC and routing seem to be very successful. Once more, making MAC and routing protocols spectrum-aware can bring relevant performance improvements in some environments.

Topology creation, control, and maintenance are very hot topics. Especially with the IEEE 802.15.4 standard, which allows creation of several types of topologies (stars, mesh, trees), these issues play a very significant role.

One more general paradigm that will be applied to the field of WSN is *cooperation*, defined as the ability of individual entities or objects (that could be sensors, controllers or actuators) to use communication as well as dynamic and loose federation to jointly strive to reach a common goal while taking care not to overtax their available resources [36].

### 1.5.2 Market- and Application-Driven Research Trends

The development of WSN solutions requires significant efforts in terms of tailoring of the available HW/SW platforms to the specific needs of the applications. Therefore, the development/deployment costs can be very high, if the market size is small. As a consequence, the most successful applications of WSN technology will be those oriented to applications including large number of nodes.

Large number of nodes require the presence of many contexts of similar nature where the same technology/application can be deployed. As examples, for this discussion, we consider buildings, humans, and vehicles: all these "contexts" exist in the current world in large numbers and, as such, they represent huge potential markets.

- Buildings: monitoring and control of building energy efficiency is one of the most relevant applications for WSNs in the short term. The sensing of energy consumption in residential buildings requires the installation of sensor nodes in each electric appliance, counting up to tens of devices per residential unit. In a large building, hundreds or even thousands of nodes might be deployed, and interference and network management issues might be based on complex approaches. Scale factors might be introduced in the market because of these large numbers, bringing to significant cost decreases. In this scenario, part of the nodes might be networked through wired connections using the electric grid, and part by means of the technology of WSNs. The application of cognitive radio networking concepts, mentioned above, might be very useful in such dynamic and unpredictable interference contexts.
- Humans: body area networks for health applications are one of the emerging markets for WSNs. However, other applications might be considered dealing with humans: inter-body communication networks might be of interest in several fields of applications, ranging from mood-based services for human networking to specific professional contexts. This type of application requires the exchange of personal body-generated information among humans that can interact for few seconds while walking or moving. Therefore, many challenges raise from the viewpoint of the ability of the WSNs to fast react to new requests of association, etc..
- Vehicles: in this context, two types of applications might be considered. First, the applications where vehicles download data from fixed on-board sensors used for transportation purposes. Second, the applications where fixed sensors upload the sensed data on vehicles passing by, with the opportunistic goal of having such information carried to the final destination exploiting the partially predictable movements of vehicles. The latter concept is sometimes mentioned as *opportunistic networking* and is one of the emerging paradigms of future wireless networks. In these scenarios, only delay-tolerant applications can be run, with the advantage of the efficient exploitation of all types of resources available in the environment where the WSN is deployed.

## 1.6 Further Readings

The increasing interest for WSNs is shown by the large amount of works, which one can found in the literature. In particular, several books and tutorial papers have been published in the last few years [1–4, 6, 8]. Moreover, the introduction of the IEEE 802.15.4 and Zigbee standards has growth the interest for this technology also in the industry. More information can be found on the respective websites, http://www.ieee802.org/15/pub/TG4.html and http://www.zigbee.org

In the recent literature, different works addressed the estimation of a scalar field using random WSNs. As an example, [24] presents a distributed algorithm able to estimate the gradient of a generic smooth physical process (energy constraints and nodes failure are not considered there). In [25], the relationship between the random topology of a sensor network and the quality of the reconstructed field is investigated and some guidelines on how sensors should be deployed over a spatial area for efficient data acquisition and reconstruction are derived. Distributed source coding techniques can be successfully exploited to reduce the amount of data to be transmitted and, hence, to improve the network energy efficiency [37].

## References

1. I. Akyildiz, W. Su, Y. Sankarasubramaniam, E. Cayirci, A survey on sensor networks. IEEE Commun. Mag. **40**(8), 102–114 (2002)
2. M. Tubaishat, S. Madria, Sensor networks: an overview. IEEE Potentials **22**(2), 20–30 (2003)
3. A. Hac, *Wireless Sensor Network Designs* (Wiley, Chichester, 2003)
4. C. Raghavendra, K.M. Sivalingam, T. Znati, *Wireless Sensor Networks* (Kluwer Academic Publishers, Boston, 2004)
5. K. Sohrabi, J. Gao, V. Ailawadhi, G. Pottie, Protocols for self-organization of a wireless sensor network. IEEE Pers. Commun. **7**(5), 16–27 (2000)
6. D. Culler, D. Estrin, M. Srivastava, Overview of sensor networks. IEEE Comput. **37**(8), 41–49 (2004)
7. V. Rajaravivarma, Y. Yang, T. Yang, An overview of wireless sensor network and applications, in *Proceedings of Southeastern Symposium on System Theory (SSST)*, Morgantown, WV, USA, 2003, pp. 432–436
8. R. Verdone, Wireless sensor networks, in *Proceedings of the 5th European Conference on Wireless Sensor Networks, EWSN 2008, Bologna, Italy* (Springer, Heidelberg, 2008)
9. R. Verdone, D. Dardari, G. Mazzini, A. Conti, *Wireless Sensor and Actuator Networks: Technologies, Analysis and Design* (Elsevier, London, 2008)
10. C.Y. Lin, Y.C. Tseng, T.H. Lai, Message-efficient in-network location management in a multi-sink wireless sensor network, in *IEEE International Conference on Sensor Networks, Ubiquitous, and Trustworthy Computing (SUTC)*, vol. 1 (Taichung, Taiwan, 2006), pp. 496–505
11. S. Basagni, M. Conti, S. Giordano, I. Stojmenovic, *Mobile Ad Hoc Networking* (Wiley Inter-Science, Hoboken, 2004)
12. W. Heinzelman, A. Chandrakasan, H.M. Balakrishnan, Energy-efficient communication protocol for wireless microsensor networks, in *Proceedings of Hawaii International Conference on System Sciences (HICSS)*. vol. 2, Maui Island, HI, USA, 2000, pp. 1–10
13. C. Merlin, W. Heinzelman, Schedule adaptation of low-power-listening protocols for wireless sensor networks. IEEE Trans. Mobile Comput. **9**(5), 672–685 (2010)

14. Z. Cheng, M. Perillo, W. Heinzelman, General network lifetime and cost models for evaluating sensor network deployment strategies. IEEE Trans. Mobile Comput. **7**(4), 484–497 (2008)
15. M. Perillo, W. Heinzelman, DAPR: a protocol for wireless sensor networks utilizing an application-based routing cost, in *Proceedings of IEEE Wireless Communications and Networking Conference (WCNC)*. vol. 3, Atlanta, GA, USA, 2004, pp. 1540–1545
16. C. Merlin, W. Heinzelman, Node synchronization for minimizing delay and energy consumption in low-power-listening MAC protocols, in *Proceedings of International Conference on Mobile Ad hoc and Sensor Systems (MASS)*, Atlanta, GA, USA, 2008, pp. 265–274
17. Texas Instruments, RF/IF and ZigBee solutions, Website: http://www.ti.com/
18. J. Ong, Y.Z. You, J. Mills-Beale, E.L. Tan, B. Pereles, K. Ghee, A wireless, passive embedded sensor for real-time monitoring of water content in civil engineering materials. IEEE Sens. J. **8**(12), 2053–2058 (2008)
19. L. Dae-Seok, L. Young-Dong, C. Wan-Young, M. Risto, Vital sign monitoring system with life emergency event detection using wireless sensor network, in *Proceedings of IEEE Conference on Sensors*, Daegu, Corea, 2006, pp. 518–521
20. J. Hao, J. Brady, B. Guenther, J. Burchett, M. Shankar, S. Feller, Human tracking with wireless distributed pyroelectric sensors. IEEE Sens. J. **6**(6), 1683–1696 (2006)
21. M. Lucchi, A. Giorgetti, M. Chiani, Cooperative diversity in wireless sensor networks, in *Proceedings of International Symposium on Wireless Personal Multimedia Communications*, Aalborg, Denmark, September 2005, pp. 1738–1742
22. T. Quek, D. Dardari, M.Z. Win, Energy efficiency of dense wireless sensor networks: to cooperate or not to cooperate. IEEE J. Sel. Areas Commun. **25**(2), 459–470 (2007)
23. S. Toriumi, Y. Sei, H. Shinichi, Energy-efficient event detection in 3D wireless sensor networks, in *IEEE Proceedings of IFIP Wireless Days (WD)*, Dubai, United Arab Emirates, 2008
24. S.N. Simic, S. Sastry, Distributed environmental monitoring using random sensor networks, in *Proceedings of International Work. on Information Processing in Sensor Networks (IPSN)*, Palo Alto, CA, USA, 2003, pp. 582–592
25. C.F. Chiasserini, A. Nordio, E. Viterbo, On data acquisition and field reconstruction in wireless sensor networks, in *Proceedings of Tyrrhenian Workshop on Digital Communications*, Sorrento, Italy, 2005
26. H. Behroozi, F. Alajaji, T. Linder, On the optimal power-distortion region for asymmetric Gaussian sensor networks with fading, in *Proceedings of IEEE Symposium on Information Theory (ISIT)*, Toronto, Canada, 2008, pp. 1538–1542
27. N. Wernersson, J. Karlsson, M. Skoglund, Distributed quantization over noisy channels. IEEE Trans. Commun. **57**(6), 1693–1700 (2009)
28. D. Dardari, F. Fabbri, Spatial field estimation through wireless sensor networks under bandwidth constraints, in *Proceedings of IEEE Global Telecommunication Conference (GLOBECOM)*, Miami, FL, USA, pp. 1–5 (2010)
29. X. Ma, W. Luo, The analysis of 6LowPAN technology, in *Proceedings of Workshop on Computational Intelligence and Industrial Application (PACIIA)*, vol. 1, Wuhan, China, 2008, pp. 963–966
30. IEEE 802.15.4 Std: Wireless Medium Access Control (MAC) and Physical Layer (PHY) Specifications for Low-Rate Wireless Personal Area Networks (LR-WPANs), IEEE Computer Society Press, pp. 1–679, (2003) ISBN: 0-7381-3677-5
31. Zigbee Alliance, Website: http://www.zigbee.org
32. C.E. Perkins, E.M. Royer, Ad hoc on demand distance vector routing, in *Proceedings of IEEE Workshop on Mobile Computing Systems and Applications*, New Orleans, LA, USA, 1999, pp. 90–100
33. P.J. Marron, D. Minder, E.W. Consortium, *Embeddeed WiseNts Reseacrh Roadmap*. (Information Society Technologies, Berlin, 2006), available: http://www.embedded-wisents.org
34. EC Project e-SENSE, FP6. See the website: http://www.ist-esense.org

35. EC Project CRUISE, FP6. See the website: http://www.ist-cruise.eu
36. P.J. Marron, *Cooperating Objects Network of Excellence* (University of Bonn, Bonn), available: http://www.cooperating-objects.eu
37. D. Dardari, A. Conti, C. Buratti, R. Verdone, Mathematical evaluation of environmental monitoring estimation error through energy-efficient wireless sensor networks. IEEE Trans. Mobile Comput. **6**(7), 790–803 (2007)

# Part II
# Distributed Processing

# Chapter 2
# Distributed Detection of Spatially Constant Phenomena

In this chapter, we analyze the problem of distributed detection of a spatially constant phenomenon in IEEE 802.15.4-based Wireless Sensor Networks (WSNs). We first present a communication-theoretic framework on distributed detection in clustered sensor networks with tree-based topologies and hierarchical multi-level fusion. The sensor nodes observe a binary phenomenon and transmit their own data to an Access Point (AP), possibly through intermediate Fusion Centers (FCs), which perform majority-like fusion strategies. Note that the AP functionality is concentrated in the Personal Area Network (PAN) coordinator and this notation is used to highlight the fact that the AP is the network collector. Moreover, the FCs correspond, according to an IEEE 802.15.4 notation, to Full Function Devices (FFDs), whereas the sensors are implemented through Reduced Function Devices (RFDs). We investigate the impact of uniform and non-uniform clustering on the system performance, evaluated in terms of probability of decision error on the phenomenon status at the AP. Our results show that, in the absence of inter-node interference (low traffic load), uniform clustering leads to minimum performance degradation, which depends only on the number of decision levels, rather than on the specific clustered topology.

Since the uniform clustering topology allows to reduce the performance loss incurred by multi-level information fusion, we then investigate the benefits, in terms of longer network lifetime, of *adaptive reclustering*. In particular, lifetime is studied under a physical layer Quality of Service QoS constraint, given by the maximum tolerable probability of decision error at the AP. On the other hand, *absence of reclustering* leads to a shorter network lifetime, and we show the impact of various clustering configurations under different QoS conditions. Our results show that, in the absence of inter-node interference, the organization of sensors in a *few big clusters* is the winning strategy to maximize network lifetime. Moreover, the observation of the phenomenon should be *frequent* in order to limit the penalties associated with the reclustering procedure.

C. Buratti et al., *Sensor Networks with IEEE 802.15.4 Systems*,
Signals and Communication Technology, DOI: 10.1007/978-3-642-17490-2_2,

Although our analysis is based on the assumption of constant Signal-to-Noise Ratio (SNR) at the sensors, we show how to extend it to sensor networks characterized by *non-constant* observation SNRs at the sensors. Furthermore, we show *how* the impact of communication noise in the links between the sensors and the AP depends on the sensor SNR profile (i.e., the spatial distribution of the observation noise). More precisely, different sensor SNR profiles are compared under two alternative assumptions: (i) common *maximum* sensor SNR or (ii) common *average* sensor SNR. Finally, we study how to combine decoding and fusion at the AP to improve the performance in scenarios where the sensors communicate to the AP through *noisy* communication links. Simple distributed channel coding strategies are considered, using either repetition coding at each sensor (i.e., multiple observations) or distributed (network-wide) systematic block channel coding. In the latter case, the use of a relay is proposed. In all cases, the system performance is analyzed *separating* or *joining* the decoding and fusion operations at the AP. Our results show that the schemes with joint decoding and fusion show a significant performance improvement with respect to that of schemes with separate decoding and fusion and the use of *multiple observations* is often the winning choice at practical values of the probability of decision error.

The analytical approach introduced in this chapter is extended to realistic sensor networks, based on commercial protocols. In particular, simulation and experimental (relative to IEEE 802.15.4-based networks) results, which confirm the analytical predictions, are presented, enriching the proposed analytical framework and showing how typical networking performance metrics (such as throughput and delay) are influenced by the probability of decision error.

This chapter is structured as follows. In Sect. 2.1, we present the analytical framework to analyze the performance of distributed detection schemes in clustered sensor networks. In Sect. 2.2, we analyze the sensor network lifetime in the presence of the proposed distributed detection strategies. In Sect. 2.3, we extend our framework to take into account possible non-constant SNR spatial distributions at the sensors. In Sect. 2.4, we extend the framework also to take into account the presence of different detection/fusion strategies. In Sect. 2.5, concluding remarks are given and, finally, a brief review of the literature is presented in Sect. 2.6.

## 2.1 Distributed Detection in Clustered Sensor Networks

### *2.1.1 Preliminaries on Distributed Binary Detection*

We consider a network scenario where $n$ sensors observe a *common binary phenomenon* whose status is defined as follows:

$$H = \begin{cases} H_0 & \text{with probability } p_0 \\ H_1 & \text{with probability } 1 - p_0 \end{cases}$$

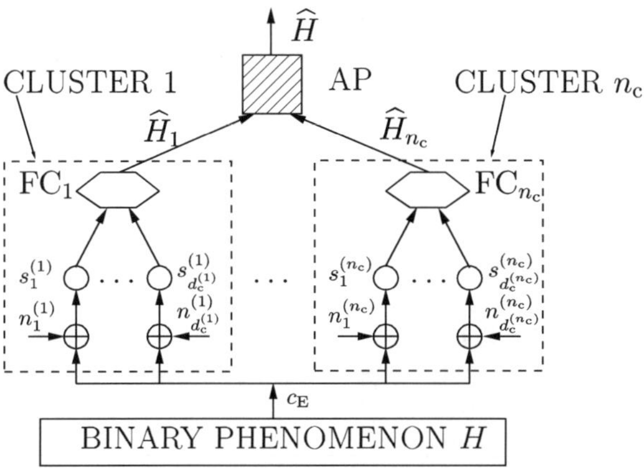


**Fig. 2.1** Block diagram of a clustered sensor network with distributed binary detection and two decision levels

where $p_0 \triangleq \mathbb{P}\{H = H_0\}$, being $\mathbb{P}\{\cdot\}$ the probability of a given event. In the remainder of this book, if no otherwise stated, we will focus on a scenario with equal a priori probabilities of the phenomenon, i.e., $p_0 = p_1 = 1/2$. However, similar results can be derived for a scenario with $p_0 \neq 1/2$. The sensors are clustered into $n_c < n$ groups, and each sensor can communicate only with its local first-level FC. The first-level FCs collect data from the sensors in their corresponding clusters and make local decisions on the status of the binary phenomenon. In a scenario with two levels of information fusion, each local FC transmits to the AP, which makes the final decision. A logical representation of this architecture is shown in Fig. 2.1.

The observed signal at the $i$th sensor[1] can be expressed as

$$r_i = c_E + n_i \quad i = 1, \ldots, n \tag{2.1}$$

where

$$c_E \triangleq \begin{cases} 0 & \text{if } H = H_0 \\ s & \text{if } H = H_1 \end{cases}$$

and $\{n_i\}$ are additive noise samples. Assuming that the noise samples $\{n_i\}$ are independent with the same Gaussian distribution $\mathcal{N}(0, \sigma^2)$, the common SNR at the sensors can be defined as follows:

$$\mathrm{SNR}_{\mathrm{sensor}} = \frac{[\mathbb{E}\{c_E|H_1\} - \mathbb{E}\{c_E|H_0\}]^2}{\sigma^2} = \frac{s^2}{\sigma^2}. \tag{2.2}$$

Each sensor makes a decision comparing its observation $r_i$ with a threshold value $\tau_i$ and computes a local decision $u_i = U(r_i - \tau_i)$, where $U(\cdot)$ is the unit step

[1] Note that in this case we do not refer, for the ease of clearness, to the specific cluster, as done in Fig. 2.1. The particular situation will clarify the ambiguity.

function. In order to optimize the system performance, the thresholds $\{\tau_i\}$ need to be optimized. Even though, in general, a common value of the decision threshold for all sensors might not be the best choice, in the remainder of this chapter we assume that all sensors use the same decision threshold $\tau$. While in a scenario with no clustering and ideal communication links between the sensors and the AP the relation between the optimized value of $\tau$ and $s$ is well known [1], in the presence of clustering it is not. In the following, the value of $\tau$ will be optimized scenario by scenario. More precisely, we consider a possible (discrete) set of values which can be assumed by $\tau : \{\tau_{\min}, \tau_{\min} + \Delta\tau, \tau_{\min} + 2\Delta\tau, \ldots, \tau_{\max}\}$. In other words, $\tau$ can assume values in $[\tau_{\min}, \tau_{\max}]$ at regular steps of (sufficiently small) width $\Delta\tau$. For a given sensor SNR, the probability of decision error is evaluated for each possible value of $\tau$, and the minimizing value is selected as threshold. In all considered cases, the optimized value of the common threshold is around $\sqrt{\mathrm{SNR}_{\mathrm{sensor}}}/2$, as already observed in [1, 2].

In a scenario with noisy communication links, modeled as Binary Symmetric Channels (BSCs), the decision $u_i$ sent by the $i$th sensor can be *flipped* with a probability corresponding to the cross-over probability of the BSC model and denoted as $p$ [3]. In general, a BSC might not be the best modelling choice for a wireless communication link, which might experience block fading [4–7]. However, in the presence of memoryless communication channels the use of a cross-over probability $p$ is accurate. More precisely, $p$ can be given a precise expression depending on the type of channel (with Additive White Gaussian Noise (AWGN) or bit-by-bit independent fading). Therefore, our simple model can give significant insights into the network behavior in many situations. The received bit at the fusion point (either an FC for clustered networks or directly the AP in the absence of clustering), referred to as $u_i^{(\mathrm{r})}$, can be expressed as

$$u_i^{(\mathrm{r})} = \begin{cases} u_i & \text{with probability } 1-p \\ 1-u_i & \text{with probability } p. \end{cases}$$

In the presence of noisy links, the value of the optimized local threshold $\tau$, fixed for all sensors, might be different from that in a scenario with ideal communication links. As for the case with ideal communication links, this optimization will be carried out, for given SNR and clustering configuration, by minimizing the probability of decision error, as outlined at the end of the previous paragraph. Note that the best strategy would consist in using a properly optimized set of decision thresholds $\{\tau_i\}$ at the sensors. In particular, in a more general scenario where the type of event perceived by the sensor might vary, a more refined per-cluster optimization of the sensor decision threshold could be considered. However, since we are interested in monitoring a spatially constant binary phenomenon, we consider a simpler optimization approach, where the same threshold is used at all sensors.

While the communication links between sensors and first level FCs can be noisy, we assume that the other communication links in the network (i.e., from each FC to higher level FCs or the AP) are ideal. The rationale behind the assumption of ideal high-level links lies in the fact that in practical sensor network design the FCs are likely to be placed relatively close to the AP. Therefore, under the assumption of a robust access control mechanism, one can assume that these links are ideal. The proposed analytical framework can be extended to encompass the presence of higher level noisy links. Moreover, realistic sensor network scenarios (with collisions) will be analyzed, through simulations and experiments, in Sect. 2.1.5.

We point out that the specific topologies of the considered networks are not explicitly taken into account. For instance, the distances between nodes are not explicitly mentioned. This corresponds to the assumption of modelling all noisy communication links as BSCs with the same cross-over probability. In order to extend our analytical framework, while still keeping the simple BSC-based link modelling, one can consider different cross-over probabilities (they could be associated with a specific network topology). This motivates the use of weighing fusion schemes, where the decisions to be fused together are weighed by the corresponding link qualities [8].

## *2.1.2 Analytical Framework*

### 2.1.2.1 Uniform Clustering

In a scenario with *uniform* clustering, the sensors are grouped into identical clusters, i.e., each of the $n_c$ clusters contains $d_c$ sensors, with $n_c \cdot d_c = n$. A pictorial description of a uniformly clustered sensor network with $n = 16$ sensors and two decision levels is shown in Fig. 2.2: there are $n_c = 4$ clusters with $d_c = 4$ sensors each.

According to the assumption of majority-like information fusion considered in this chapter, the $j$th FC ($j = 1, \ldots, n_c$) computes a local decision using the following rule:

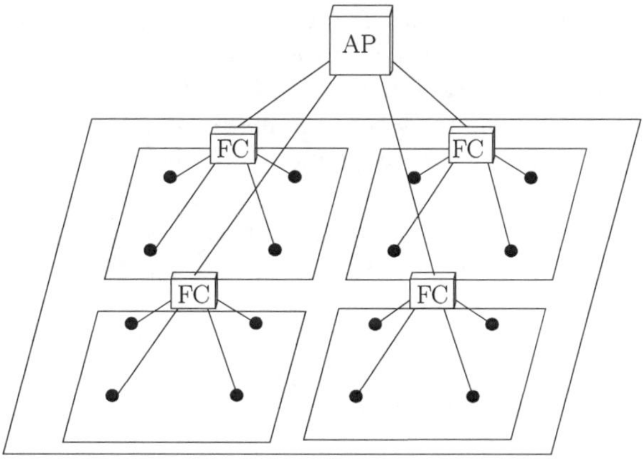


**Fig. 2.2** An example of a uniformly clustered sensor network with $n = 16$ sensors. There are $n_c = 4$ clusters with $d_c = 4$ sensors each

$$\widehat{H}_j = \Gamma\left(u_1^{(j)}, \ldots, u_{d_c}^{(j)}\right) = \begin{cases} 0 & \text{if } \sum_{m=1}^{d_c} u_m^{(j)} < k \\ 1 & \text{if } \sum_{m=1}^{d_c} u_m^{(j)} \geq k \end{cases} \tag{2.3}$$

where $u_m^{(j)}$ is the $m$th decision of a sensor in the $j$th cluster and $k$ is the FC decision threshold—since the clusters have the same dimension, the threshold $k = \lfloor d_c/2 \rfloor + 1$ is the same at all FCs. The AP decides with the following majority-like rule based on the local FC decisions $\{\widehat{H}_j\}$:

$$\widehat{H} = \Psi\left(\widehat{H}_1, \ldots, \widehat{H}_{n_c}\right) = \begin{cases} H_0 & \text{if } \sum_{j=1}^{n_c} \widehat{H}_j < k_f \\ H_1 & \text{if } \sum_{j=1}^{n_c} \widehat{H}_j \geq k_f \end{cases} \tag{2.4}$$

where $k_f = \lfloor n_c/2 \rfloor + 1$ is the fusion threshold at the AP. Using a combinatorial approach (based on the repeated trials formula [9]) and taking into account the decision rules (2.3) and (2.4), the probability of decision error at the AP can be expressed as follows:

$$\begin{aligned} P_e &= \mathbb{P}\{\widehat{H} = H_1 | H_0\}\mathbb{P}\{H_0\} + \mathbb{P}\{\widehat{H} = H_0 | H_1\}\mathbb{P}\{H_1\} \\ &= p_0 \operatorname{bin}(k_f, n, n_c, \operatorname{bin}(k, d_c, d_c, Q(\tau))) \end{aligned} \tag{2.5}$$

$$+(1 - p_0) \operatorname{bin}(0, k_f - 1, n_c, (k, d_c, d_c, Q(\tau - s))) \tag{2.6}$$

where $Q(x) \triangleq \int_x^\infty \frac{1}{\sqrt{2\pi}} \exp(-y^2/2)\, dy$ and

$$\operatorname{bin}(a, b, n, z) \triangleq \sum_{i=a}^{b} \binom{n}{i} z^i (1 - z)^{(n-i)} \tag{2.7}$$

where $a, b, n \in \mathbb{N}$ and $z \in (0, 1)$. If $n_c = k_f = 1$ and $d_c = n$, i.e., there is no clustering, and the probability of decision error (2.6) reduces to that derived in [3].

We point out that the majority fusion rule (2.3) with FC decision threshold $k = \lfloor d_c/2 \rfloor + 1$ is exact for *odd* values of $k$. For *even* values of $k$, the proposed fusion strategy tends to favor a final decision equal to '0.' For example, if $d_c = 2$, then only the received sequence 11 leads to a final decision in favor of '1.' However, since in all considered scenarios the two statuses of the binary phenomenon are equiprobable, setting $k$ to $\lfloor d_c/2 \rfloor$ would unbalance the decision towards '1,' but, *on average*, the final performance would be the same.

Although we have previously derived the probability of decision error in a scenario with uniform clustering and two levels of information fusion, this analysis can be extended to a scenario with three levels of information fusion. In Fig. 2.3c, the logical structure of a sensor network with three decision levels is illustrated. For comparison, in the same figure the schemes with (a) no clustering and (b) two decision level uniform clustering are also shown. One should note that Fig. 2.3b is logically equivalent to the network schemes shown in Figs. 2.1 and 2.2. In a three decision level scenario the probability of decision error at the AP becomes

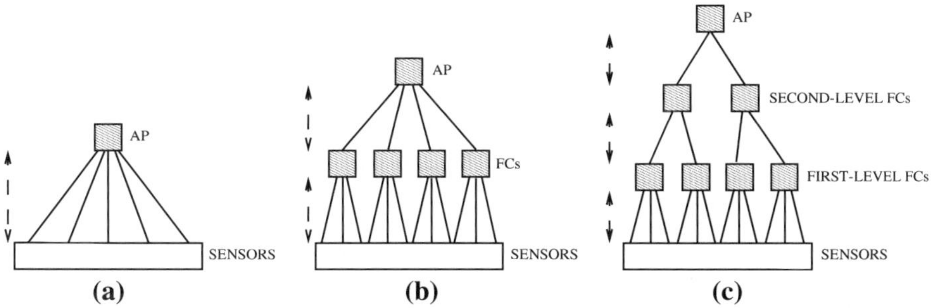


**Fig. 2.3** Basic structures for sensor networks with distributed detection. Three cases are shown: **a** absence of clustering, **b** uniform clustering with two levels of information fusion, and **c** uniform clustering with three levels of information fusion

$$
\begin{aligned}
P_{\mathrm{e}} &= p_0 \operatorname{bin}(k_{\mathrm{f}}, n_{\mathrm{c}_2}, n_{\mathrm{c}_2}, \operatorname{bin}(k_2, d_{\mathrm{c}_2}, d_{\mathrm{c}_2}, \operatorname{bin}(k_1, d_{\mathrm{c}_1}, d_{\mathrm{c}_1}, Q(\tau)))) \\
&\quad + (1 - p_0) \operatorname{bin}(0, k_{\mathrm{f}} - 1, n_{\mathrm{c}_2}, \operatorname{bin}(k_2, d_{\mathrm{c}_2}, d_{\mathrm{c}_2}, \operatorname{bin}(k_1, d_{\mathrm{c}_1}, d_{\mathrm{c}_1}, Q(\tau - s))))
\end{aligned}
\tag{2.8}
$$

where $n_{\mathrm{c}_i}$ and $d_{\mathrm{c}_i}$ denote the number of clusters and nodes per cluster at the $i$th level ($i = 1, 2$), whereas $k_i$ ($i = 1, 2$) is the majority-like fusion threshold at the $i$th level.

We remark that the above derivation can be straightforwardly extended to a scenario with a generic number of fusion levels. As for the scenario with uniform clustering and one decision level, the thresholds $\{k_i\}$ can be optimized by minimizing the probability of decision error at the AP.

#### 2.1.2.2 Non-Uniform Clustering

Assuming a two-level sensor network topology, the probability of decision error in a generic scenario with non-uniform clustering can be evaluated as follows. Define the cluster size vector $\mathscr{D} \triangleq \{d_{\mathrm{c}}^{(1)}, d_{\mathrm{c}}^{(2)}, \ldots, d_{\mathrm{c}}^{(n_{\mathrm{c}})}\}$, where $d_{\mathrm{c}}^{(i)}$ is the number of sensors in the $i$th cluster ($i = 1, \ldots, n_{\mathrm{c}}$) and $\sum_{i=1}^{n_{\mathrm{c}}} d_{\mathrm{c}}^{(i)} = n$. Furthermore, define also the following two probability vectors:

$$
\mathscr{P}^{\mathbf{1|1}} \triangleq \left\{p_1^{1|1}, p_2^{1|1}, \ldots, p_{n_{\mathrm{c}}}^{1|1}\right\} \quad \mathscr{P}^{\mathbf{1|0}} \triangleq \left\{p_1^{1|0}, p_2^{1|0}, \ldots, p_{n_{\mathrm{c}}}^{1|0}\right\}
$$

where $p_\ell^{1|1}$ ($p_\ell^{1|0}$, respectively) is the probability that the $\ell$th FC decides for $H_1$ when $H_1$ ($H_0$, respectively) has happened. We still consider the use of a common threshold $\tau$ at the sensors, and its value is optimized as described in Sect. 2.1.1. The elements of $\mathscr{P}^{\mathbf{1|1}}$ (equivalently, the elements of $\mathscr{P}^{\mathbf{1|0}}$) are, in general, different from each other and depend on the particular distribution of the sensors among the clusters. In [8], it is shown that the probability of decision error can be expressed as follows:

$$P_e = p_0 \sum_{i=k_f}^{n} \sum_{j=1}^{\binom{n_c}{i}} \prod_{\ell=1}^{n} \left\{ c_{i,j}(\ell) p_\ell^{1|0} + (1 - c_{i,j}(\ell))(1 - p_\ell^{1|0}) \right\}$$

$$+ (1 - p_0) \sum_{i=0}^{k-1} \sum_{j=1}^{\binom{n_c}{i}} \prod_{\ell=1}^{n} \left\{ c_{i,j}(\ell) p_\ell^{1|1} + (1 - c_{i,j}(\ell))(1 - p_\ell^{1|1}) \right\} \quad (2.9)$$

where $\boldsymbol{c}_{i,j} = (c_{i,j}(1), \ldots, c_{i,j}(n_c))$ is a vector which designates the $j$th configuration of the decisions from the first-level FCs in a case with $i$ '1's (and, obviously, $n_c - i$ '0's). In Table 2.1, the possible configurations of $\boldsymbol{c}_{i,j}$ are shown in the presence of $n_c = 3$ clusters. For example, $\boldsymbol{c}_{1,2}$ is the second possible configuration with one '1' (and two '0'): the '1' is the decision of the second FC.

A scenario with *uniform* clustering can be interpreted as a special case of a generic non-uniform scenario. In this case, in fact, the elements of the three vectors $\mathcal{D}$, $\boldsymbol{\mathcal{P}}^{1|1}$, and $\boldsymbol{\mathcal{P}}^{1|0}$, become equal, i.e.:

$$\begin{cases} d_c^{(i)} & = & d_c \\ p_i^{1|1} & = & \text{bin}(k, d_c, d_c, Q(\tau - s)) \\ p_i^{1|0} & = & \text{bin}(k, d_c, d_c, Q(\tau)) \end{cases}$$

$\forall i = 1, \ldots, n_c$. It can be shown that (2.9) reduces to (2.6) in the presence of uniform clustering.

#### 2.1.2.3 Scenarios with Noisy Communication Links

In a scenario with non-uniform clustering and two decision levels, the probability of decision error can be derived from (2.9), by replacing the probabilities $\{p_\ell^{1|i}\}_{\ell=1,\ldots,n_c}^{i=0,1}$ with the probabilities $\{p_{\ell,\text{noisy}}^{1|i}\}_{\ell=1,\ldots,n_c}^{i=0,1}$, which take into account the noise in the communication links between sensors and FCs and are defined as follows:

**Table 2.1** Possible configurations of $\boldsymbol{c}_{i,j}$ in a scenario with $n_c = 3$ clusters

| $i$ | $j$ | $\boldsymbol{c}_{i,j}$ |
|---|---|---|
| 0 | 1 | 000 |
| | 1 | 100 |
| 1 | 2 | 010 |
| | 3 | 001 |
| | 1 | 110 |
| 2 | 2 | 101 |
| | 3 | 011 |
| 3 | 1 | 111 |

$$p^{1|0}_{\ell,\text{noisy}}\left(d^{(\ell)}_{\text{c}}\right) = \sum_{m=k_\ell}^{d^{(\ell)}_{\text{c}}} \binom{d^{(\ell)}_{\text{c}}}{m} P^m_{10} P^{d^{(\ell)}_{\text{c}}-m}_{00} \tag{2.10}$$

$$p^{1|1}_{\ell,\text{noisy}}\left(d^{(\ell)}_{\text{c}}\right) = \sum_{m=k_\ell}^{d^{(\ell)}_{\text{c}}} \binom{d^{(\ell)}_{\text{c}}}{m} P^m_{11} P^{d^{(\ell)}_{\text{c}}-m}_{01} \tag{2.11}$$

where $k_\ell$ depends on the number of packets received at the $\ell$th FC. Since the same majority-like fusion rule of the AP is applied to each FC, the same considerations given above for $k_{\text{f}}$ still apply here for the value of $k_\ell$.

In (2.10), $P_{10} = 1 - P_{00}$ is the probability that a sensor decision sent to an FC is in favor of $H_1$ when $H_0$ has happened and can be expressed, according to the BSC model for a noisy communication link, as

$$P_{10} = Q(\tau)(1-p) + [1 - Q(\tau)]p. \tag{2.12}$$

In fact, the first term at the right-hand side is obtained when there is an observation error but error-free communications, whereas the second term is obtained when there are error-free observations but communication errors. Similarly, in (2.11) $P_{11} = 1 - P_{01}$ represents the probability that a decision sent by a sensor to an FC is in favor of $H_1$ when $H_1$ has happened and can be given the following expression:

$$P_{11} = Q(\tau - s)(1-p) + [1 - Q(\tau - s)]p. \tag{2.13}$$

Finally, the probability of decision error in a scenario with noisy communication links becomes

$$\begin{aligned} P_{\text{e}} = p_0 \sum_{i=k_{\text{f}}}^{n} \sum_{j=1}^{\binom{n_{\text{c}}}{i}} \prod_{\ell=1}^{n} \left\{ c_{i,j}(\ell) p^{1|0}_{\ell,\text{noisy}} + (1 - c_{i,j}(\ell))(1 - p^{1|0}_{\ell,\text{noisy}}) \right\} \\ + (1-p_0) \sum_{i=0}^{k_{\text{f}}-1} \sum_{j=1}^{\binom{n_{\text{c}}}{i}} \prod_{\ell=1}^{n_{\text{c}}} \left\{ c_{i,j}(\ell) p^{1|1}_{\ell,} + (1 - c_{i,j}(\ell))(1 - p^{1|1}_{\ell,\text{noisy}}) \right\}. \end{aligned} \tag{2.14}$$

## *2.1.3 Communication-Theoretic Characterization*

### 2.1.3.1 Ideal Communication Links

The analytical framework presented in Sect. 2.1.2 leads to a communication-theoretic characterization of the network performance in terms of probability of decision error at the AP as a function of the sensor SNR and the communication noise level.

In Fig. 2.4, the probability of decision error is shown, as a function of the sensor SNR, in the case with $n = 16$ sensors, considering two and three decision levels.

In the scenario with two decision levels, the following topologies are possible:

- 8-8 (2 clusters with 8 sensors each);
- 4-4-4-4 (4 clusters with 4 sensors each);
- 2-2-2-2-2-2-2-2 (8 clusters with 2 sensors each).

For a three decision level scenario, the following topologies are considered:

- 4-4-4-4/2-2 (4 first-level FCs, each connected with 4 sensors, and 2 second-level FCs, each connected with 2 first-level FCs);
- 2-2-2-2-2-2-2-2/4-4 (8 first-level FCs, each connected with 2 sensors, and 2 second-level FCs, each connected with 4 first-level FCs);
- 2-2-2-2-2-2-2-2/2-2-2-2 (8 first-level FCs, each connected with 2 sensors, and 4 second-level FCs, each connected with 2 first-level FCs).

Lines and symbols (circles, triangles, and stars) correspond to analytical and simulation results, respectively. For comparison, the probability of decision error with no clustering is also shown. We point out that the simulation results shown in Fig. 2.4 and those shown, in the following, in Fig. 2.5 are meant to verify the correctness of the analytical framework. In other words, these results are obtained by simulating systems which are identical to those behind the analytical models. Obviously, the agreement between analysis and simulations is perfect, as no approximations were included. In Sect. 2.1.5, instead, the presented simulation results will refer to realistic IEEE 802.15.4 networks.

In Fig. 2.4, only one curve is shown for the scenario with two levels of information fusion, since the performance curves associated with all possible configurations (i.e., 8-8, 4-4-4-4, 2-2-2-2-2-2-2-2) overlap. This implies that one can choose between a uniform network topology with a small number of large clusters and a uniform network topology with a large number of small clusters, still

**Fig. 2.4** Probability of decision error, as a function of the sensor SNR, in a scenario with $n = 16$ sensors and uniform clustering

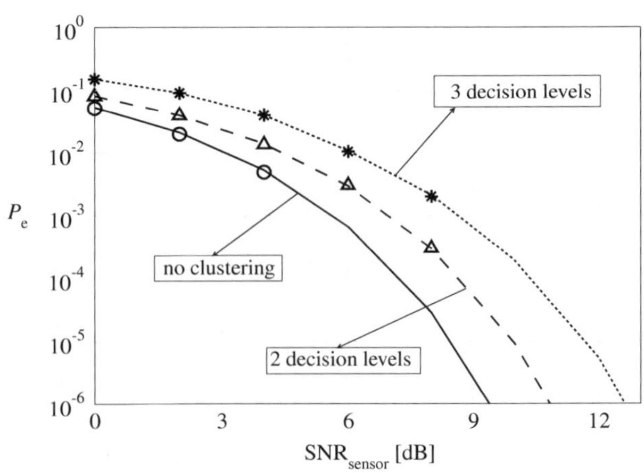

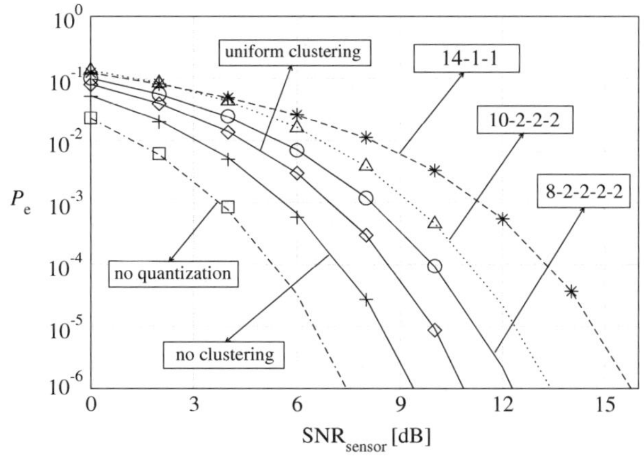


**Fig. 2.5** Probability of decision error, as a function of the sensor SNR, in a scenario with $n = 16$ sensors. Various configurations are considered

guaranteeing the same performance level. The intuition behind this result is the following.

- If one considers an architecture with small clusters, then fusion at the first-level FCs is not effective. However, many local cluster decisions are then fused together, and this allows to recover (partially) the first-level information loss.
- On the other hand, considering large clusters leads to more reliable local first-level decisions. However, a few of them are then fused together, so that the supplementary (higher-level) refinement is not relevant.

Similar considerations also hold for a three decision level scenario. We point out that in Fig. 2.4 the obtained analytical expressions of the probability of decision error are numerically evaluated and verified through simulations.

Comparing the performance in the absence of clustering with that in the presence of uniform clustering (with either two or three decision levels), one can conclude that the larger is the number of decision levels, the worse is the performance. This is intuitive, since a larger number of decision levels corresponds to a larger number of partial information losses in correspondence to the fusion operations. However, this holds in scenarios with ideal communication links. In a wireless communication scenario, where some links may be completely obstructed, a sensor network with multiple communication layers might not yield the worst performance.

Although the analytical framework derived in the previous subsections is general, the presented results refer to networks with a (relatively) small number of sensors. However, our framework can be extended to scenarios with a large number of sensors. To this regard, in [8] a simple, yet very accurate, approximation of the derived framework, based on the application of the De-Moivre Laplace (DML) theorem, is proposed.

In order to evaluate the impact of non-uniform clustering, we consider a scenario with $n = 16$ sensors and various non-uniform network topologies. In Fig. 2.5, the probability of decision error is shown, as a function of the sensor SNR, considering no clustering, two level uniform clustering, and various configurations with two decision levels and non-uniform clustering (explicitly indicated). For comparison, the curve in the absence of quantization at the sensors is also shown. The lines correspond to analytical results, whereas symbols are associated with simulations. In the scenarios with non-uniform clustering, the considered configurations are 8-2-2-2-2 (5 clusters, out of which 4 contain 2 nodes each and 1 contains 8 nodes), 10-2-2-2, and 14-1-1. As one can see from Fig. 2.5, in the presence of majority-like information fusion the higher is the non-uniformity degree among the clusters, i.e., the more unbalanced is clustering, the worse is system performance. Consequently, a sensor network designer should avoid non-uniform configurations with one big cluster and remaining small clusters. In general, a two-level uniformly clustered scenario is desirable, since it guarantees the smallest energetic loss with respect to a network with no clustering. However, uniform clustering in a realistic scenario might not be possible (e.g., in environmental monitoring applications). In fact, the area over which the sensors are distributed could be irregular and, therefore, uniform clustering of the sensors could not be feasible. An interesting application of our framework could consist in the identification of non-uniform clustering "classes," with similar performance per class. This could help significantly a network designer in predicting, for example, the performance degradation caused by the loss of some sensors (e.g., for battery exhaustion).

The above analysis in non-uniformly clustered scenarios applies to situations where the AP does not know the exact distribution of the sensors among the clusters. This is meaningful, for instance, in large networks where only local topology knowledge is possible. If, on the other hand, the distribution is very unbalanced (e.g., 14-1-1 with $n = 16$ sensors) and the AP knows the exact topology, the less reliable decisions originated by small clusters can be ignored. In a scenario with $n = 16$ sensors and the considered 14-1-1 topology, at $P_e = 10^{-4}$ a sensor SNR gain equal to 5.47 dB can be obtained without using, at the AP, the decisions associated with the smaller clusters—this corresponds to the performance of a sensor network with $n = 14$ sensors and no clustering. Therefore, knowledge of the clustering configuration at the AP allows to obtain a performance very close to that in the absence of clustering. In particular, in the previous case with $n = 16$ sensors and 14-1-1 configuration, the sensor SNR loss (with respect to a scenario with no clustering) can be reduced to 0.77 dB by using only the decision sent by the 14-sensor cluster. Our current focus, however, is on the comparison of clustering topologies when the AP gives the same weight to all received decisions. This is meaningful for a *dynamic* sensor network scenario, where sensors might die and sensors clusters might become unbalanced. In this case, intelligent reclustering techniques can be used to improve the system performance, as it will be shown in Sect. 2.2.

In Fig. 2.6, the probability of decision error is shown, as a function of the sensor SNR, for different values of the number of sensors $n$ in a scenario with uniform clustering. In particular, the considered values for $n$ are 16, 20, 32, 40, and 64. Observe that only one curve is associated with each value of $n$, since we have

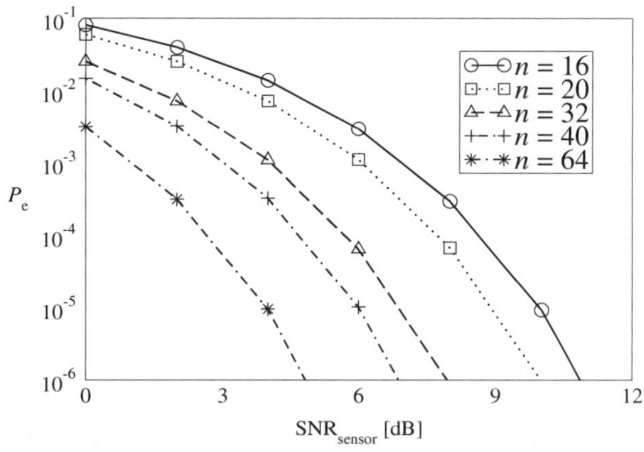


**Fig. 2.6** Probability of decision error, as a function of the sensor SNR, in a scenario with uniform clustering. Different values of the number of sensors are considered

previously shown that the performance does not depend on the number of clusters (for a given $n$), as long as clustering is uniform. Obviously, the performance improves (i.e., the probability of decision error decreases) when the number of sensors in the network becomes larger. The results in Fig. 2.6 will be used in Sect. 2.2.1 to compute the sensor network lifetime under a QoS condition on the maximum acceptable probability of decision error.

#### 2.1.3.2 Noisy Communication Links

While in Sect. 2.1.3.1 the performance in scenarios with ideal communication links has been analyzed, we now turn our attention to scenarios with noisy communication links. In particular, it is interesting to investigate how the probability of decision error behaves as a function of the communication noise level, i.e., the cross-over probability $p$. To this end, we introduce a communication-theoretic QoS condition, in terms of the maximum tolerable probability of decision error, denoted as $P_e^*$, at the AP. A physical layer-oriented QoS condition can be written as

$$P_e \leq P_e^*. \qquad (2.15)$$

Since the probability of decision error is a monotonically decreasing function of the sensor SNR, the QoS condition (2.15) can be equivalently rewritten as

$$\mathrm{SNR}_{\mathrm{sensor}} \geq \mathrm{SNR}_{\mathrm{sensor}}^*$$

where $\mathrm{SNR}_{\mathrm{sensor}}^*$ depends on $P_e^*$. It is then possible to evaluate the performance under a desired QoS constraint, given by the maximum tolerable probability of decision error $P_e^*$.

In Fig. 2.7, the value of the minimum sensor SNR required to guarantee $P_e^*$, i.e., $\mathrm{SNR}_{\mathrm{sensor}}^*$, is shown, as a function of the cross-over probability $p$, in scenarios (i) without clustering and (ii) with clustering and two decision levels, respectively. Two possible values for $P_e^*$ are considered: (i) $10^{-3}$ (curves with circles) and (ii) $10^{-4}$ (curves with triangles). As expected, when the noise level increases, the minimum

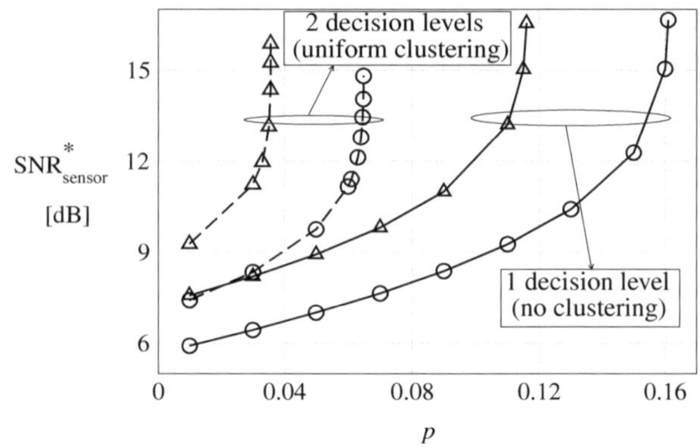


**Fig. 2.7** Minimum sensor SNR required to obtain a desired QoS, in scenarios with noisy communication links in the cases (i) without clustering and (ii) with uniform clustering and two decision levels. Two possible QoS conditions are considered: (i) $P_e^* = 10^{-3}$ (*lines with circles*) and (ii) $P_e^* = 10^{-4}$ (*lines with triangles*)

sensor SNR required to guarantee the desired network performance also increases. In fact, since communications become less reliable, a higher accuracy in the observation phase is needed in order to maintain the same overall performance. Besides, one can observe that there exists a vertical asymptote in each curve in Fig. 2.7. In other words, there exists a critical value $p_{\text{crit}}$ of the noise level, such that: (i) for $p < p_{\text{crit}}$, the sensor network is operative, i.e., there exists a finite value of the sensor SNR which satisfies the desired QoS condition (2.15); (ii) for $p > p_{\text{crit}}$, instead, the network is not operative, i.e., it is not possible to achieve the desired performance level, *regardless* of the value of the sensor SNR. One could equivalently describe this behavior as *bimodal*. This is a typical behavior of distributed communication networks, such as the bimodal connectivity behavior in ad hoc wireless networks [10–14]. Proper operation of the considered sensor networks with distributed detection can be equivalently interpreted as a symptom of network connectedness. In Fig. 2.7, this bimodal behavior is also confirmed in a scenario with uniform clustering and two decision levels. However, in the latter case the impact of the communication noise is stronger with respect to a scenario with no clustering, i.e., the network looses connectivity for smaller values of $p$. Consequently, the larger is the number of decision levels in the network, the lower is the maximum tolerable communication noise level.

### 2.1.4 Joint Communication/Information-Theoretic Characterization

The considered sensor network schemes can be modeled as "black boxes" with a binary input (the phenomenon $H$) and a binary output (the decision $\widehat{H}$ at the AP). Using the model in Fig. 2.1, the final decision $\widehat{H}$ can be described as a binary random variable[2] with $P_0 \triangleq P(\widehat{H} = H_0)$. In a scenario with two-level uniform

[2] Note that the definition of $P_0 = \mathbb{P}\{\widehat{H} = H_0\}$ (relative to the decision $\widehat{H}$) is different from that given for the *a priori* probability of the phenomenon $p_0 = \mathbb{P}\{H = H_0\}$ given in Sect. 2.1.1.

clustering and ideal communication links, the parameter $P_0$ can be rewritten (using the results in Sect. 2.1.1) as

$$P_0 = p_0 \operatorname{bin}(0, k_f - 1, n_c, \operatorname{bin}(k, d_c, d_c, Q(\tau))) \\ + (1 - p_0) \operatorname{bin}(0, k_f - 1, n_c, \operatorname{bin}(k, d_c, d_c, Q(\tau - s))). \quad (2.16)$$

We remark that Eq. 2.16 may look identical to (2.6). In (2.16), however, the term on the right-hand side in the first row corresponds to $\mathbb{P}\{\widehat{H} = H_0|H_0\}$, whereas in (2.6) it is given by $\mathbb{P}\{\widehat{H} = H_1|H_0\}$—the second parameter of the function "bin" is, in fact, different in the two cases.

The mutual information of the Binary Input Binary Output (BIBO) sensor network can then be written as [15, Chap. 2]

$$I(H; \widehat{H}) = H_e(\widehat{H}) - H_e(\widehat{H}|H)$$

where $H_e(\cdot)$ is the entropy of a random variable; in particular, $H_e(\widehat{H}|H)$ is the conditional entropy of $\widehat{H}$ given $H$ [15]. After a few manipulations, the mutual information becomes

$$I(H; \widehat{H}) = H_e(p_0(1 - p_{10}) + (1 - p_0)p_{01}) - p_0 H_e(p_{10}) - (1 - p_0)H_e(p_{01}) \quad (2.17)$$

where $p_{ij} \triangleq \mathbb{P}\{\widehat{H} = H_i|H_j\}$, $i, j = 0, 1$.

In Fig. 2.8, the probability of decision error is shown, as a function of the *mutual information*, for the same scenario considered in Fig. 2.5, i.e., with no clustering (circles), uniform clustering (triangles), and non-uniform clustering (pluses, 14-1-1 configuration), respectively. The communication links are ideal. The curves considered in this figure are parameterized curves, obtained by

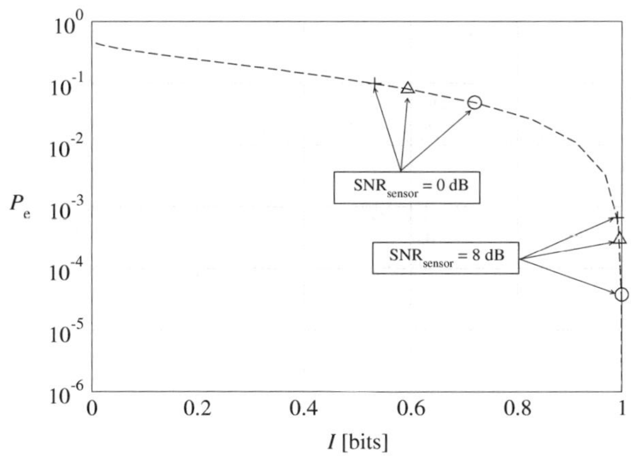


**Fig. 2.8** Probability of decision error, as a function of the mutual information, in a scenario with $n = 16$ sensors. The operating points for various clustering configurations and two sensor SNRs are shown

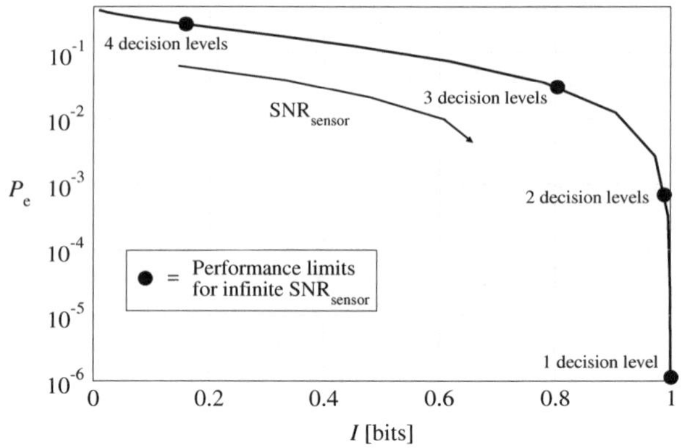


**Fig. 2.9** Probability of decision error, as a function of the mutual information, in a scenario with $n = 16$ sensors, uniform clustering, and noisy communication links ($p = 0.05$)

combining probability of decision error curves with mutual information curves, through the common parameter given by the sensor SNR. As one can see, the curves associated with different sensor network topologies overlap. In other words, for a given value of the mutual information, the probability of decision error is fixed. Note, however, that a specific mutual information is obtained in clustered (for example, 4-4-4-4 or 2-2-2-2-2-2-2-2) and non-clustered scenarios for different values of the sensor SNR (in the figure, a few representative points associated with two SNRs are highlighted). This means that, for a given mutual information, the presence of clustering leads to an *energetic loss* at the sensors (in the observation phase). The loss with non-uniform clustering is higher than with uniform clustering. Similar curves can be derived for the other scenarios considered in this chapter, e.g., for a large number of sensors, with more than two decision levels, and in the presence of noisy communication links between sensors and first-level FCs (with sufficiently low values of the noise level $p$). However, the network behavior does not change, i.e., for a fixed value of the mutual information, the probability of decision error is uniquely determined.

In Fig. 2.9, the probability of decision error is shown, as a function of the mutual information, in a scenario with $n = 16$ sensors and uniform clustering. Communication links between sensors and first-level FCs are noisy, with cross-over probability $p = 0.05$. The limiting (for $\text{SNR}_{\text{sensor}} \rightarrow \infty$) operating points over the $P_e - I$ curve of a BIBO sensor network, corresponding to all possible numbers of decision levels (1–4, respectively), are shown. For a given number of decision levels, the system operating point moves from the position corresponding to $I = 0$ (for very low values of $\text{SNR}_{\text{sensor}}$) to the corresponding limiting point, which is asymptotically approached for $\text{SNR}_{\text{sensor}} \rightarrow \infty$. As one can see, the presence of noise over the communication links limits the maximum achievable mutual information, i.e., the maximum information transfer rate across the network.

In [16], possible simplified expressions for the probability of decision error (as a function of the mutual information) are presented. In particular, (i) polynomial approximations, (ii) asymptotic (for sufficiently large sensor SNR) analytical expressions, and (iii) *bimodal* approximations (valid for all sensor SNRs) are derived.

## *2.1.5 Realistic Clustered Networks with Data Fusion*

In this subsection, we present *simulation* and *experimental* results which validate our analytical framework on distributed detection in practical sensor networking scenarios, where nodes comply with the IEEE 802.15.4 standard.

### 2.1.5.1 Simulations

The simulations have been carried out with the Opnet Modeler simulator [17] and a built-in IEEE 802.15.4 network model designed at the National Institute of Standards and Technologies (NIST) [18]. This model provides only the first two layers of the ISO/OSI stack, and we have extended it with a simple Opnet model for a FC, which, in addition to providing relaying functionalities, implements the intermediate data fusion mechanisms described in the previous subsections. Our Opnet model assumes strong line-of-sight communications between the sensors and the FCs, and between the FCs and the coordinator.

According to the theoretical analysis, the sensors make a noisy observation (affected by AWGN) of a randomly generated binary phenomenon $H$ and make local decisions on the status of the phenomenon. Subsequently, the sensors embed their decisions into proper data packets of length 216 bits,[3] which are sent either to the coordinator (in the absence of clustering) or to the first-level FCs (in the presence of clustering). The decisions are assumed to be either 0 (no phenomenon) or 1 (presence of the phenomenon). Obviously, if some packets are lost due to medium access collisions, decisions (either at the FCs or at the AP) are made only on the basis of the received packets (this leads to a reduced reliability of the decisions). If all the packets related to a set of observations of the same phenomenon are lost, instead, the final binary decision is random. Finally, if half of the decisions are in favor of one phenomenon status and the other half are in favor of the other, the coordinator decides for the presence of the phenomenon. More details about the implementation of the data fusion mechanism in Opnet can be found in [19].

[3] This length corresponds to a payload of 96 bits and a header of 120 bits introduced by physical and MAC layers.

In both scenarios, it is possible to evaluate, by simulation, the probability of decision error. Together with the probability of decision error, the simulator allows to evaluate (i) the *packet delivery fraction*, denoted as $\xi$ and defined as the ratio between the number of packets correctly delivered at the coordinator and the number of packets sent by the sensors, and (ii) the *delay*, defined as the time interval between the transmission instant and the reception instant of a generic packet. Results about the aggregate throughput [dimension: (pck/s)], defined as $S_{agg} = n \cdot g \cdot \xi$, where $n$ is the number of transmitting sensors and $g$ is the packet generation rate (set to 2 pck/s in all simulation results presented in the remainder of this subsection), can be found in [19]. Moreover, no acknowledgement (ACK) messages are used to confirm successful transmissions. In order to eliminate possible statistical fluctuations, each simulation performance point is obtained by averaging the results of ten Opnet simulation runs.

In Fig. 2.10, the packet delivery fraction and the delay are shown as functions of the number $n$ of transmitting sensors. These curves are obtained considering a fixed observation SNR at the sensors (equal to 0 dB). Our results, however, show that the packet delivery fraction and the delay are not affected by the value of the observation SNR at the sensors. We consider, in fact, ideal communication channels, so that only the observations at sensors are noisy, whereas the packets sent from the sensors to either an FC (clustered schemes) or the coordinator (non-clustered schemes) are received without errors. Consequently, the performance does not depend on the considered SNR, since packet delivery fraction and delay are network performance indicators and do not depend on the observation reliability. The packet delivery fraction (solid line with circles) decreases monotonically. In particular, for small values of $n$, it remains close to 1. When the number of transmitting nodes increases, instead, the number of collisions in the channel increases as well, and the packet delivery fraction reduces. In the same figure, the delay (dotted line with diamonds) is also shown. As the intuition suggests, the delay is short for small values of $n$. When the traffic

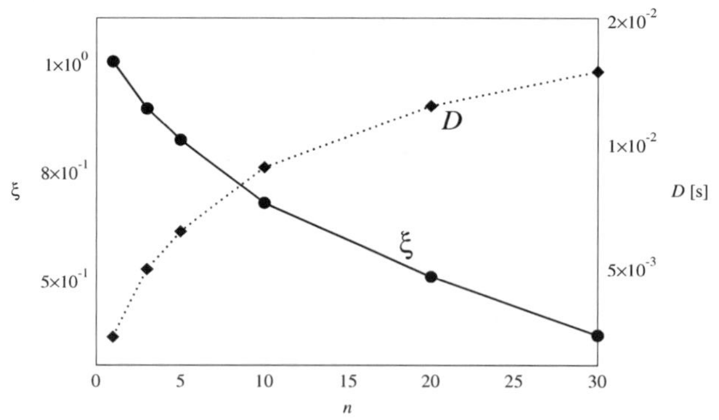


**Fig. 2.10** Performance analysis in a scenario without clustering: packet delivery fraction and delay performance as functions of the number $n$ of transmitting sensors

increases, instead, due to a larger number of collisions, the delay is longer, since the channel is busy for a longer period of time and the probability of finding the channel idle reduces. Finally, for large values of $n$, the delay seems to start saturating to a maximum value. In this case, in fact, due to the increased offered traffic, at least one sensor is likely to be ready to send its packet as soon as the channel becomes idle.

In Fig. 2.11, we analyze the impact of non-uniform clustering on the probability of decision error—as a performance benchmark, the probability of decision error in the case with uniform clustering is also shown. We consider scenarios with $n = 16$ sensors and the following network configurations: (i) no clustering, (ii) 8-8, (iii) 4-4-4-4 FCs, (iv) 14-1-1, (v) 10-2-2-2, and (vi) 8-2-2-2-2. According to the results in Fig. 2.11, the best performance is obtained in the absence of clustering, whereas the worst performance is obtained in the 14-1-1 scenario, i.e., with 3 FCs and non-uniform clustering. From Fig. 2.11, one can conclude that, in the presence of non-uniform clustering, the performance improves for relatively balanced clusters (as also predicted by the analytical framework). In this case, in fact, decisions made by intermediate FCs are more reliable, so the final decision made by the coordinator is more likely to be correct. In the case of uniform clustering, instead, the probability of decision error is *not* affected by the number of clusters in the network, as long as the total number of sensors remains the same: observing Fig. 2.11, one can note that the curves relative to the scenarios with four 4-sensor clusters and two 8-sensor clusters are almost overlapped. This is due to the fact that a smaller number of clusters is compensated by a higher quality of the intermediate decisions. This result is in agreement with the theoretical conclusions reached in Sect. 2.1.3.1. However, note that the performance in IEEE 802.15.4 scenarios worsens with respect to the analytical case, because the simulator takes into account the losses due to collisions. Since some packets may be lost, the probability of decision error is influenced by the collisions.

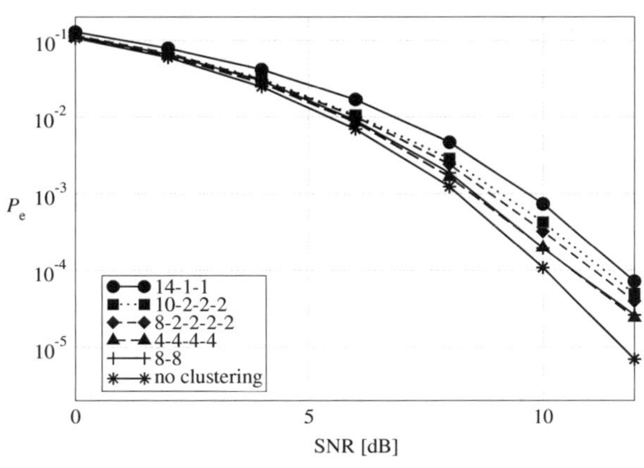


**Fig. 2.11** BER performance in scenarios with $n = 16$ sensors both in the case of uniform and non-uniform clustering. Various topologies (indicated in the figure) are considered

### 2.1.5.2 Experiments

In order to verify the predictions of the theoretical framework from an experimental perspective, we consider a networking set-up formed by MicaZ nodes [20]. MicaZ platforms include an ATmega128L 7.3 MHz micro-controller [21], FLASH and EEPROM memories, and a 2.4 GHz IEEE 802.15.4 Chipcon CC2420 radio-frequency transceiver [22]. The nodes' operating system is TinyOS. The experimental set-up is characterized by $n = 16$ nodes, organized in uniform clusters, with two and three decision levels, respectively. In our implementation, each node observes a "0" phenomenon and adds a Gaussian observation noise generated through the function "random" available in the TinyOS environment. According to the local decision threshold, each source node makes a decision on the observed phenomenon and embeds it in a packet to be transmitted. Since each TinyOS packet is formed by a payload of 30 bytes (the first byte contains the dimension and the following 29 the information data), we embed in each packet $29 \times 8 = 232$ consecutive binary decisions. This corresponds to 232 consecutive (time-wise) realizations of the observed binary phenomenon. The packets originated by the source nodes are then transmitted, through the intermediate FCs, to the AP. Note that a packet duration is of the order of 1 ms, and consecutive packet transmissions are separated by approximately 0.1 s. The transmit power is set to $-25$ dBm and the sensitivity threshold at the receivers is $-100$ dBm. The distances between communicating nodes (of the order of 2 m) are such that the received power is significantly higher than the sensitivity threshold. The data fusion mechanisms at the intermediate FCs and at the AP follows the majority decision rules described in the analytical framework.

The experimental Bit Error Rate (BER) performance is shown in Fig. 2.12. In the same figure, for comparison, we also show the corresponding theoretical results extracted from Fig. 2.4. As one can see, the experimental results are slightly worse than the theoretical ones (as observed also, in Sect. 2.1.5.1 for

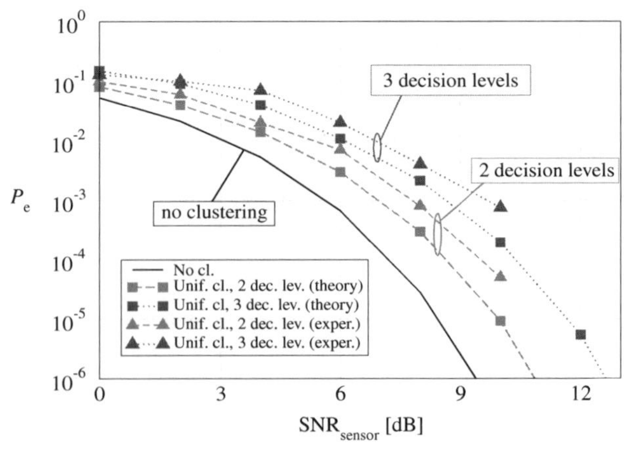


**Fig. 2.12** Experimental BER performance in scenarios with $n = 16$ sensors and uniform clustering. Two and three decision levels are considered

simulation results), but confirm the trend. This discrepancy is due to the more realistic experimental scenario, where some packets may get lost because of the wireless communication links. Since the decision rules at the FCs and at the AP do not adapt to the number of received observations, this explains the performance degradation. We point out that in our experiments the packet losses are typically *not* due to collisions, i.e., the traffic load of the considered network scenarios is too low to create problems at the access level. On the opposite, the performance degradation is due to losses of packets due to propagation reasons.

## 2.2 Extending the Lifetime of Clustered Sensor Networks

### *2.2.1 Sensor Network Lifetime under a Physical Layer QoS Condition*

In order to evaluate the sensor network lifetime, one needs first to define when the network has to be considered "alive." We assume that the network is "alive" until the QoS condition in (2.15) is satisfied. When a sensor in the network dies (e.g., there is a hardware failure or its battery exhausts), the probability of decision error increases since a smaller number of sensors is alive (see, for instance, Fig. 2.6). Moreover, the presence of a specific clustering configuration might make the process of network death faster. More precisely, the network dies when the desired QoS condition (2.15) is no longer satisfied, as a consequence of the death of a *critical sensor*. Therefore, the network lifetime corresponds to the lifetime of this critical sensor. Obviously, the criticality of a sensor's death depends on the particular sequence of previous sensors' deaths.

On the basis of the considerations in the previous section, in order to estimate the *network* lifetime one, first, needs to consider a reasonable model for the *sensor* lifetime. We denote by $F(t) \triangleq \mathbb{P}\{T_{\text{sensor}} \leq t\}$ the Cumulative Distribution Function (CDF) of a sensor's lifetime $T_{\text{sensor}}$ (the same for all sensors) and we consider the following exponential distribution as representative:

$$F(t) = \left[1 - e^{-t/\mu}\right] U(t) \tag{2.18}$$

where $\mu$ is the mean of the exponential distribution, the time $t$ is measured in arbitrary units [dimension: (aU)]. We have chosen the distribution in (2.18) as a good model for a sensor lifetime [23, Ch. 8]. Note that the results presented here for an exponential distributions also hold for other sensor lifetime distributions [24].

As mentioned before, we are interested in analyzing the network behavior when the QoS condition (2.15) is satisfied. More precisely, in the following subsections we evaluate the sensor network lifetime in scenarios with (A) ideal reclustering and (B) no reclustering. The obtained results are then commented.

#### 2.2.1.1 Analysis with Ideal Reclustering

In the case of *ideal reclustering*, the network dynamically reconfigures its topology, immediately after a sensor death, in order to recreate a uniform configuration. Obviously, the time needed for rearranging the network topology depends on the specific strategy chosen in order to reconfigure correctly (according to the updated network configuration) the connections between the sensors and the FCs and those between the FCs and the AP. In Sect. 2.2.2, a simple reconfiguration strategy will be proposed.

Given a maximum tolerable probability of decision error $P_e^*$, one can determine the smallest number of sensors, denoted as $n_{\min}$, required to satisfy the desired QoS condition. For instance, considering Fig. 2.6 and fixing a maximum tolerable value $P_e^*$, one can observe that, for decreasing numbers of sensors, at some point the actual probability of decision error $P_e$ becomes higher than $P_e^*$. In other words, the probability of decision error is lower than $P_e^*$ if *at least*$n_{\min}$ sensors are alive or, equivalently, until $n_{\text{crit}} = n - n_{\min} + 1$ sensors die. Therefore, denoting as $T_{\text{net}}$ the network lifetime, one can write:

$$\mathbb{P}\{T_{\text{net}} \leq t\} = \mathbb{P}\{\text{at least } n_{\text{crit}} \text{ sensors have } T_{\text{sensor}} < t\}$$

where $T_{\text{sensor}}$ is the sensor lifetime (recall that this random variable has the same distribution for all sensors) with CDF $F(t)$. Since the lifetimes of different sensors are assumed independent, using the repeated trials formula, one obtains

$$\mathbb{P}\{T_{\text{net}} \leq t\} = \sum_{i=n_{\text{crit}}}^{n} \binom{n}{n_{\text{crit}}} [F(t)]^i [1 - F(t)]^{n-i}.$$

#### 2.2.1.2 Absence of Reclustering

In the previous subsection, we have analyzed the network evolution in an ideal scenario where the topology is dynamically reconfigured in response to a sensor death (e.g., because of the depletion of its battery or hardware failure). However, it might happen that the initial clustered configuration is fixed, i.e., the connections between sensors, FCs, and AP cannot be modified after a sensor death. In this case, the following question is relevant: is there an optimum initial topology which leads to longest network lifetime? In order to answer this question, we will analyze the network evolution in scenarios where there is no reclustering. The network is still considered dead when the QoS condition (2.15) is no longer satisfied.

In the absence of ideal reclustering, an analytical performance evaluation is not feasible, i.e., there does not exist a closed-form expression for the CDF of the network lifetime. In fact, the CDF depends on the particular network evolution, i.e., it depends on how the sensors die among the clusters in the network. Therefore, each sequence of sensors' deaths is characterized by a specific lifetime, and one needs to resort to simulations in order to extrapolate an average statistical characterization. The simulations are performed according to the following steps.

1. The lifetimes of all $n$ sensors are generated according to the chosen distribution and the sensors are randomly assigned to the clusters.
2. The sensors' lifetimes are ordered in an increasing manner.
3. After a sensor death, the network topology is updated.
4. The probability of decision error is computed in correspondence to the surviving topology determined at the previous point: if the QoS condition (2.15) is satisfied, then the evolution of the network continues from step 3, otherwise, step 5 applies.
5. The network lifetime corresponds to the lifetime of the last dead sensor.

In Fig. 2.13, the CDF of the network lifetime is shown, as a function of time, in a scenario with $n = 32$ sensors grouped, respectively, in 2, 4, and 8 clusters. The sensor SNR is set to 5 dB and the maximum tolerable probability of decision error is $P_e^* = 10^{-3}$. For comparison, the curve associated with ideal reclustering is also shown. One can observe that the larger is the number of clusters, the worse is the performance, i.e., the higher is the probability of network death. Moreover, the curve associated with two clusters is very close to that relative to ideal reclustering. In fact, in a scenario with only two clusters, the average number of sensors which die in each cluster is approximately the same and, consequently, the topology remains approximatively uniform.

In Table 2.2, the network lifetime corresponding to a CDF equal to 0.9 (i.e., an outage probability of 90%) is shown, assuming an *exponential* sensor lifetime (with $\mu = 1$ aU), for various clustering configurations and values of the maximum tolerable probability of decision error $P_e^*$. The number of sensors is $n = 64$. For comparison, the network lifetime with ideal reclustering is also shown. From the results in Table 2.2, the following observations can be carried out.

- For a small number of clusters (2 or 4), the lifetime reduction, with respect to a scenario with ideal reclustering, is negligible. This is to be expected from the results in Fig. 2.13 and is due to the fact that the sensors die "more or less"

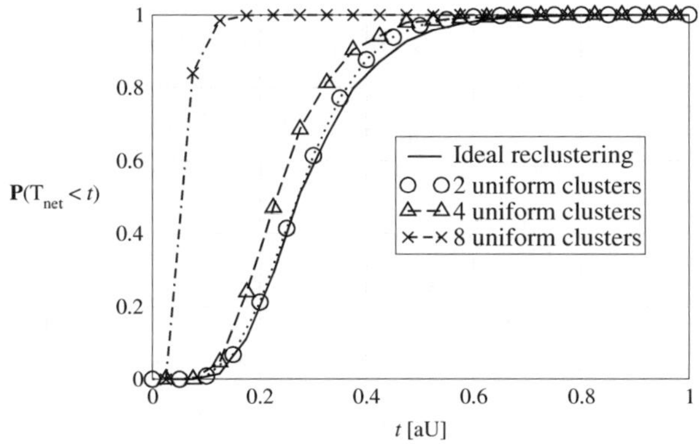


**Fig. 2.13** CDF of the network lifetime, as a function of time, in a scenario with $n = 32$ sensors, uniform clustering (with, respectively, 2, 4, and 8 clusters), and *absence of reclustering* (simulation results). The sensor SNR is set to 5 dB and the maximum tolerable probability of decision error is $P_e^* = 10^{-3}$. For comparison, the curve associated with ideal reclustering (analytical results) is also shown

**Table 2.2** Sensor network lifetime corresponding to an outage probability equal to 90% in a scenario with $n = 64$ sensors and $\mathrm{SNR}_{\mathrm{sensor}} = 5$ dB. Three values for the maximum tolerable probability of decision error $P_e^*$ are considered: (i) $10^{-2}$, (ii) $10^{-3}$, and (iii) $10^{-4}$. The mean parameter of the exponential distribution is $\mu = 1$ aU. All time values in the table entries are expressed in aU

| $P_e^*$ | Ideal reclustering | No reclustering (2 clusters) | No reclustering (4 clusters) | No reclustering (8 clusters) |
|---|---|---|---|---|
| $10^{-2}$ | 2.1 | 2.1 | 2.0 | 1.68 |
| $10^{-3}$ | 1.3 | 1.3 | 1.2 | 1.012 |
| $10^{-4}$ | 0.78 | 0.78 | 0.74 | 0.725 |

uniformly in all clusters. When the number of clusters increases beyond 4, the network lifetime starts to reduce appreciably. Therefore, our results show that, in the *absence of ideal reclustering*, the winning strategy to prolong network lifetime is to *form a few large clusters.*

- The impact of the QoS condition is very strong. In fact, when the QoS condition becomes more stringent (i.e., $P_e^*$ decreases), the network lifetime shortens, since a smaller number of sensor deaths is sufficient to violate this condition. On the other hand, if the QoS condition is less stringent, then a larger number of sensors have to die in order to violate it.
- The impact of the number of nodes on the network lifetime has not been directly analyzed. However, since the performance improves when the number of sensors increases (as shown in Fig. 2.6), one can conclude that, for a fixed QoS condition, a network with a larger number of sensors will satisfy the QoS condition for a longer time and, therefore, the network lifetime will be prolonged. Equivalently, one can impose a stronger QoS condition (a lower value of $P_e^*$), still guaranteeing the same network lifetime.

### *2.2.2 Analytical Computation of Network Lifetime*

In Sect. 2.2.1, we have analyzed the network performance without taking into account the *cost* of reclustering. In this subsection, instead, we investigate, from an analytical viewpoint, the cost of the used reclustering protocol in terms of its impact on the sensor network lifetime. In order to evaluate the cost of reclustering, one first needs to detail a reclustering protocol. We remark that we limit ourselves mainly (but not only) to scenarios with two (big) clusters, since they are associated with the minimum loss, in terms of probability of decision error at the AP, with respect to the scenario with the absence of clustering.

The reclustering protocol which will be used can be characterized as follows.

1. When an FC senses that a sensor belonging to its cluster is dead, e.g., when it does not receive packets from this sensor, it sends a control message, referred to as "ALERT," to the AP.

2. Assuming that the AP is aware of the current network topology, when it receives an ALERT message, it decides if reclustering has to be carried out. If so, the optimized network topology is determined.
3. If no reclustering is required, the AP sends to both FCs an "OK" message to confirm the current topology. On the other hand, if reclustering has to be carried out, another message, referred to as "CHANGE" and containing the new topology information, is sent to the FCs. In the latter case, the FCs send the CHANGE message also to sensors in order to allow them to communicate with the correct FC from then on.
4. If reclustering has happened, the sensors retransmit their previous packet to the FCs according to the new topology and a new data fusion is carried out at the AP.

In Fig. 2.14, the behavior of this simple protocol is pictured in an illustrative scenario with $n = 11$ sensors and two clusters (with 6 and 5 sensors, respectively). The control messages associated with solid lines are exchanged in the absence of reclustering, whereas the messages associated with dashed lines are exchanged in the presence of reclustering.

In order to derive a simple analytical framework for evaluating the sensor network lifetime, the following assumptions are expedient.

(a) The observation frequency, referred to as $f_{obs}$, is sufficiently low to allow regular transmissions from the sensors to the AP and, if necessary, the applicability of the reclustering protocol (this is reasonable for scenarios where the status of the observed phenomenon does not change rapidly).
(b) Transmissions between sensors and FCs and between FCs and AP are supposed instantaneous (this is reasonable, for example, if FCs and AP are connected through wired links or very reliable wireless links).
(c) Data processing and topology reconfiguration are instantaneous (this is reasonable if the processing power at the AP is sufficiently high).
(d) There is perfect synchronization among all nodes in the network (this is a reasonable assumption if nodes are equipped with synchronization devices, e.g., global positioning system).

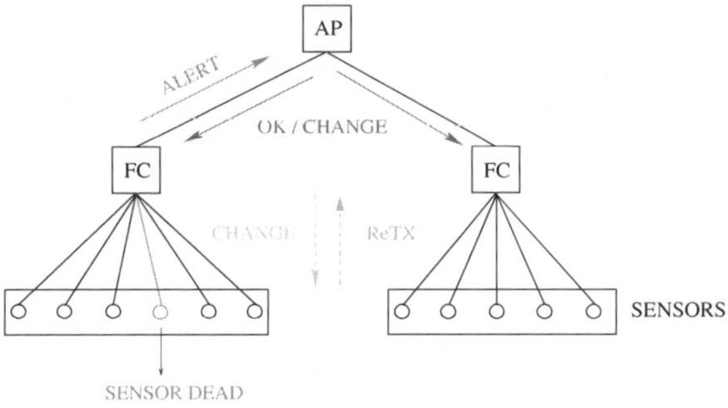


**Fig. 2.14** Message exchange in the proposed reclustering protocol. A network scenario with $n = 11$ sensors and two clusters (with 6 and 5 sensors, respectively) is considered. The control messages evolution follows the death of a sensor

The proposed reclustering algorithm and the assumptions above might look too simplistic for a realistic wireless sensor network scenario. However, they allow to obtain significant insights about the cost, in terms of network lifetime, of adaptive reclustering.

We preliminary assume that the duration of a data packet transmission has no influence on the lifetime of a single sensor—a more accurate analysis, which takes properly into account the actual duration of a data transmission, will be proposed in Sect. 2.2.4. In this case, the network lifetime can be written as

$$T_{\text{net}} = \sum_{i=1}^{n_{\text{crit}}} T_{\text{d},i}$$

where $n_{\text{crit}}$ has been introduced in Sect. 2.2.1.1 and $T_{\text{d},i}$ is the time interval between the $(i-1)$th sensor death and the $i$th sensor death. Obviously, $T_{\text{d},1}$ is the time interval until the death of the first sensor and can be written as

$$T_{\text{d},1} = \min_{j=1,\dots,n} \{T_j\} \tag{2.19}$$

where $T_j$ is the lifetime of the $j$th sensor. Since $T_{\text{net}}$ is a r.v., one could determine its statistics (e.g., the CDF). However, in order to concisely characterize the impact of reclustering, it is of interest to evaluate its average value, i.e.,

$$\mathbb{E}[T_{\text{net}}] = \mathbb{E}\left[\sum_{i=1}^{n_{\text{crit}}} T_{\text{d},i}\right]. \tag{2.20}$$

#### 2.2.2.1 Absence of Reclustering

In this case, $n_{\text{crit}}$ and $\{T_{\text{d},i}\}$ in (2.20) are independent r.v.s. In fact, they depend on the sensors' lifetime distribution and on the particular evolution (due to the nodes' deaths) of the network topology. Therefore, the sum in (2.20) is a stochastic sum. Using the conditional expectation theorem and the fundamental theorem of probability [9], one can write

$$\mathbb{E}\left[\sum_{i=1}^{n_{\text{crit}}} T_{\text{d},i}\right] = \sum_{j=1}^{n} \mathbb{P}\{n_{\text{crit}} = j\} \sum_{i=1}^{j} \mathbb{E}\left[T_{\text{d},i}\right].$$

At this point, one needs to resort to simulations to compute the probabilities $\{P(n_{\text{crit}} = j)\}$. In fact, they strongly depend on the particular network evolution before its death.

#### 2.2.2.2 Ideal Reclustering

In Sect. 2.2.1, we have shown that the presence of ideal reclustering leads to an upper bound on the network lifetime, i.e., it tolerates the maximum number of sensors' deaths before the network dies. This bound can be analytically evaluated using (2.20) and replacing $n_{\mathrm{crit}}$ with the value $n_{\mathrm{crit}}^{\mathrm{R}}$ defined as follows:

$$n_{\mathrm{crit}}^{\mathrm{R}} = \min_{n_{\mathrm{crit}}^{*}=1,\ldots,n} \{P_{\mathrm{e}}(\text{after } n_{\mathrm{crit}}^{*} \text{ sensors' deaths}) \geq P_{\mathrm{e}}^{*}\}.$$

The value of $n_{\mathrm{crit}}^{\mathrm{R}}$ can be determined by numerical inversion of the QoS condition. Therefore, an upper bound for the network lifetime can be expressed as

$$\mathrm{UB}_{T_{\mathrm{net}}} \triangleq \mathbb{E}\left[T_{\mathrm{net}} | n_{\mathrm{crit}} = n_{\mathrm{crit}}^{\mathrm{R}}\right] = \sum_{i=1}^{n_{\mathrm{crit}}^{\mathrm{R}}} \mathbb{E}\left[T_{\mathrm{d},i}\right]. \tag{2.21}$$

In this case, one can observe that the sum in (2.21) is deterministic and, therefore, can be analytically evaluated through the computation of $\{\mathbb{E}\left[T_{\mathrm{d},i}\right]\}$. In [24], it is shown that this upper bound is equal to

$$\mathrm{UB}_{T_{\mathrm{net}}} = \frac{\mu}{n} + \sum_{i=2}^{n^{\mathrm{R}}} \mu \frac{n-i}{(n-i+1)^2}. \tag{2.22}$$

Similarly, we can derive a lower bound on the network lifetime. This bound, for a fixed number of sensors, is obtained when all sensors' deaths occur in the same cluster. In this way, for a fixed topology, the highest possible probability of decision error is obtained at each instant and, consequently, the corresponding network lifetime is the shortest possible. This bound can be expressed as

$$\mathrm{LB}_{T_{\mathrm{net}}} \triangleq \mathbb{E}\left[T | n_{\mathrm{crit}} = n_{\mathrm{crit}}^{\mathrm{LB}}\right] = \frac{\mu}{n} + \sum_{i=2}^{n_{\mathrm{crit}}^{\mathrm{LB}}} \mu \frac{n-i}{(n-i+1)^2}. \tag{2.23}$$

Expression (2.23) for $\mathrm{LB}_{T_{\mathrm{net}}}$ is derived from (2.22) by replacing $n_{\mathrm{crit}}^{\mathrm{R}}$ with $n_{\mathrm{crit}}^{\mathrm{LB}}$, which is obtained through simulations, since it depends on the network evolution. The value of $\mathrm{LB}_{T_{\mathrm{net}}}$ is smaller than that of $\mathrm{UB}_{T_{\mathrm{net}}}$, since $n_{\mathrm{crit}}^{\mathrm{R}} > n_{\mathrm{crit}}^{\mathrm{LB}}$. As previously mentioned, we consider an initial topology with two big clusters. In fact, this scenario allows to obtain the lowest probability of decision error at each instant, because the network topology is less unbalanced than in scenarios with a higher number of clusters, e.g., 8. Therefore, evolution of the lower bound (2.23) in correspondence to a scenario with two clusters leads to the tightest possible lower bound with respect to a scenario with no reclustering.

Finally, one needs to evaluate the extra time required by the application of the reclustering procedure. We will refer to this quantity as $T_{\mathrm{R}}$. Under the given assumptions and since the probability that reclustering has happened is equal to 1/2 (the derivation of this probability is summarized in [24]), $T_{\mathrm{R}}$ can be expressed as

$$T_{\mathrm{R}} = (n_{\mathrm{crit}}^{\mathrm{R}} - 1) T_{\mathrm{RECL}}$$

where $T_{\text{RECL}}$ represents the time required by a single reclustering operation. The duration of this time interval cannot be a priori specified, since it depends on the dimensions of the OK, CHANGE, and ALERT messages, the data-rate, and other network parameters. It is reasonable to assume that the longer is the average sensor lifetime $\mu$, the shorter should be (proportionally) $T_{\text{RECL}}$. In other words, one could assume $T_{\text{RECL}} = c \cdot \mu$, where $c$ is small if $\mu$ is large and vice versa. In general, $c$ can be chosen to model accurately the situation of interest.

Finally, one can define a *time penalty* as the ratio between the time necessary for the application of the reclustering protocol and the total time, given by the sum of reclustering and "useful" times (i.e., the time spent for data transmission). It follows that:

$$P^{\text{time}} = \frac{T_{\text{R}}}{T + \mathbb{E}[T_{\text{net}}]} = \frac{(n_{\text{crit}} - 1)T_{\text{RECL}}}{(n^{\text{R}}_{\text{crit}} - 1)T_{\text{RECL}} + \frac{\mu}{n} + \sum_{i=2}^{n^{\text{R}}_{\text{crit}}} \mu \frac{n-i}{(n-i+1)^2}}. \tag{2.24}$$

After a few mathematical passages, from (2.24) one obtains

$$P^{\text{time}} \gtrsim \frac{(n - k^* - 1)c}{(n - k^* - 1)c + \frac{1}{n} + \ln(n-2) - \ln(k^* - 1)} \tag{2.25}$$

where we have used the fact that $\sum_{i=1}^{m} 1/i \simeq \ln m + 0.577$ [25].

From (2.25) and owing to the fact that $k^*$ is approximately constant, one can analytically show that

$$\lim_{n\to\infty} P^{\text{time}} \simeq 1 \quad \forall c.$$

In other words, if the number of sensors is large, for a fixed value of $c$ the proposed reclustering algorithm does not guarantee a limited time penalty. Similarly, one can show that

$$\lim_{c\to 0} P^{\text{time}} \simeq 0 \quad \forall n.$$

In other words, for a fixed number of nodes the reclustering protocol is effective, using the algorithm proposed in Sect. 2.2.2, *provided that* the duration of a single reclustering operation is sufficiently short (e.g., very small control packets are used).

### *2.2.3 Numerical Results*

In Fig. 2.15, numerical results based on the application of the analytical framework derived in Sects. 2.2.2.1 and 2.2.2.2 are shown. In particular, (a) the average network lifetime $\mathbb{E}[T_{\text{net}}]$ and (b) the critical number of deaths $n_{\text{crit}}$ are shown as functions of the number of sensors $n$. The average network lifetime in a scenario with no reclustering (for various numbers of clusters) is compared with the upper and lower bounds derived in Sect. 2.2.2.2. The QoS condition is associated with $P^*_{\text{e}} = 10^{-3}$

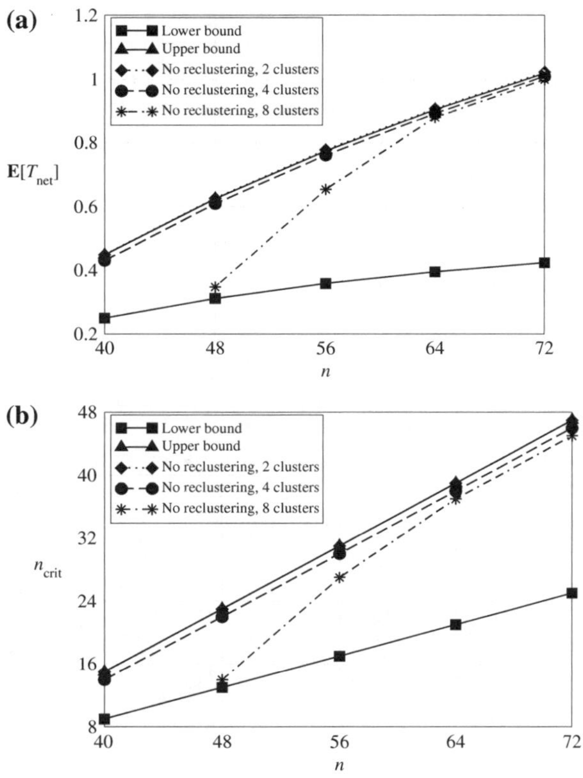


**Fig. 2.15** Sensor network performance using the proposed reclustering algorithm: **a** network lifetime and **b** critical number of deaths, as functions of the number of sensors. The performance in the absence of reclustering (with 2, 4, and 8 clusters, respectively) is compared with the proposed upper bound $\mathrm{UB}_{T_{net}}$ and lower bound $\mathrm{LB}_{T_{net}}$. The QoS condition is $P_e^* = 10^{-3}$ and the sensor SNR is set to 5 dB. The average sensor lifetime is $\mu = 1$

and the sensor SNR is set to 5 dB. In order to compare these results with those in Sect. 2.2.1.2, the distribution of the sensors' lifetime is assumed to be exponential with $\mu = 1$ aU. From the results in Fig. 2.15a, one can observe that, when the number of sensors increases, also the network lifetime becomes longer, since a larger number of sensors' deaths have to occur in order to violate the QoS condition. This is confirmed in Fig. 2.15b, where the critical number of sensors' deaths is shown as a function of the number of sensors. Moreover, as expected, the sensor network lifetime in the absence of reclustering is shorter than in the presence of ideal reclustering (with the proposed reclustering protocol), since the network topology becomes less and less uniform and, therefore, the probability of decision error becomes higher and higher. As previously shown in Fig. 2.13, when the initial number of clusters is equal to two, the network lifetime with no reclustering is very close to that corresponding to the application of the reclustering protocol. This is due to the fact that the sensors' deaths are, on average, equally distributed among the two clusters, i.e., there is a sort of "natural" reclustering. Finally, one can observe that when the number of clusters in the initial topology increases (e.g., is equal to 8) the network lifetime drastically reduces for *small* values of the number of sensors, since it is more difficult to satisfy the QoS condition. However, it is interesting to observe that for sufficiently large values of $n$, the lifetime penalty incurred by the presence of a large number of clusters is negligible, suggesting that there may exist a minimum cluster dimension which guarantees acceptable performance. This is probably due to the fact that when the number of sensors is sufficiently large, the cluster dimension is also sufficiently large and, consequently, its lifetime is longer. Therefore, the lifetime of the entire sensor network is longer, since the network topology is less unbalanced.

## 2.2.4 *Energy Budget*

The analysis of the reclustering cost provided in Sect. 2.2.2 is ideal, since it does not consider the energy spent by the nodes in the network. Although this assumption is reasonable for the FCs and the AP,[4] this is not realistic for remote nodes (sensors) which need to rely on batteries with limited energy. Moreover, there is a delay associated with a packet transmission. In this subsection, the realistic network energy consumption is evaluated in the presence of ideal reclustering, using the reclustering protocol proposed in Sect. 2.2.2. In order to analyze this energy consumption, we will refer to a commercial WSN with a communication protocol based on the IEEE 802.15.4 standard (also considered in Sect. 2.2.6) [26]. In particular, while the first analysis does not take into account the energy of the sensor battery, we then show the impact of a limited battery energy at the sensors.

### 2.2.4.1 Analysis with Infinite Energy Battery at the Sensors

The energetic cost, for a single sensor, of the application of our reclustering algorithm can be written as

$$C_{\text{tot}}^{\text{en}} = P_{\text{t}}\, C_{\text{tot}}^{\text{time}} \tag{2.26}$$

where $C_{\text{tot}}^{\text{en}}$ is the total cost in terms of energy spent by a sensor, $P_{\text{t}}$ is the transmit power at each sensor, and $C_{\text{tot}}^{\text{time}}$ is the total time cost associated with packet transmission. After a few manipulations, the total energetic cost can be written as [24]

$$C_{\text{tot}}^{\text{en}} = P_{\text{t}} \Bigg\{ \underbrace{\frac{1}{2}\left[\frac{L_{\text{cont}} + L_{\text{data}}}{R_{\text{b}}}\right]\left(n_{\text{crit}}^{\text{R}} - 1\right)}_{\substack{\text{Costfortransmissionof} \\ \text{controlpackets}\,:\, C_{\text{R}}^{\text{time}}}} + \underbrace{\frac{L_{\text{data}}}{R_{\text{b}}} f_{\text{obs}} \sum_{i=1}^{n_{\text{crit}}^{\text{R}}} \mathbb{E}\left[\mathbb{T}_{\text{d},\beth}\right]}_{\substack{\text{Costfortransmission} \\ \text{ofdatapackets}\,:\, C_{\text{data}}^{\text{time}}}} \Bigg\} \tag{2.27}$$

---

[4] In fact, they may be placed by the network designer so that they can be power-supplied.

where $R_b$ is the data-rate [dimension: (bit/s)], $L_{cont}$ and $L_{data}$ are, respectively, the length of a control packet and data packet [dimension: (b/pck)], and $f_{obs}$ is the observation frequency. Expression (2.27) for the energetic cost represents the total energy spent by any of the $n - n_{crit}^{R}$ surviving sensors after the network death. Obviously, this energetic cost represents a worst case, since there are $n_{crit}^{R}$ nodes (i.e., those which die while the network is still alive) which spend a smaller amount of energy in their shorter lifetimes. An average cost per sensor can be easily computed using the same approach proposed above. In [24], the following expression for the average energy cost is derived:

$$\begin{aligned}\overline{C}_{tot}^{en} &= P_t(\overline{C}_{R}^{time} + \overline{C}_{data}^{time}) \\ &= P_t\left\{\frac{(n_{crit}^{R} - 1)(L_{data} + L_{cont})}{4R_b} + \frac{L_{data}f_{obs}}{R_b n}\sum_{i=1}^{n_{crit}^{R}}\left((n - n^{R})\mathbb{E}\left[T_{d,i}\right] + \sum_{j=1}^{i}\mathbb{E}\left[T_{d,j}\right]\right)\right\}.\end{aligned} \tag{2.28}$$

Similarly to (2.24), we define the following *energy penalties*:

$$P^{en-1} \triangleq \frac{C_{R}^{en}}{C_{tot}^{en}} = \frac{C_{R}^{time}}{C_{R}^{time} + C_{data}^{time}} \tag{2.29}$$

$$P^{en-2} \triangleq \frac{\overline{C}_{R}^{en}}{\overline{C}_{tot}^{en}} = \frac{\overline{C}_{R}^{time}}{\overline{C}_{R}^{time} + \overline{C}_{data}^{time}} \tag{2.30}$$

where $P^{en-1}$ is the *worst-case* penalty (associated with a sensor which survives until the end) and $P^{en-2}$ is the *average-case* penalty (associated with the average energetic costs among all sensors in the network). As mentioned before, the energy penalties (2.29) and (2.30) take into account, with respect to (2.24), realistic network parameters, such as $L_{data}$, $f_{obs}$, $R_b$, and $P_t$.

In Fig. 2.16, the energy penalty is shown, as a function of the number of sensors $n$ and the observation frequency $f_{obs}$, in the two cases previously highlighted: (a) *worst-case* energy consumption (obtained by using expression (2.29)) and (b) *average-case* energy consumption (obtained by using expression (2.30)). In order to compare the results in Fig. 2.16 with the results given in the previous subsections, we have set $P_e^* = 10^{-3}$ and $SNR_{sensor} = 5$ dB. Realistic values for the network parameters, provided by the IEEE 802.15.4 standard, correspond to $P_t = 1$ mW, $R_b = 250$ Kbit/s, $L_{data} = 1024$ bit/pck, and $L_{cont} = 80$ bit/pck.[5] One can note that for small values of the observation frequency (*rare observations*), the performance worsens since the network spends more time in reclustering than in

[5] In our analysis, we use the maximum possible data-rate allowed by the IEEE 802.15.4 standard, i.e., $R_b = 250$ Kbit/s. However, our experimental results show that only a maximum value $R_b = 25$ Kbit/s can be achieved by practical sensor networks [27]. Moreover, the length of data packets is the maximum allowed by the standard.

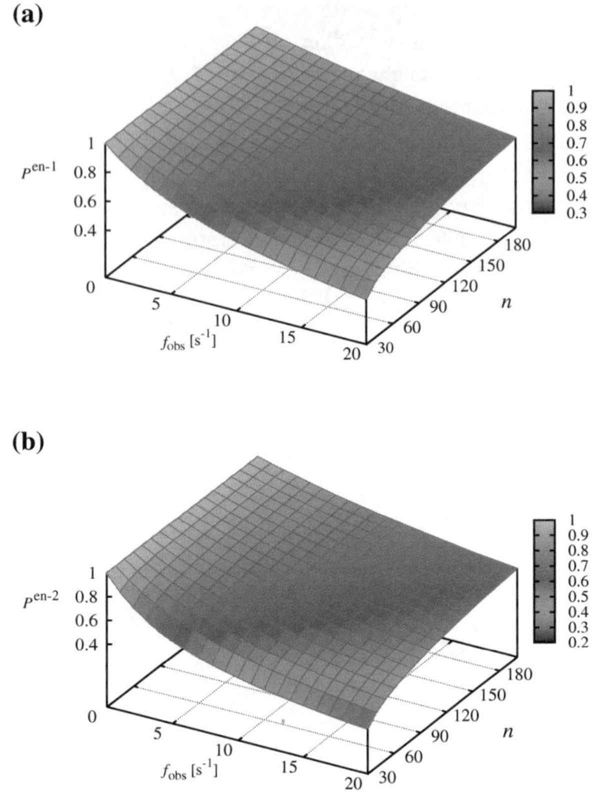


**Fig. 2.16** Energy penalty, associated with the reclustering protocol, as a function of both the observation frequency $f_{\text{obs}}$ and the number of sensors $n$. Two possible cases are considered: **a** *maximum* penalty (associated with a sensor which survives until the end) and **b** *average* penalty (among all the sensors in the network)

transmitting useful data. For a fixed value of the number of sensors $n$, the following limits hold:

$$\lim_{f_{\text{obs}} \to 0} P^{\text{en}-1} = \frac{C_{\text{R}}^{\text{en}}}{C_{\text{R}}^{\text{en}}} = 1 \qquad \lim_{f_{\text{obs}} \to 0} P^{\text{en}-2} = \frac{\overline{C}_{\text{R}}^{\text{en}}}{\overline{C}_{\text{R}}^{\text{en}}} = 1.$$

Besides, one can observe that for increasing values of the observation frequency (*frequent observations*), the performance is better. In fact, for a fixed number of sensors, there is a larger number of data transmissions from the sensors to the AP and the value of $D_{\text{R}}^{\text{en}}$ becomes increasingly negligible with respect to the value of $D_{\text{data}}^{\text{en}}$. Analytically, one can write

$$\lim_{f_{\text{obs}} \to \infty} P^{\text{en}-1} = \frac{1}{C_{\text{data}}^{\text{en}}} = 0 \qquad \lim_{f_{\text{obs}} \to \infty} P^{\text{en}-2} = \frac{1}{\overline{C}_{\text{data}}^{\text{en}}} = 0.$$

Note that a high value of the observation frequency might not be admissible. In fact, in Sect. 2.2.2 we have supposed that the inverse of the observation frequency is much smaller than the time necessary to complete a transmission to the AP and, eventually, the reclustering protocol (hypothesis (a) in Sect. 2.2.2).

#### 2.2.4.2 Analysis with Energy-Limited Battery at the Sensors

In the previous derivations, the proposed framework and the presented results have used arbitrary time units. However, it is of interest to map these arbitrary time units into realistic units. In order to do so, we assume that a node is equipped with a limited-energy battery with initial energy $E_{\text{battery}}$ [dimension: (J)]. When a sensor battery energy exhausts, the sensor dies and, consequently, the network is closer to breaking the QoS condition. The average sensor lifetime [dimension: (s)] can be expressed as

$$\mathbb{E}[T_{\text{sensor}}] = \frac{E_{\text{battery}}}{\overline{P}}$$

where $\overline{P}$ is the average power depleted at the node [dimension: (W)]. In a realistic wireless sensor network (e.g., IEEE 802.15.4 wireless sensor networks [26]), four *states* are admissible at the node: (1) *transmission*, (2) *reception*, (3) *idle*, and (4) *sleep*. In this case, the average power depleted at the node is given by

$$\overline{P} = \sum_{i=1}^{4} P_i p_i \tag{2.31}$$

where $P_i$ and $p_i$ $(i = 1, 2, 3, 4)$ are, respectively, the power consumption in the $i$th state and the probability that the sensor is in the $i$th state—note that $P_1 = P_t$. Typically, in a IEEE 802.15.4 wireless sensor network $P_4 \ll 1$ and $p_2 \ll p_3, p_1$ [28]. Therefore, the average depleted power in (2.31) can be written as

$$\overline{P} \simeq P_1 p_1 + P_2 p_2$$

where $p_2 = 1 - p_1$ and $P_1 = P_2 = P_t$ [28]. Therefore, the average consumed power in (2.31) becomes

$$\overline{P} = P_t$$

and it follows that

$$\mathbb{E}[T_{\text{sensor}}] = \frac{E_{\text{battery}}}{P_t}. \tag{2.32}$$

Using the value of $\mathbb{E}[T_{\text{sensor}}]$ given in (2.32) for the computation of $C_{\text{tot}}^{\text{time}}$ according to the framework derived in Sect. 2.2.4.1, the lifetime of a realistic IEEE 802.15.4 wireless sensor network, with the parameters used to derive the results in Fig. 2.16, can be obtained. The sensor network lifetime values, associated with different battery energies at the sensors (typical for practical applications), are summarized in Table 2.3. In particular, a scenario with $n = 64$ sensors, $P_t = 1$ mW, and $f_{\text{obs}} = 20\,\text{s}^{-1}$ is considered. One can observe that the theoretical results given in Sect. 2.2.3 are confirmed also in a more realistic IEEE 802.15.4 WSN. However, note that for $n = 64$ sensors the network lifetime in the ideal

**Table 2.3** Sensor network lifetime for a realistic IEEE 802.15.4 wireless sensor network in a scenario with $n = 64$ sensors, $P_t = 1$ mW, and $f_{obs} = 20\,s^{-1}$. The IEEE 802.15.4 parameters are the same considered in Fig. 2.16. Different values of the battery energy at a sensor are considered

| Battery energy $E_{battery}$ (kJ) | Average sensor lifetime $\mathbb{E}[T_{sensor}]$ (days) | Sensor network lifetime $C_{tot}^{time}$ (days) |
|---|---|---|
| 12.96 (400 mAh, 9 V) | 150 | 196 |
| 19.44 (600 mAh, 9 V) | 224 | 294 |
| 31.68 | 365 | 480 |
| 32.4 (1 Ah, 9 V) | 375 | 491 |

scenario is shorter than $\mathbb{E}[T_{sensor}]$, whereas it is longer in a realistic scenario. This behavior is due to the fact that our theoretical framework does not consider the delay associated with packet transmissions, as considered, instead, in the performance analysis for an IEEE 802.15.4 network.

### *2.2.5 Noisy Communication Links*

The analysis of the sensor network lifetime proposed in Sect. 2.2.2 is quite general and, in particular, no assumption has been made on the communication links. However, the results presented in Sect. 2.2.3 are obtained under the assumption of *ideal* communication links. In a scenario with noisy communication links, two main differences, with respect to a scenario with ideal communication links, can be observed:

- for a given value of the sensor SNR, the presence of noisy communication links leads to a performance loss (i.e., higher probability of decision error);
- a probability of decision error floor can be visualized for high values of the sensor SNR.

These differences between the scenarios with ideal communication links and those with noisy communication links imply that the network lifetime will be shorter, since the QoS condition will be satisfied for a shorter time. Moreover, the presence of a probability of decision error floor implies that, for a given value of the sensor SNR, the QoS condition might never be satisfied. These considerations suggest that the QoS condition and the operating sensor SNR, for a given value of the number of sensors $n$, have to be properly chosen.

In Fig. 2.17, the CDF of the network lifetime is shown, as a function of time,[6] in a scenario with $n = 64$ sensors, uniform clustering, and noisy communication links.

[6] We recall that the time is measured, here, in arbitrary units. For more realistic scenarios, see the considerations at the end of Sect. 2.2.4.

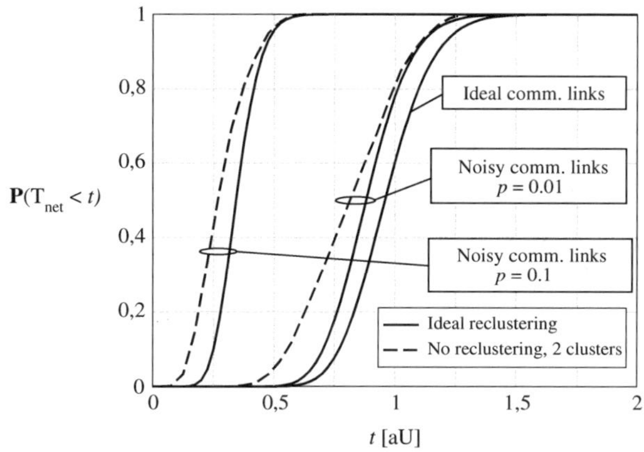


**Fig. 2.17** CDF of the network lifetime, as a function of time, in a scenario with $n = 64$ sensors, uniform clustering, and noisy communication links. Two possible values for the cross-over probability are considered: (i) $p = 0.1$ and (ii) $p = 0.01$. The sensor SNR is set to 5 dB and the maximum tolerable probability of decision error is $P_e^* = 10^{-3}$

Two possible values for the cross-over probability are considered: (i) $p = 0.1$ and (ii) $p = 0.01$. For comparison, the curve associated with ideal communication links is also shown. The sensor SNR is set to 5 dB and the maximum tolerable probability of decision error is $P_e^* = 10^{-3}$. One can observe that the higher the noise intensity in the communication links is, the higher the CDF of the network lifetime becomes. In fact, in this case the transfer of information from the sensors to the AP is less reliable and, consequently, the probability of decision error becomes higher and higher and the QoS condition can be guaranteed for a shorter time. As in a scenario with ideal communication links, the presence of reclustering prolongs the network lifetime with respect to a scenario with no reclustering. Obviously, for a given reclustering strategy a scenario with ideal communication links corresponds to a longer network lifetime, since the probability of decision error is the lowest possible.

### *2.2.6 Throughput and Delay with Varying Sensor Network Lifetime*

In this subsection, we evaluate the performance of a realistic IEEE 802.15.4 WSN subject to nodes' failures. In order to carry out this analysis, we resort, as in Sect. 2.1.5, to simulations using Opnet Modeler 11.5 [17] and a built-in model for IEEE 802.15.4 networks, provided by the NIST [18]. The network performance (in terms of number of transmitted packets, throughput, and delay) is analyzed in scenarios with no clustering (and, therefore, no reclustering). The goal of this subsection is to show the impact of different QoS conditions (given in terms of the required percentage of nodes' deaths which makes the network die) on different network performance indicators (e.g., throughput and delay). For the performance in the presence of relaying, the reader is referred to [29]. As discussed in Sect. 2.2.1, the performance of sensor networks with no clustering can be

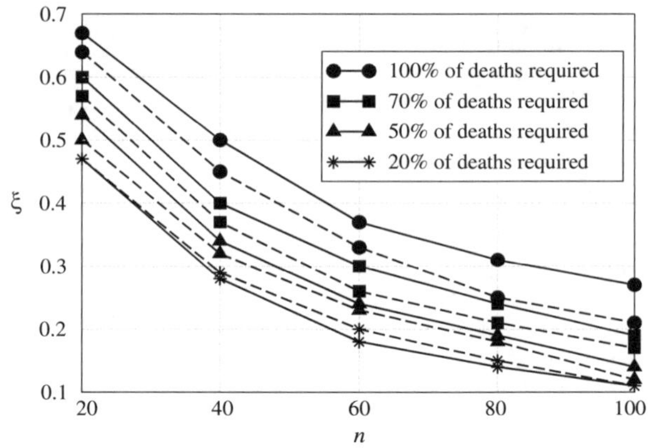


**Fig. 2.18** Packet delivery fraction, as a function of the number of sensors $n$, in an IEEE 802.15.4 WSN with nodes' failures. Two possible distributions for a single sensor lifetime are considered: (**a**) exponential with $\mu = 300$ s (*solid lines*) and (**b**) uniform with $t_{max} =$ 600 s (*dashed lines*)

considered, from a network lifetime viewpoint, as a lower bound, since the probability of decision error is lower than in scenarios with clustering. In the simulations, the following parameters are considered: $R_b = 250$ Kbit/s, $L_{data} = 994$ bit/pck, and $g = 0.236$ s, where $g$ is the packet interarrival time at the sensors. Moreover, no transmission of ACK packets is considered from the AP to the remote nodes. In all presented results, four QoS conditions will be considered: (i) network death corresponds to 100% of sensors' deaths (i.e., the network survives until there is a single sensor alive), (ii) network death corresponds to 70% of sensors' deaths, (iii) network death corresponds to 50% of sensors' deaths, and (iv) network death corresponds to 20% of sensors' deaths.

In Fig. 2.18, the packet delivery fraction is shown, as a function of the number of sensors $n$, for two possible distributions of a single sensor lifetime: (a) exponential with $\mu = 300$ s (solid lines) and (b) uniform with $t_{max} = 600$ s (dashed lines). First, one can observe that the more stringent is the QoS condition, the lower is the throughput. In fact, a smaller number of transmissions is possible (since the network lifetime is shorter) and a larger number of collisions happens, because there is a large number of sensors that try to transmit to the AP and a larger number of packets is lost. Moreover, a scenario with uniform distribution of the sensors' lifetime has a lower throughput with respect to a scenario with exponential distribution, since more packets are lost due to the collisions.

In Fig. 2.19, the average MAC delay[7] $D$ over all the received packets is shown, as a function of the number of sensors $n$, for two possible distributions of a single sensor lifetime: (a) exponential with $\mu = 300$ s (solid lines) and (b) uniform with $t_{max} = 600$ s (dashed lines). Similarly to what happens to the throughput in Fig. 2.18, a larger number of collisions also causes a higher delay in receiving the packets. Therefore, scenarios with a uniform distribution of the sensors' lifetimes are characterized by a higher delay with respect to scenarios with an exponential

[7] The average MAC delay corresponds to the delay averaged over all packets which are correctly received at the MAC level during the Opnet simulations.

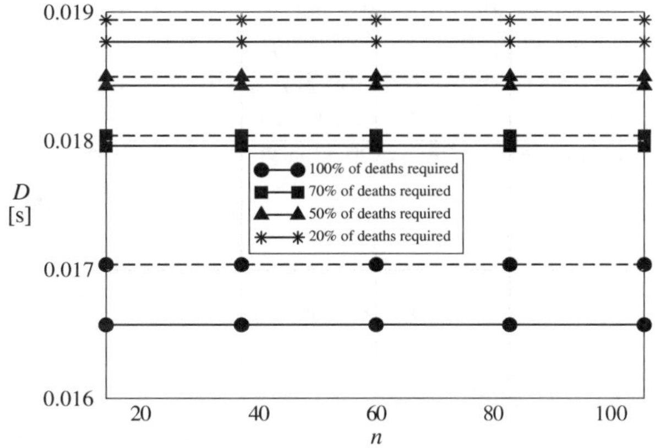


**Fig. 2.19** Average MAC delay $D$, as a function of the number of sensors $n$, in an IEEE 802.15.4 WSN with nodes' failures. Two possible distributions for a single sensor lifetime are considered: **a** exponential with $\mu = 300$ s (*solid lines*) and **b** uniform with $t_{\max} = 600$ s (*dashed lines*)

distribution. In this case as well, however, the more stringent is the QoS condition, the higher is the average MAC delay. Finally, the average MAC delay does not depend on the number of sensors, for a fixed QoS condition, since the number of surviving sensors is (almost) the same and, therefore, the average delay in the packet transmissions is constant.

## 2.3 Impact of Different SNRs at the Sensors

Consider now a generic scenario with different SNRs at the sensors. For the sake of simplicity, we consider a scenario with no clustering, i.e., direct communications between the sensors and the AP. In this case, a decision based on the majority-like fusion rule might not be the best choice. In fact, if a sensor is very noisy (i.e., its observation SNR is very small), its decision should be taken into account with a low level of reliability in the fusion process at the AP. Therefore, it would be reasonable to assign each sensor a weight proportional to its own SNR—this approach is similar to that proposed in [7], where the weights are assigned according to the link qualities. The AP could then make a final decision taking into account the weights assigned to the sensors. Note that the improvement, in terms of probability of decision error, comes at the price of a non-optimal network energy efficiency, since all sensors, even those with low SNR, have to send their decisions to the AP and waste the same amount of energy.

In the following, we consider a system where the AP takes into account the $n$ local sensor decisions with the same weight, i.e., without considering their SNRs, and adopts a majority-like decision rule. In order to take into consideration the sensor SNR profile, the threshold for local decision at each sensor is properly optimized, as explained in detail in Sect. 2.3.1.2.

We now derive analytical expressions for the probability of decision error, distinguishing between a scenario with *ideal* communication links and a scenario with *noisy* communication links. In [30], the reader might found an analytical expression for the probability of decision error also in the case when no quantization is carried out at the sensors, i.e., when sensors transmit their local likelihood values.

## 2.3.1 Ideal Communication Links

### 2.3.1.1 Probability of Decision Error

Consider the first conditional probability at the right-hand side of (2.5) and define the threshold value $k$ in the majority-like decision rule. There is an error, i.e., $\widehat{H} = H_1$ given that $H = H_0$, if $i \geq k$ sensors decide for $H_1$ when $H_0$ has happened. In this case, there can be $\binom{n}{i}$ combinations of sensors deciding for $H_1$. We denote as $\Omega_i(j)$ the $j$th possible combination ($j = 1, \ldots, \binom{n}{i}$) in a scenario where $i$ sensors are in error.[8] Therefore, the conditional probability of interest can be expressed as follows:

$$\mathbb{P}\{\widehat{H} = H_1|H_0\} = \sum_{i=k}^{n} \sum_{j=1}^{\binom{n}{i}} \left\{ \prod_{\ell=1}^{i} \mathbb{P}\left\{u_\ell^{(\Omega_i(j))} = H_1|H_0\right) \prod_{m=i+1}^{n} P\left(u_m^{(\Omega_i(j))} = H_0|H_0\right\} \right\} \tag{2.33}$$

where $\mathbb{P}\{u_\ell^{(\Omega_i(j))} = H_1|H_0\}$ is the probability that at the $\ell$th sensor, in the $\Omega_i(j)$th combination (out of the $\binom{n}{i}$ possible ones), a wrong decision is made when $H_0$ has happened.

Similarly, the second conditional probability at the right-hand side of (2.5) can be expressed as

$$\mathbb{P}\{\widehat{H} = H_0|H_1\} = \sum_{i=0}^{k-1} \sum_{j=1}^{\binom{n}{i}} \left\{ \prod_{\ell=1}^{i} \mathbb{P}\{u_\ell^{(\Omega_i(j))} = H_1|H_1\} \prod_{m=i+1}^{n} \mathbb{P}\{u_m^{(\Omega_i(j))} = H_0|H_1\} \right\} \tag{2.34}$$

---

[8] Note that $\Omega_i(j)$ depends also on $n$. However, for the sake of notational simplicity, this dependence is not explicitly indicated. The context should eliminate any ambiguity.

where $\mathbb{P}\{u_\ell^{(\Omega_i(j))} = H_1|H_1\}$ is the probability that at the $\ell$th sensor, in the $\Omega_i(j)$th combination, a correct decision is made when $H_1$ has happened.

### 2.3.1.2 Decision Threshold Selection at the Sensors

In the literature, it is shown that using the same threshold at every sensor is an asymptotically optimal solution *if and only if* the SNR at the sensors is constant [31]. In the currently considered scenario (with different SNRs at the sensors), it is not reasonable to use the same threshold at all sensors. Therefore, one needs to choose another criterion for local decisions at the sensors.

In this subsection, we consider a *locally optimal* decision scheme.[9] In other words, each sensor makes a binary decision which minimizes, for the corresponding SNR, its probability of (local) error—this corresponds to a *Person-By-Person Optimization (PBPO)* approach to distributed detection [32]. The optimal value for the threshold $\tau_i$ is such that

$$p(\tau_i|H_1)\mathbb{P}\{H_1\} = p(\tau_i|H_0)\mathbb{P}\{H_0\}. \tag{2.35}$$

In general, the computation of the probability of decision error, based on the evaluation of (2.33) and (2.34), depends on (i) the chosen value for $k$, (ii) the sequence of the detected phenomenon amplitudes $\{s_i\}$ at the sensors, (iii) the sequence of noise variances $\{\sigma_i\}$, and (iv) the sequence of thresholds $\{\tau_i\}$. Recalling the Gaussian model for the observable in (2.1), one can obtain [30]

$$\begin{aligned}\mathbb{P}\{u_\ell = H_1|H\} &= 1 - Q\left(\frac{\tau_\ell - s_\ell \cdot H}{\sigma_\ell}\right)\\ &= 1 - Q\left(\frac{1}{2}\sqrt{\mathrm{SNR}_{\mathrm{sensor}}^{(\ell)}} + \frac{1}{\sqrt{\mathrm{SNR}_{\mathrm{sensor}}^{(\ell)}}}\ln\frac{p_0}{1-p_0} - \sqrt{\mathrm{SNR}_{\mathrm{sensor}}^{(\ell)}}H\right).\end{aligned}$$

As expected, the probability of decision error does not depend on the sequences $\{s_i\}$ and $\{\sigma_i\}$ separately but, rather, only on the sequence of ratios $\{s_i/\sigma_i\}$, i.e., on the sequence of sensor SNRs. In other words, the probability of decision error depends on the sensor *SNR profile* $\{\mathrm{SNR}_{\mathrm{sensor}}^{(i)}\}$. Therefore, evaluating the system performance of the sensor network as a function of the sensor SNR profile is a meaningful problem.

## *2.3.2 Noisy Communication Links*

Let us denote by $p$ the cross-over probability of the BSCs (the same for all noisy communication links). In this case, the decision made at the $\ell$th sensor, i.e., $u_\ell$,

[9] We are implicitly assuming that each sensor estimates its own observation SNR.

might be "flipped," with probability $p$, by the communication link. In particular, the component conditional probabilities in (2.5) depend on $p$. For instance, the conditional probability (2.33) has to be modified by replacing the decisions made locally by the sensors with the corresponding *received* decisions:

$$\mathbb{P}\{\widehat{H} = H_1|H_0\} = \sum_{i=k}^{n}\sum_{j=1}^{\binom{n}{i}}\left\{\prod_{\ell=1}^{i}\mathbb{P}\{u_\ell^{(\Omega_i(j))-\mathrm{rec}} = H_1|H_0\}\prod_{m=i+1}^{n}\mathbb{P}\{u_m^{(\Omega_i(j))-\mathrm{rec}} = H_0|H_0\}\right\} \tag{2.36}$$

where $u_\ell^{(\Omega_i(j))-\mathrm{rec}}$ and $u_m^{(\Omega_i(j))-\mathrm{rec}}$ are the received versions of the local decisions $u_\ell^{(\Omega_i(j))}$ and $u_m^{(\Omega_i(j))}$, respectively. The conditional probability (2.34) has to be modified similarly. A generic term in (2.36) can then be expressed as follows:

$$\mathbb{P}\{u_\ell^{\mathrm{rec}} = H_1|H_0\} = (1-p)Q\left(\frac{\tau_l}{\sigma_l}\right) + p\left[1 - Q\left(\frac{\tau_l}{\sigma_l}\right)\right]. \tag{2.37}$$

Since we are considering locally optimal selection of the decision thresholds at the sensors, there is no difference (in terms of the decision strategy at the sensors) between a scenario with ideal communication links and a scenario with noisy communication links. Therefore, the derivation considered in Sect. 2.3.1.2 for sensor threshold selection holds in this case as well.

### 2.3.3 Sensor SNR Profiles

As observed in Sect. 2.3.1.2, the probability of decision error ultimately depends on the *sensor SNR profile* $\{\mathrm{SNR}_{\mathrm{sensor}}^{(i)}\}$. A generic example of sensor SNR profile is shown in Fig. 2.20a: the sensor SNRs are generally not monotonically ordered. However, since it is always possible to reorder the sensor SNRs from highest to lowest, as shown in Fig. 2.20b, without loss of generality, one can restrict his/her attention to a scenario where the sensor SNR profile is *non-increasing*.

Based on the observation in the previous subsection, in order to characterize non-increasing sensor SNR profiles we consider four possible cases (the SNRs are expressed in dB):

$$\begin{aligned}
&\text{Linearprofile:} \quad \mathrm{SNR}_i = \mathrm{SNR}_0 - c\cdot i\\
&\text{Quadratic profile:} \quad \mathrm{SNR}_i = \mathrm{SNR}_0 - c\cdot i^2\\
&\text{Cubic profile:} \quad \mathrm{SNR}_i = \mathrm{SNR}_0 - c\cdot i^3\\
&\text{Hyperbolicprofile:} \quad \mathrm{SNR}_i = \frac{\mathrm{SNR}_0}{1+c\cdot i}
\end{aligned} \tag{2.38}$$

where: $i = 0,\ldots,n-1$; $n$ is the number of sensors; $\mathrm{SNR}_0$ is the highest sensor SNR; and $c$ is a suitable constant which uniquely characterizes the sensor SNR

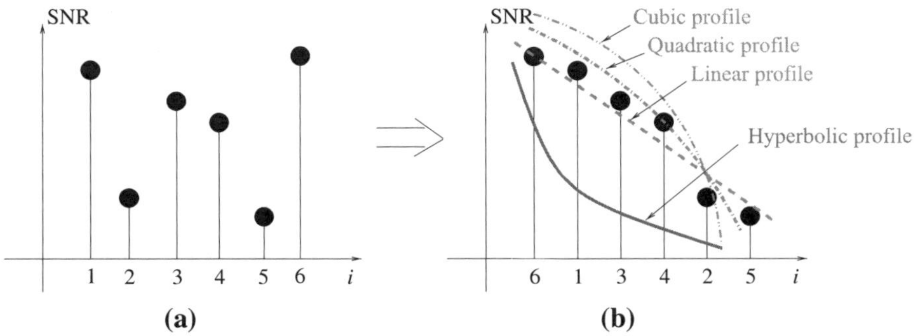


**Fig. 2.20** Illustrative sensor SNR profile: **a** realistic and **b** reordered with non-increasing values of the SNRs. In particular, in (**b**) four possible interpolating profiles (linear, quadratic, cubic, and hyperbolic) are shown

profile *slope*. For this reason, we denote $c$ as *slope coefficient*. A large value of $c$ corresponds to a scenario where the sensor SNRs decrease rapidly (i.e., the corresponding realistic non-ordered sensor SNR profile is highly varying), whereas a small value of $c$ corresponds to a scenario where the sensor SNRs are similar (i.e., the corresponding realistic non-ordered sensor SNR profile is almost constant). If $c = 0$, all profiles degenerate into a constant profile, i.e., $\mathrm{SNR}_i = \mathrm{SNR}_0, \forall i$. In Fig. 2.20b, illustrative graphical examples of the four profiles are shown. In the following, we will restrict our attention to scenarios with convex SNR profiles (linear, quadratic, and cubic), since concave profiles (e.g., hyperbolic) can be shown to lead to worse performance [30]. As one can see, by suitably setting the values of $\mathrm{SNR}_0$ and $c$, a large number of realistic sensor SNR profiles can be characterized. This underlines the applicability of our framework. In Sect. 2.3.5, we will propose a simple experiment to characterize a realistic sensor SNR profile.

In (2.38), we have assumed that the maximum SNR and the slope coefficient $c$ are the same for all profiles. However, in this case the winning profile is always the linear, since the sensor SNR at any position is higher than the corresponding one in any other profile. In order to obtain a "fair" comparison between the various profiles, one can impose that all the SNR profiles have the same average value, denoted as $\overline{\mathrm{SNR}}$.

- By imposing that the slope coefficient $c$ is the same for all profiles, after a few manipulations one obtains that the maximum SNRs in the various cases need to be set as follows:

$$\begin{aligned}
\mathrm{SNR}_{0,\mathrm{l}} &= \overline{\mathrm{SNR}} + c\frac{n-1}{2} \\
\mathrm{SNR}_{0,\mathrm{q}} &= \overline{\mathrm{SNR}} + c\frac{(n-1)(2n-1)}{6} \qquad (2.39) \\
\mathrm{SNR}_{0,\mathrm{c}} &= \overline{\mathrm{SNR}} + c\frac{n(n-1)^2}{4}.
\end{aligned}$$

• Specularly, imposing that the maximum SNR is the same for all the sensors, the slope coefficient in the four considered cases need to be set in the following way:

$$c_l = (\mathrm{SNR}_0 - \overline{\mathrm{SNR}})\frac{2}{n-1}$$
$$c_q = (\mathrm{SNR}_0 - \overline{\mathrm{SNR}})\frac{6}{(n-1)(2n-1)} \tag{2.40}$$
$$c_c = (\mathrm{SNR}_0 - \overline{\mathrm{SNR}})\frac{4}{n(n-1)^2}.$$

Finally, one should observe that in (2.40) it must hold that $\mathrm{SNR}_0 - \overline{\mathrm{SNR}} \geq 0$.

We point out that throughout this subsection we make the implicit assumption that the SNR profiles are perfectly known and available at the AP. This is expedient for performance analysis. However, in a realistic scenario, the mechanisms to collect SNR values from the resource-constrained sensors may not be very accurate, and relying too much on them may not be helpful. Collecting the values accurately is a challenging problem, which needs further investigation. For example, the SNR values could be collected during a *training phase*, when each sensor computes its local SNR and send it to the AP. In Sect. 2.3.5, we propose a simple experimental validation of our theoretical assumptions.

## 2.3.4 Numerical Results

### 2.3.4.1 Ideal Communication Links

Let us first consider a sensor network with ideal communication links from the sensors to the AP. Moreover, the a priori probabilities of the phenomenon are such that $\mathbb{P}\{H_0\} = 10\mathbb{P}\{H_1\}$: this is meaningful for situations where a phenomenon is rare (e.g., the phenomenon under observation is an unusually high humidity level).

The following question is meaningful: for a given value of $\mathrm{SNR}_0$, what are the conditions under which the use of a limited number of sensors (lower, for instance, than $n$) is the winning strategy? In order to answer this question, in Fig. 2.21 the probability of decision error is shown, as a function of the coefficient $c$, in a scenario with linear SNR profile. The lines correspond to analytical results, whereas the symbols are associated with Monte Carlo simulation results. Two possible values for the highest sensor SNR, i.e., $\mathrm{SNR}_0$, are considered: 12 and 16 dB, respectively. For each value of the sensor SNR, various numbers of sensors are considered. Obviously, the curves corresponding to scenarios with only $n = 1$ sensor are constant with respect to $c$. The impacts of the parameters $c$ and $\mathrm{SNR}_0$ can be characterized as follows.

- For *small* values of $c$, i.e., in a scenario with almost constant SNR profile, the best performance is obtained using *all sensors*, regardless of the value of $\mathrm{SNR}_0$.

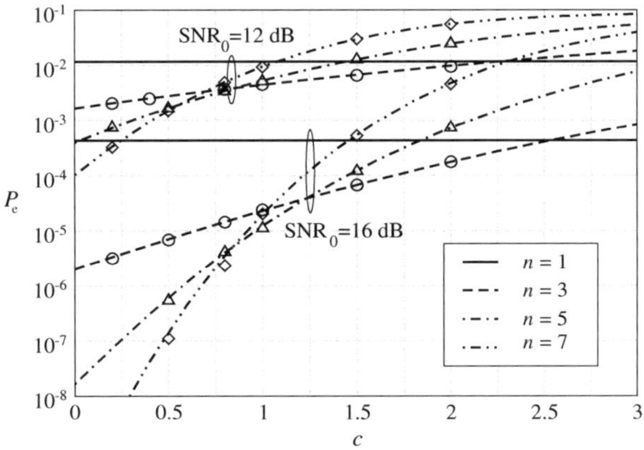


**Fig. 2.21** Probability of decision error, as a function of the coefficient $c$, with $SNR_0$ equal to 12 and 16 dB, respectively. Various values of the number of sensors $n$ are considered, in a scenario with *linear* sensor SNR profile. The *lines* correspond to analytical results, whereas the symbols are associated with simulation results

For *large* values of $c$ (i.e., irregular sensor SNR profile before monotonic reordering), the best performance is obtained using only the sensors with *highest SNRs*. Note that the best asymptotic performance ($c \to \infty$) is obtained using only *the* sensor with highest SNR ($SNR_0$): however, the probability of decision error might be intolerably high.

- For low values of $SNR_0$, the impact of $c$ is "mild," whereas for high values of $SNR_0$ the impact of $c$ is relatively stronger. This behavior can be interpreted as follows. If *at least* one sensor is highly accurate, i.e., $SNR_0$ is high, then in order to optimize the network performance the right subset of sensors should be carefully chosen. In other words, the higher is the sensitivity of at least one sensor in observing the phenomenon, the more accurate the selection of a suitable subset of sensors has to be carried out.

As one can observe from Fig. 2.21, for a given value of $c$, the best performance is obtained selecting a specific number of sensors—those with highest SNRs, starting from the one with $SNR_0$. In order to characterize this behavior in more detail, in Fig. 2.22 the optimal value of the number of sensors to be selected is shown, as a function of $c$, for various values of $SNR_0$. The results in Fig. 2.22 show that (i) the optimal number of sensors is a decreasing function of $c$ and (ii) the lower $SNR_0$ is, the faster the optimal number of sensors decreases for increasing values of $c$. A careful reader might wonder, at this point, why the optimal number of sensors does not reduce by one in correspondence with each vertical (decreasing) step. This behavior is due to the fact that the decision threshold $\tau_i$ at the $i$th sensor is computed according to (2.35), which represents a *locally optimal* threshold selection strategy. Therefore, one can conclude that such a threshold selection strategy is not *globally optimal* (from the entire distributed decision process), as observed in [33]. The individuation of globally optimal decision thresholds at the sensors in a scenario with non-constant sensor SNR profile goes beyond the scope of this book.

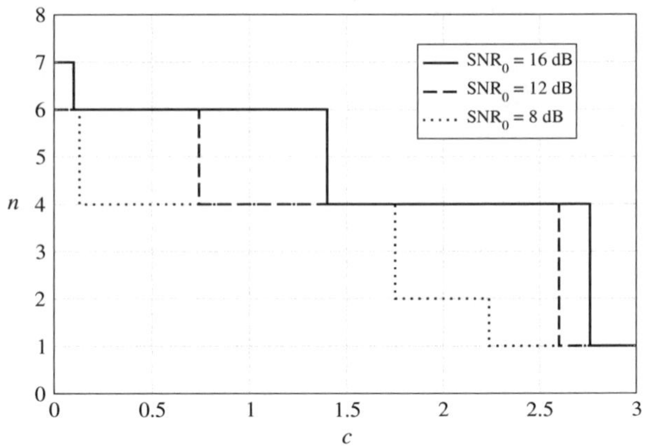


**Fig. 2.22** "Optimal" number of sensors (for minimizing the probability of decision error), as a function of the coefficient $c$, in a scenario with *linear* sensor SNR profile and $\mathbb{P}\{H_0\} = 10\,\mathbb{P}\{H_1\}$. Three values for $\text{SNR}_0$ are considered

#### 2.3.4.2 Noisy Communication Links

While in the previous section we have considered a scenario with *ideal* communication links, we now extend the previous analysis in order to evaluate the impact of the sensor SNR profile in the presence of noisy communication links. More precisely, in a simple network scenario with $n = 3$ sensors, we compare directly the performance with linear, quadratic, and cubic sensor SNR profiles. We do not consider the hyperbolic profile, since we have shown in Sect. 2.3.4.1 that the overall performance with this profile is worse than that with the other profiles—in fact, in the presence of a hyperbolic profile the average sensor SNR has to be very high in order to obtain an acceptable performance level. We evaluate the probability of decision error in a scenario with *all noisy* communication links (considering two values for the cross-over probability $p$, equal to $10^{-3}$ and $10^{-1}$, respectively) and, for comparison, in a scenario with all ideal links.

In Fig. 2.23, the probability of decision error is shown, as a function of the slope coefficient $c$, in various scenarios with $\text{SNR}_0 = 16$ dB and $\mathbb{P}\{H_0\} = 10\mathbb{P}\{H_1\}$.

In Fig. 2.24, the same sensor network scenario is considered, but the *average* sensor SNR is kept constant to $\overline{\text{SNR}} = 16$ dB—for each value of $c$, the corresponding value of $\text{SNR}_0$ is determined according to (2.39). On the basis of the results shown in Figs. 2.23 and 2.24, it is possible to characterize, performance-wise, the interaction between the sensor SNR profile and the communication noise as follows.

- In a scenario with a *common* value of $\text{SNR}_0$, the impact of the sensor SNR profile is very similar in scenarios with ideal communication links and with noisy communication links. For the same value of $c$, the probability of decision error increases if the profile changes from linear to cubic. Obviously, for $c = 0$ the performance with the three profiles coincides. Moreover, asymptotically (for large values of $c$) the probability of decision error is the same regardless of the profile. Therefore, it is possible to identify a critical value of $c$ beyond which the impact of the sensor SNR profile is the highest.

**Fig. 2.23** Probability of decision error, as a function of the coefficient $c$, in a scenario with $n = 3$ sensors. The common value of the maximum sensor SNR is $\text{SNR}_0 = 16$ dB. Three possible scenarios are considered: (i) all ideal links ($p = 0$), and all noisy links with (ii) $p = 10^{-3}$ and (iii) $p = 10^{-1}$, respectively. For comparison, the performance with $n = 1$ sensor is also shown (*horizontal solid line*)

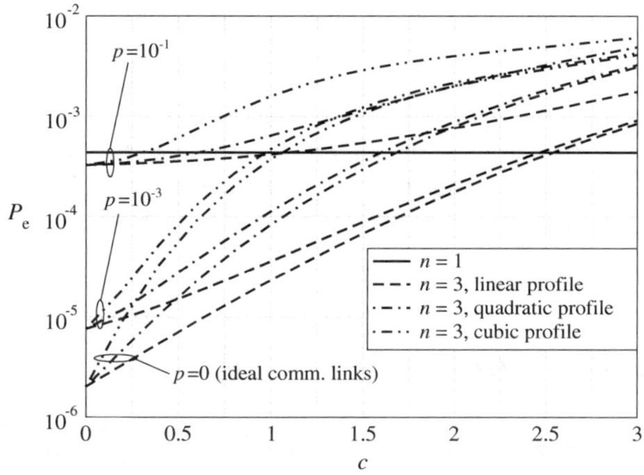


**Fig. 2.24** Probability of decision error, as a function of the coefficient $c$, for the same scenario of Fig. 2.23 and a common *average* value of the sensor SNR equal to $\overline{\text{SNR}} = 16$ dB

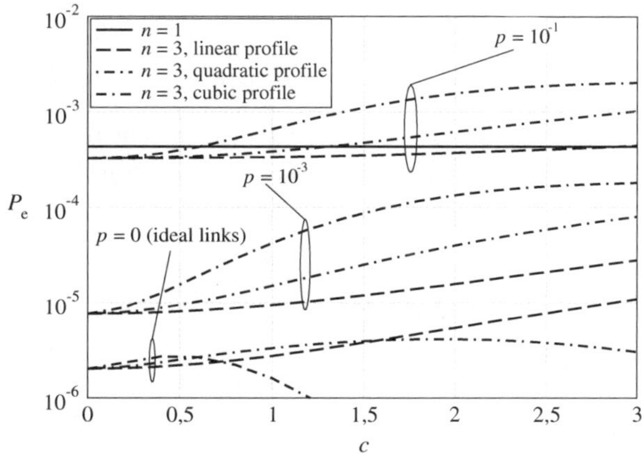


The impact of the noise is strong for small values of $c$, whereas it becomes negligible for large values of $c$. In fact, for any given profile, the curves associated with ideal links and those associated with noisy links tend to coincide for increasing values of $c$. In other words, the less regular is the sensor SNR profile (i.e., the larger is $c$), the milder is the impact of the noise in the communication links. On the other hand, if the sensor SNR is very similar across the sensors, then the noise in the communication links has a severe impact of the network performance. This latter scenario is analyzed in detail in [3].

- In a scenario with a common value of $\overline{\text{SNR}}$, rather than a common maximum sensor SNR, the $P_e - c$ curves do not tend to coincide for large values of the slope coefficient $c$. In other words, the impact of the value of $c$ in a scenario with common $\overline{\text{SNR}}$ is stronger than in a scenario with common $\text{SNR}_0$. On the other

hand, for small values of the slope coefficient $c$, the performance in a scenario with common $\overline{\text{SNR}}$ is similar to that in a scenario with common $\text{SNR}_0$.
From the results in Fig. 2.24, one can also make another observation. In the presence of ideal communication links, for increasing values of $c$ the best performance is obtained by quadratic and cubic profiles. On the opposite, in the presence of noisy communication links, for increasing values of $c$ the best performance is given by a linear sensor SNR profile.

### 2.3.5 Experimental Validation

In this subsection, we show experimental results relative to the SNRs measured at the sensors, in order to validate the theoretical models proposed in this section. In particular, we evaluate the Received Signal Strength Indication (RSSI) in order to obtain *sensor SNR-like* profiles. Equivalently to the RSSI, one could also use the *Path Loss* indicator. In fact, the following equation (in logarithmic scale) holds:

$$P_{\mathrm{t}} = \mathrm{RSSI} + \mathrm{PathLoss}$$

where $P_{\mathrm{t}}$ is the transmit power [dimension: (dBm)] and Path Loss is the power reduction incurred by propagation [dimension: (dB)]. Since in our experiments we set $P_{\mathrm{t}} = 0$ dBm, one easily obtains:

$$\mathrm{RSSI} = -\mathrm{PathLoss}.$$

The main idea of our experiments is the following. A mobile mote periodically sends a message, called *beacon*, whereas $n$ remote nodes, at fixed positions with respect to the mobile mote, receive the beacon and store the received power. Finally, a vector of $n$ power levels is obtained, and an SNR-like profile can be derived. The experimental set-up[10] is schematically shown in Fig. 2.25, from (a) practical and (b) logical viewpoints, respectively. We deploy four MicaZ nodes at the vertices of a square area of $90 \times 90\,\mathrm{cm}^2$, and the remaining mobile (beacon) mote acts as the event "generator" and is denoted as firing mote (fm). As shown in Fig. 2.25, four nodes are placed at the vertices of the network surface. The fm moves inside the network, sending messages to the fixed nodes. Note that in the considered experimental set-up, the observed phenomenon corresponds to the message sent by the mobile node. In order to replicate the theoretical analysis,

[10] Since our experiments are developed in a laboratory environment, there is furniture all around the square area where the sensors are deployed. However, we can consider the reflected signals negligible.

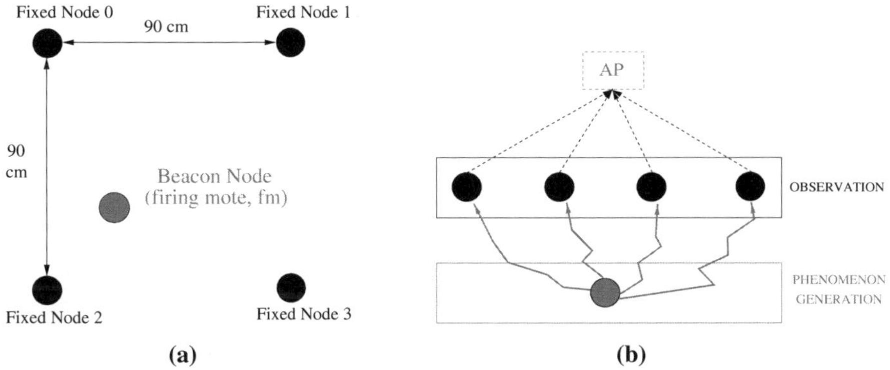


**Fig. 2.25** Experimental set-up: **a** practical scheme with five motes (one "firing/beacon node" and four fixed nodes), deployed over a square network surface with area equal to $90 \times 90\,\text{cm}^2$, and **b** its corresponding logical scheme. The considered platforms are constituted by MicaZ motes using a communication protocol compliant with the IEEE 802.15.4 standard

after receiving the message from the fm, the four fixed nodes should take a decision (e.g., based on the received power), and send their decisions to an AP. Since our goal, in this subsection, is to characterize the sensor SNR profile, we do not consider the communication phase from the sensors to the AP.

Two experiments have been run:

- the fm, which sends the beacon, is very close to one of the remote (fixed) nodes;
- the fm is in the middle between the network center and one of the four vertices of the square network surface, i.e., a fixed node.

In Fig. 2.26, the Path Loss is shown, as a function of the remote node IDs (indicated in Fig. 2.25a), in two different scenarios: (a) the fm is very close to one of the fixed nodes, and (b) the fm is in the middle between the network center and one of the fixed nodes. As one can see from Fig. 2.26a, the lowest Path Loss is obtained, as expected, in correspondence to the nearest remote node. In this case, the profile described is a *heavyside-like* function, since only the fixed node closest to the fm senses a high RSSI (or, equivalently, a low Path Loss), while the others do the opposite. In Fig. 2.26b, the fm is in a more central region and, therefore, the measured power profile is, as expected, smoother than that observed in Fig. 2.26a.

Rearranging the values in Fig. 2.26b in an increasing order, one can obtain a decreasing profile, as described in the previous subsections, of Path Loss or RSSI measurements. In Fig. 2.27, the *Path Loss* profile is shown, as a function of the mote ID, for the four different cases (relative to the position of the mobile mote) considered in Fig. 2.26b. As one can observe, on the average, the profile is approximately linear.

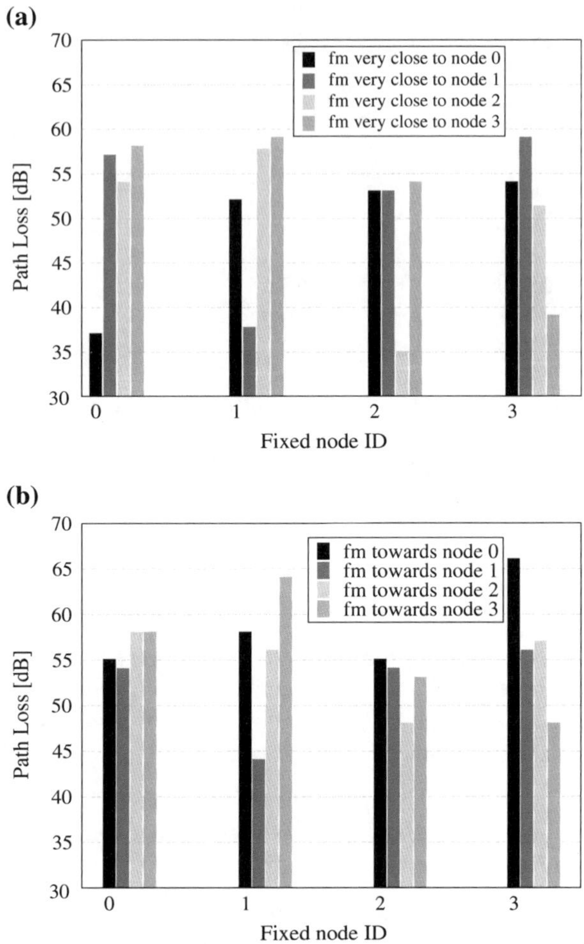


**Fig. 2.26** Path Loss profiles in the presence of four MicaZ motes sensing a fm. The fm is placed either **a** very close to one of the vertices or **b** between the center of the area and one of the vertices

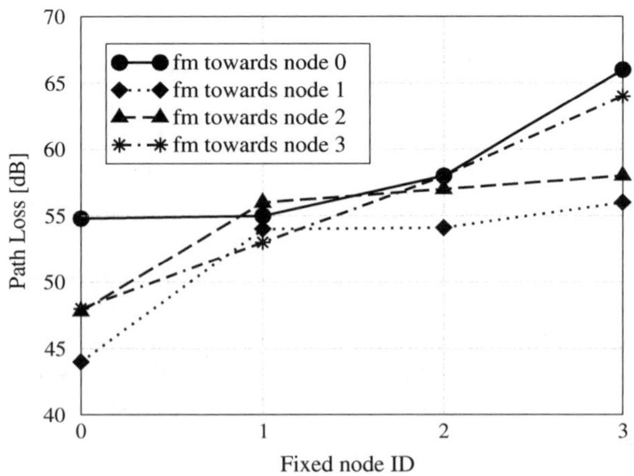


**Fig. 2.27** Reordered Path Loss profiles in the scenarios considered in Fig. 2.26b

## 2.4 On the Interplay Between Decoding and Fusion

### 2.4.1 Distributed Channel Coding and Detection/Decoding/Fusion Strategies

In Fig. 2.28, a pictorial description of the considered sensor network model is shown. There are source nodes (the sensors), which observe (in a noisy manner) a spatially constant phenomenon and send their decisions to the AP, possibly using channel coding. The presence of a relay is also considered and a simple relaying strategy is proposed. The impact of multiple access interference is not investigated here: in other words, we assume orthogonal transmissions to the AP (e.g., perfect transmission scheduling between the sensors and, if present, the relay). The AP performs the following operations:

- *detection* of the observables, taking into account their statistical characterization;
- *decoding* of the embedded error correction code (when used);
- *fusion* of the decoded data to estimate the status of the phenomenon under observation.

Note that some of the elements in Fig. 2.28 are present only in specific scenarios—for instance, the relay node and the decoding block in the AP appear only in coded scenarios.

#### 2.4.1.1 Repetition Coded Sensor Network

A sensor network with multiple observations ($M$ consecutive and independent observations of the same phenomenon) can be interpreted as a system embedding a *repetition code* (with code rate $1/M$) at each sensor. In this case, redundant

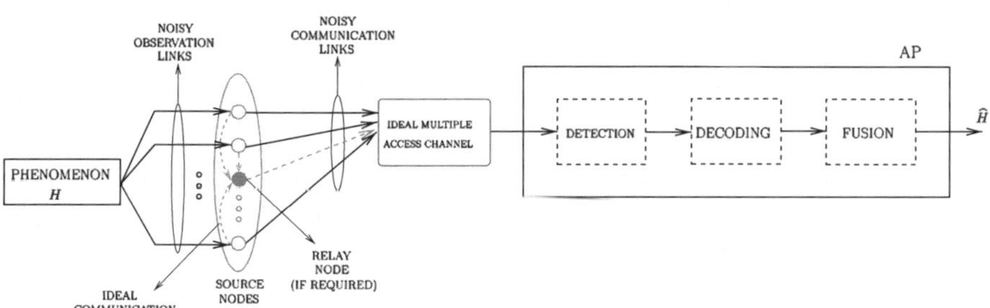


**Fig. 2.28** Pictorial description of the considered sensor network schemes. *Solid lines* are associated with *mandatory* elements (either blocks or connections), whereas *dashed lines* are associated with *optional* elements

information is not sent by a relay, but from the sensors themselves through $M$ consecutive transmission acts per sensor.

#### 2.4.1.2 Systematic Block Coded Sensor Network

In order to embed a *systematic* block channel code into a sensor network, we propose a simple relaying strategy. More precisely, we assume that each sensor transmits its (uncoded) decision to the AP and, owing to the broadcast nature of the wireless medium, also to the relay. Upon reception of the decisions from the sources, the relay, by using a systematic block code, generates parity bits and sends them to the AP. For example, a $(n_{\rm cod}, n) = (7, 4)$ systematic Hamming code [34, p. 562] can be embedded into a sensor network with $n = 4$ sensors and one relay, which generates $n_{\rm cod} - n = 3$ bits according to the parity-check equations of the Hamming code. Assuming (as mentioned) that each sensor can reach both the AP and the relay in a single transmission act, the total number of transmission acts in the proposed sensor network is $n_{\rm cod}$. The equivalent code rate of this distributed coded scheme is $R_{\rm c} = n/n_{\rm cod} = 4/7$. Note, however, that the connections between the sensors and the relay have to be *ideal* (i.e., with no communication noise) in order for the proposed schemes to be applicable. This assumption is reasonable provided that, for example, the relay is relatively closer to the sensors than the AP is. In Sect. 2.4.4, we will comment on the impact of the noise in the communication links from the sensors to the relay.

With a slight abuse of notation, in the following we will denote a scenario as "coded" only if a block channel code is embedded into the network structure, in order to distinguish it from a scenario with multiple observations (i.e., repetition coded).

#### 2.4.1.3 Communication Schemes

In a coded scenario with Binary Phase Shift Keying (BPSK) and Rayleigh faded links, the observable at the output of the communication channel can be expressed as

$$r_i = f_i(2c_i - 1)\sqrt{E_{\rm c}} + w_i \quad i = 1, \ldots, n_{\rm cod} \tag{2.41}$$

where $c_i \in \{0, 1\}$ is the symbol transmitted from either a sensor ($c_i$ is an information bit, $i = 1, \ldots, n$) or the relay ($c_i$ is a parity bit, $i = n + 1, \ldots, n_{\rm cod}$), $\{w_i\}$ are statistically independent AWGN samples with the same distribution $\mathcal{N}(0, N_0/2)$, $N_0$ being the single-sided noise power spectral density, $E_{\rm c} \triangleq R_{\rm c} E_{\rm b}$ is the energy per coded bit, $E_{\rm b}$ being the energy per information bit, and $f_i$ is a random variable with Rayleigh distribution—perfectly coherent demodulation is considered. Under the assumptions of independence between consecutive fading samples (e.g., through the use of channel interleaving) and that $\mathbb{E}[|f_i|^2] = 1$, the BER at the output of the detector at the AP is [35]

$$p^{\text{Rayleigh}} = \frac{1}{2}\left[1 - \sqrt{\frac{R_c \gamma_b}{1 + R_c \gamma_b}}\right] \tag{2.42}$$

$\gamma_b \triangleq E_b/N_0$ is the SNR at the AP. A scenario with AWGN communication links can be modeled using (2.41), by imposing $f_i = 1$ $(i = 1, \ldots, n_{\text{cod}})$. In this case, the BER at the output of the detector at the AP can be written as [35]

$$p^{\text{AWGN}} = Q\left(\sqrt{2R_c \gamma_b}\right). \tag{2.43}$$

In general, one can denote as $p$ the BER at the output of the detector, where $p$ has a specific expression (either (2.43) or (2.42)), depending on the communication channel and the detection strategy (this is also compliant with the initial approach proposed in Sect. 2.1). For simplicity, we assume that $p$ is the same for all sensor-AP links.

In all above communication schemes, the probability of decision error at the AP can be evaluated by computing the conditional probabilities $\mathbb{P}\{\widehat{H} = H_i | H = H_j\}$ in (2.5) $(i, j = 0, 1, i \neq j)$. These values depend on the presence/absence of channel coding and on the detection/decoding/fusion strategy at the AP, as will be described in the following subsections, distinguishing on the basis of the observations at the sensors.

## 2.4.2 *Ideal Observations at the Sensors*

In order to obtain performance benchmarks, we first consider scenarios where the spatially constant phenomenon $H$ is detected by the sensors ideally. In this case, we distinguish between AP structures where the decoding and fusion operations are either separate or joint.

### 2.4.2.1 Separate Decoding and Fusion

When the decoding and fusion operations are separate, assuming majority-like fusion the conditional probabilities at the right-hand side of (2.5) can be computed as follows:

$$\mathbb{P}\{\widehat{H} = H_1 | H = H_0\} = \sum_{i=k}^{n} \binom{n}{i} (p_{\text{ch}}^{\text{ideal}})^i (1 - p_{\text{ch}}^{\text{ideal}})^{n-i} \tag{2.44}$$

$$\mathbb{P}\{\widehat{H} = H_0 | H = H_1\} = \sum_{i=0}^{k-1} \binom{n}{i} (1 - p_{\text{ch}}^{\text{ideal}})^i (p_{\text{ch}}^{\text{ideal}})^{n-i} \tag{2.45}$$

where the repeated trials formula has been used [9], $k$ (i.e., the majority decision threshold) is $\lfloor \frac{n}{2} \rfloor + 1$, and the probability $p_{\text{ch}}^{\text{ideal}}$ depends on the noisy communication link model and the specific distributed channel coding strategy. Note that the upper index of the sum in (2.44) is $n$ (and not $n_{\text{cod}}$) also in coded scenarios, since the information from the relay (i.e., the parity bits) is not used in the *fusion* process (only the systematic bits are used). The parity bits are used only in the *detection/decoding* process.

Since the local sensors' decisions are error-free, $p_{\text{ch}}^{\text{ideal}}$ and $1 - p_{\text{ch}}^{\text{ideal}}$ in (2.44) and (2.45) correspond to the probabilities of error and correct decision at the detector output, respectively. In an "uncoded scenario" (i.e., $n_{\text{cod}} = n$), it holds that $p_{\text{ch}}^{\text{ideal}} = p$. In a scenario with multiple observations, the AP preliminary decides for the phenomenon status at each sensor through a majority fusion rule over the $M$ consecutive decisions sent by that sensor. In this case, $p_{\text{ch}}^{\text{ideal}}$ can be expressed, similarly to (2.44), as

$$p_{\text{ch}}^{\text{ideal}} = \sum_{i=k_{\text{NC}}}^{M} \binom{M}{i} p^i (1-p)^{M-i} \tag{2.46}$$

where $k_{\text{NC}} \triangleq \lfloor \frac{M}{2} \rfloor + 1$. In a coded scenario and for sufficiently small values of $p$, the following approximation holds [34]:

$$p_{\text{ch}}^{\text{ideal}} \simeq \binom{n_{\text{cod}} - 1}{t} p^{t+1}$$

where $t = (d_{\min} - 1)/2$ is the number of errors which can be corrected by a code with minimum distance $d_{\min}$ [34, 36]. We point out that, provided that $1/M = n/n_{\text{cod}}$, the comparison between coded schemes and schemes with multiple observations is consistent from an energetic viewpoint.

#### 2.4.2.2 Joint Decoding and Fusion

In a scenario with multiple ($M$) independent observations at the sensors, joining the decoding and fusion operations consists in adopting a majority fusion rule over all the $n \times M$ bits sent from the sensors to the AP. In this case, the probability of decision error becomes

$$P_{\text{e}}^{\text{mult.obs.}} = \frac{1}{2} \left[ \sum_{i=k_{\text{M}}}^{n \times M} \binom{n \times M}{i} p^i (1-p)^{n \times M - i} + \sum_{i=0}^{k_{\text{M}}-1} \binom{n \times M}{i} (1-p)^i p^{n \times M - i} \right] \tag{2.47}$$

where $k_{\text{M}} \triangleq \lfloor \frac{n \times M}{2} \rfloor + 1$ is the majority decision threshold.

In a coded scenario, the receiver with joint decoding and fusion can be designed as follows. Since the considered sensor networks embed *systematic* codes, we denote as $[u_1^{(j)}, \ldots, u_n^{(j)}, b_1^{(j)}, \ldots, b_{n_{\text{cod}}-n}^{(j)}]$ $(j = 0, 1)$ the entire sequence of bits transmitted by the sensors ($u_i^{(j)}$ from sensor $i$) and the relay ($\{b_i\}_{i=1}^{n_{\text{cod}}-n}$ from the relay) in correspondence to the phenomenon status $H_j$. Note that in the current case with a spatially constant binary phenomenon and ideal observations at the sensors, $(u_1, \ldots, u_n)$ is either $(0, \ldots, 0)$ or $(1, \ldots, 1)$. In other words, in the presence of *ideal* observations, only two codewords, denoted as $\boldsymbol{c}^{(0)}$ and $\boldsymbol{c}^{(1)}$, are allowed—this does not hold with noisy observations, as will be shown in Sect. 2.4.3. In particular, $\boldsymbol{c}^{(0)} = (0, \ldots, 0)$. In all cases considered in this subsection, it will also hold that $\boldsymbol{c}^{(1)} = (1, \ldots, 1)$.

Given that decoding and fusion are joint, two possible detection strategies at the AP can be devised:

- hard-output detection is followed by (hard-input) joint decoding/fusion;
- detection, decoding, and fusion are all joined together.

In the former case, the Maximum A posteriori Probability (MAP) joint decoding/fusion strategy can be formalized as

$$\widehat{H} = \underset{j=0,1}{\operatorname{argmax}}\, \mathbb{P}\{\boldsymbol{c}^{(j)}|\boldsymbol{c}_{\text{rx}}\} = \underset{j=0,1}{\operatorname{argmax}}\, \mathbb{P}\{\boldsymbol{c}|\boldsymbol{c}^{(j)}\}\mathbb{P}\{\boldsymbol{c}^{(j)}\} \tag{2.48}$$

where $\boldsymbol{c}_{\text{rx}}$ is the codeword at the output of the detector at the AP. Since only two codewords $\boldsymbol{c}^{(0)}$ and $\boldsymbol{c}^{(1)}$ are used, the a priori probability of the sequence $\boldsymbol{c}^{(j)}$ is equal to the a priori probability of the phenomenon status $H_j$, i.e., $\mathbb{P}\{\boldsymbol{c}^{(j)}\} = p_j = 1/2$. Owing to the independence of the communication channels (conditionally on the transmitted bits), the MAP decoding/fusion strategy in (2.48) can be rewritten as

$$\widehat{H} = \underset{j=0,1}{\operatorname{argmax}}\, p_j \prod_{i=1}^{n_{\text{cod}}} \mathbb{P}\{c_{i,\text{rx}}|c_i^{(j)}\}. \tag{2.49}$$

After a few manipulations, the MAP decoding/fusion strategy in (2.49) can be finally formulated as

$$\left(\frac{1-p}{p}\right)^{2\vartheta(1,\boldsymbol{c}_{\text{rx}})-n_{\text{cod}}} \underset{H_1}{\overset{H_0}{\gtrless}} 1. \tag{2.50}$$

where $\vartheta(1, \boldsymbol{c}_{\text{rx}})$ is the number of 0's in $\boldsymbol{c}_{\text{rx}}$.

At this point, one can evaluate the probability of decision error in (2.5). In particular, the terms $\left\{\mathbb{P}\{\widehat{H} = H_i | H = H_j\}\right\}$ $(i,j = 0, 1,\ i \neq j)$ can be computed from the decision rule (2.50). After a few manipulations, one obtains:

$$P_{\text{e}} = \frac{1}{2}\left[\sum_{k=k^*}^{n_{\text{cod}}} \binom{n_{\text{cod}}}{k} p^k (1-p)^{n_{\text{cod}}-k} + \sum_{k=0}^{k^*-1} \binom{n_{\text{cod}}}{k} (1-p)^k p^{n_{\text{cod}}-k}\right]$$

where we have used the fact that $\vartheta(1, \boldsymbol{c}_{\text{rx}})$ is a binomial random variable with parameters $n_{\text{cod}}$ and $p$, $\boldsymbol{c}^{(1)} = \mathbf{1}$, and $k^*$ is defined as follows:

$$k^* = \min\{1, \ldots, n_{\text{cod}}\}$$
$$\text{s.t.} \left(\frac{1-p}{p}\right)^{2k^* - n_{\text{cod}}} > 1.$$

In the case with joint detection/decoding/fusion, we first consider a scenario with Rayleigh faded links, and we denote by $\boldsymbol{f} = [f_1, \ldots, f_{n_{\text{cod}}}]$ the fading samples and by $\boldsymbol{r} = [r_1, \ldots, r_{n_{\text{cod}}}]$ the observables at the output of the communication links. Under the assumption of perfect channel state information at the AP, the MAP detection/decoding/fusion strategy can be formulated as[11] [36]

$$\widehat{H} = \underset{j=0,1}{\operatorname{argmax}}\, p\left(\boldsymbol{r}|\boldsymbol{c}^{(j)}, \boldsymbol{f}\right)\mathbb{P}\{\boldsymbol{c}^{(j)}|\boldsymbol{f}\} = \underset{j=0,1}{\operatorname{argmax}}\, p_j \prod_{i=1}^{n} p\left(r_i|c_i^{(j)}, f_i\right) \tag{2.51}$$

where we have used the facts that the observables are conditionally independent given $\{c_i^{(j)}\}$ and the coded bit $c_i^{(j)}$ is independent of the fading sample $f_i$. Discarding $p_j = 1/2$, from (2.51) one can derive, after a few manipulations, the following decision rule:

$$\sum_{i=1}^{n_{\text{cod}}} r_i f_i c_i^{(1)} \underset{H_0}{\overset{H_1}{\gtrless}} 0. \tag{2.52}$$

On the basis of (2.52) and recalling that a linear combination of Gaussian random variables is still a Gaussian random variable [9], after a few manipulations the probability of decision error at the AP (2.5) becomes

$$P_e = \frac{1}{2}\left[Q\left(2\frac{\sqrt{R_c E_b}\sum_{i=1}^{n_{\text{cod}}} f_i c_i^{(1)}}{\sqrt{N_0 \sum_{i=1}^{n_{\text{cod}}} f_i^2 (c_i^{(1)})^2}}\right) + \Phi\left(-2\frac{\sqrt{R_c E_b}\sum_{i=1}^{n_{\text{cod}}} f_i (2c_i^{(1)} - 1) c_i^{(1)}}{\sqrt{N_0 \sum_{i=1}^{n_{\text{cod}}} f_i^2 (c_i^{(1)})^2}}\right)\right] \tag{2.53}$$

where $\Phi(x) \triangleq 1 - Q(x)$. Observe that (2.53) depends on the particular sequence of fading samples $\{f_i\}$.

An expression for the probability of decision error in the case of AWGN links can be directly obtained from (2.53) by imposing $f_i = 1$ $(i = 1, \ldots, n_{\text{cod}})$. In particular, in the presence of a code with $\boldsymbol{c}^{(1)} = \mathbf{1}$ (recall that, in all cases, $\boldsymbol{c}^{(0)} = \mathbf{0}$) it can be shown that

$$P_e = Q\left(\sqrt{2 n_{\text{cod}} R_c \gamma_b}\right) = Q\left(\sqrt{2n\gamma_b}\right).$$

[11] In (2.51) and in the remainder of this subsection, the uppercase $\mathbb{P}$ is used to denote the probability of an event, whereas the lowercase $p$ is used to denote the conditional probability density function (p.d.f.) of a random variable.

## 2.4.3 Noisy Observations at the Sensors

We now extend the derivation presented in Sect. 2.4.2 to encompass the presence of observation noise.

### 2.4.3.1 Separate Decoding and Fusion

In the case of separate decoding and fusion, only the expression of the probability $p_{\mathrm{ch}}^{\mathrm{ideal}}$ in (2.44) and (2.45) need to be modified. In particular, by using the total probability theorem [9], one can write

$$\begin{aligned} p_{\mathrm{ch}}^{\mathrm{noisy}} &= \mathbb{P}\{c_{i,\mathrm{rx}} = 1 | H_\ell\} \quad i = 1, \ldots, n, \ \ell = 0, 1 \\ &= p_{\mathrm{ch}}^{\mathrm{ideal}}[1 - Q(\tau - s \cdot \ell)] + (1 - p_{\mathrm{ch}}^{\mathrm{ideal}})Q(\tau - s \cdot \ell) \end{aligned}$$

where the sensors' decisions $\{c_i^{(\ell)}\}$ are done as outlined in Sect. 2.1.1 and $p_{\mathrm{ch}}^{\mathrm{ideal}}$ is the final BER, which depends on the presence/absence of distributed channel coding, as shown in Sect. 2.4.2.

In a scenario with $M$ observations at each sensor, expression (2.46) for $p_{\mathrm{ch}}^{\mathrm{ideal}}$ has to be similarly modified. In particular, one obtains:

$$p_{\mathrm{ch}}^{\mathrm{noisy}} = \sum_{i=k_{\mathrm{NC}}}^{M} \binom{M}{i} [g(p,\ell)]^i [1 - g(p,\ell)]^{M-i} \tag{2.54}$$

where $g(p,\ell) \triangleq p[1 - Q(\tau - s \cdot \ell)] + (1 - p)Q(\tau - s \cdot \ell)$.

### 2.4.3.2 Joint Decoding and Fusion

In the case of hard-output detection followed by joint decoding/fusion, expression (2.48) for the phenomenon estimate in a scenario with multiple observations at the sensors has to be modified, similarly to (2.54), as follows:

$$\begin{aligned} P_{\mathrm{e,noisy}}^{\mathrm{mult.obs.}} = & \frac{1}{2} \sum_{i=k_{\mathrm{M}}}^{n \times M} \binom{n \times M}{i} [g(p,1)]^i [1 - g(p,1)]^{n \times M - i} \\ & + \frac{1}{2} \sum_{i=0}^{k_{\mathrm{M}}-1} \binom{n \times M}{i} [1 - g(p,0)]^i [g(p,0)]^{n \times M - i}. \end{aligned} \tag{2.55}$$

We now derive the MAP decoding/fusion strategy for the coded scenarios in the presence of noisy observations at the sensors. In the case of hard-output detection followed by (hard-input) joint decoding/fusion, in order to take into account the observation noise statistics expression (2.48) has to be modified as follows:

$$\widehat{H} = \underset{j=0,1}{\operatorname{argmax}}\, \mathbb{P}\{H_j | \boldsymbol{c}_{\mathrm{rx}}\} = \underset{j=0,1}{\operatorname{argmax}} \prod_{i=1}^{n_{\mathrm{cod}}} \mathbb{P}\{c_{i,\mathrm{rx}} | H_j\}$$

where the irrelevant term $P(H_j) = p_j = 1/2$ has been discarded and the probability $P(c_{i,\mathrm{rx}}|H_j)$ can be written, after a few manipulations, as

$$\mathbb{P}\{c_{i,\mathrm{rx}}|H_j\} = \begin{cases} (1-p)[1-Q(\tau - s\cdot j)] + pQ(\tau - s\cdot j) & \text{if } c_{i,\mathrm{rx}} = 0 \\ p[1-Q(\tau - s\cdot j)] + (1-p)Q(\tau - s\cdot j) & \text{if } c_{i,\mathrm{rx}} = 1. \end{cases}$$

In a coded scenario with joint detection/decoding/fusion, the MAP estimation strategy (2.51) has to be modified as follows:

$$\widehat{H} = \underset{j=0,1}{\operatorname{argmax}} \prod_{i=1}^{n_{\mathrm{cod}}} \mathbb{P}\{r_i|H_j,f_i\}$$

which can be rewritten, after a few manipulations, as

$$\frac{\prod_{i=1}^{n_{\mathrm{cod}}} \Upsilon(0, r_i, f_i)}{\prod_{i=1}^{n_{\mathrm{cod}}} \Upsilon(1, r_i, f_i)} \underset{H_1}{\overset{H_0}{\gtrless}} 1$$

where

$$\Upsilon(m, r_i, f_i) \triangleq \Phi(\tau - m\cdot s)\exp\left(-2\frac{r_i f_i \sqrt{E_c}}{N_0}\right) + [1-\Phi(\tau - m\cdot s)]\exp\left(2\frac{r_i f_i\sqrt{E_c}}{N_0}\right).$$

## *2.4.4 Impact of Noisy Communication Links Towards the Relay*

The previous derivations in coded scenarios are based on the assumption of ideal communication links between the sensors and the relay. In this subsection, we briefly discuss on the impact of *noisy* communication links between the sensors and the relay. Neither analytical derivation nor numerical results will be presented. The considerations which will be carried out are simply meant to give some guidelines on the benefits brought by the distributed use of properly designed block error correction codes.

We first consider the case with *ideal* observations at the sensors. In Fig. 2.29, we give a pictorial description of how the communication noise influences data transmission to the relay. As previously seen, two possible codewords are selected at the sensors and relay, namely $\boldsymbol{c}^{(0)}$ and $\boldsymbol{c}^{(1)}$, which are shown, in Fig. 2.29, as a filled circle and an empty circle, respectively.

- In the scenario with no communication noise between the sensors and the relay (case (a)), we denote the Hamming distance between the two codewords as $d$. If $\boldsymbol{c}^{(0)} = \mathbf{0}$ and $\boldsymbol{c}^{(1)} = \mathbf{1}$, then $d = n_{\mathrm{cod}}$. The presence of noisy communication links from the sensors and the relay to the AP is such that the word $\boldsymbol{c}_{\mathrm{rx}}$ (one of the $2^{n_{\mathrm{cod}}}$ possible binary sequences of length $n_{\mathrm{cod}}$) received at the AP may be different from the codeword transmitted by the sensors and the relay. In particular, $\boldsymbol{c}_{\mathrm{rx}}$

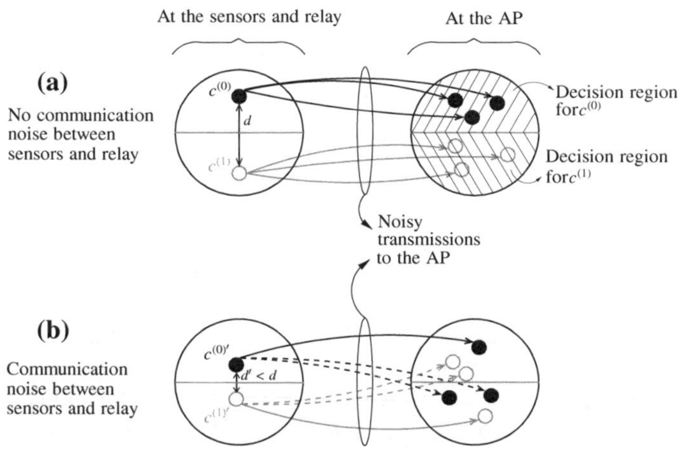


**Fig. 2.29** Codebook perspective on the considered distributed detection schemes: **a** ideal communication links between sensors and relay and **b** noisy communication links. In each case, on the left the two possible codewords at sensors and relay are shown, whereas on the right possible received words at the AP are shown

may not even be a codeword. Decoding and fusion at the AP corresponds to associating the received word to one of the information sequences **0** or **1**. It is intuitive that the larger is $d$, the more robust is the system against communication noise in the links to the AP.

- In the presence of communication noise between the sensors and the relay (case b), the latter may receive a sequence of bits which differs from that sent by the sensors. Therefore, the parity bits generated by the relay may lead to the association of $H_0$ and $H_1$ to two codewords $\boldsymbol{c}^{(0)'}$ and $\boldsymbol{c}^{(1)'}$ which are at a distance $d' < d$. As a consequence of this decreased distance, the system performance will be worse than in the previous scenario, since the probability of associating (through decoding and fusion) the received word to the wrong phenomenon status will increase. This can be understood from the codebook scenario at the AP, where the received word at the AP might belong to the portion of the signal space which is associated (by decoding and fusion) to the wrong phenomenon status.

The presence of *noisy* observations may lead to the association of the phenomenon statuses $H_0$ and $H_1$ to two codewords $\boldsymbol{c}^{(0)''}$ and $\boldsymbol{c}^{(1)''}$ at a distance smaller than $d$. In particular, in the presence of both (i) observation noise and (ii) communication noise from the sensors to the relay, when the intensities of these two noises are sufficiently small, their negative effects tend to add, so that the distance $d''$ between $\boldsymbol{c}^{(0)''}$ and $\boldsymbol{c}^{(1)''}$ might be even smaller than $d'$.

Obviously, an open problem is to quantify precisely the decrease of the error correction capability $t$ of the code in the presence of noisy communication links between the sensors and the relay. In fact, the parameter $t$ depends on the particular structure (codebook) of the considered error correction code. An interesting research direction is the design of robust (fault tolerant) error correcting codes for the proposed distributed detection schemes.

### 2.4.5 Numerical Results

We resort to Monte Carlo simulations to evaluate the probability of decision error with the devised MAP detection/decoding/fusion strategies presented in Sects. 2.4.2 and 2.4.3.

In order to make the detection/decoding process at the input of the AP more effective, *soft-input* decoding/fusion (either separate or joint), rather than hard-input decoding/fusion, can be considered. In Fig. 2.30, the probability of decision error is shown, as a function of the SNR at the AP, in a scenario with $n = 16$ sensors, AWGN communication links (similar results can be obtained in scenarios with Rayleigh faded communication links), and *error-free* observations at the sensors. Six coding strategies are considered: (i) uncoded, (ii) (31,16) BCH [36, p. 438] (the corresponding BCH code has $t = 3$)[12] coded with hard-input and separate decoding/fusion, (iii) (31,16) BCH coded with soft-input and joint decoding/fusion, (iv) with $M = 2$ observations and joint decoding/fusion, (v) with $M = 3$ observations and separate decoding/fusion, and (vi) with $M = 3$ observations and joint decoding/fusion. One can observe that the probability of decision error in coded scenarios shows a "waterfall" behavior, which is due to the concatenation of the decoding and fusion operations. However, the improvement brought by the presence of distributed channel coding, with respect to schemes with multiple observations, becomes apparent at very low probabilities of decision error, which may not be of practical interest. One can observe that the coded network with soft-input and joint decoding/fusion at the AP has a performance significantly better than that associated with the schemes with hard-input and separate decoding/fusion. This is to be expected, since in a scenario with soft-input decoding no information is lost upon reception of the observables from the communication links. Note, however, that in this case as well the proposed coded scheme outperforms a scheme with multiple observations only at very low values of the probability of decision error.

In Fig. 2.31, the probability of decision error is shown, as a function of the BER $p$ at the output of the AP detector, in a scenario with $n = 16$ sensors and *noisy* phenomenon observations. Two values for the observation SNR are considered: (a) 20 dB and (b) 10 dB. The performance is evaluated with six sensor network architectures: (i) uncoded, (ii) (31,16) BCH coded with separate decoding/fusion, (iii) (31,16) BCH coded with joint decoding/fusion, (iv) with $M = 2$ observations and joint decoding/fusion, (v) with $M = 3$ observations and separate

[12] We remark that the BCH is one of the block channel codes that it is possible to consider. However, the same results would be asymptotically obtained with any code with $t = 3$.

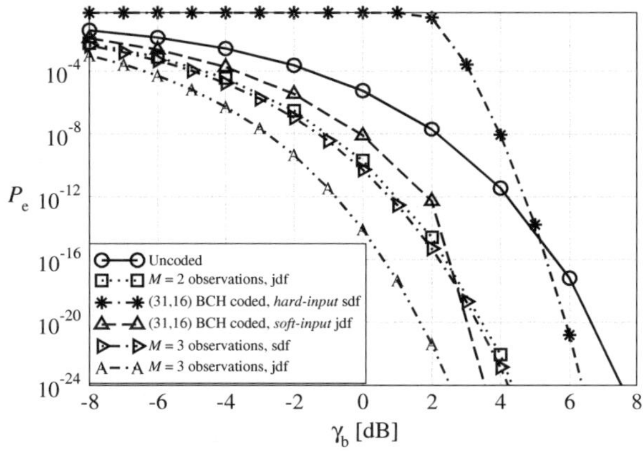


**Fig. 2.30** Probability of decision error, as a function of the SNR at the AP, in a scenario with $n = 16$ sensors, AWGN communication links, and *error-free* phenomenon observations. Various coding strategies are considered

decoding/fusion, and (vi) with $M = 3$ observations and joint decoding/fusion. In the case of high observation SNR (e.g., in Fig. 2.31a), the phenomenon observations at the sensors are practically error-free and, therefore, the network performance is similar to that in Fig. 2.30. When the observation SNR decreases (e.g., in Fig. 2.31b), instead, the proposed detection/decoding/fusion strategies are not effective, since the quality of the sensors' observations heavily affects the system performance, and this is more pronounced in the presence of joint decoding/fusion. One can observe that the probability of decision error curve reaches a floor, due to the observation noise (which is independent of the communication noise). As before, the schemes with multiple observations at the sensors outperform those with block channel coding.

Finally, we investigate the performance of the proposed distributed schemes in *large scale* sensor networks, by using an Low-Density Parity-Check (LDPC) code and the sum–product (SP) decoding algorithm. In particular, we consider a (3,6) regular and systematic LDPC code: the systematic bits of the codeword correspond to the $n$ decisions sent by the sensors, whereas the $n_{\text{cod}} - n$ parity bits are generated by the relay node. The LDPC code is constructed in a *random* fashion, according to an algorithm, which exploits an idea similar to the progressive edge growth (PEG) algorithm presented in [37]. In Fig. 2.32, the probability of decision error is shown, as a function of the SNR at the AP, in a scenario with $n = 100$ sensors, AWGN communication links, and *noisy* phenomenon observations. Three sensor network architectures are considered: (i) LDPC coded with standard SP decoding [38, 39], (ii) LDPC coded with *enhanced* (as described in the following) channel Logarithmic Likelihood Ratios (LLRs), and (iii) with $M = 2$ observations and separate decoding/fusion. Two values for the observation SNR are considered: (i) 10 dB (dashed lines) and (ii) 20 dB (solid lines). While in the LDPC coded case with standard SP decoding the channel LLRs (input at the variable nodes of the LDPC bipartite graph) do not take into account the observation noise, in the enhanced SP decoding case the channel LLRs are modified by properly taking into

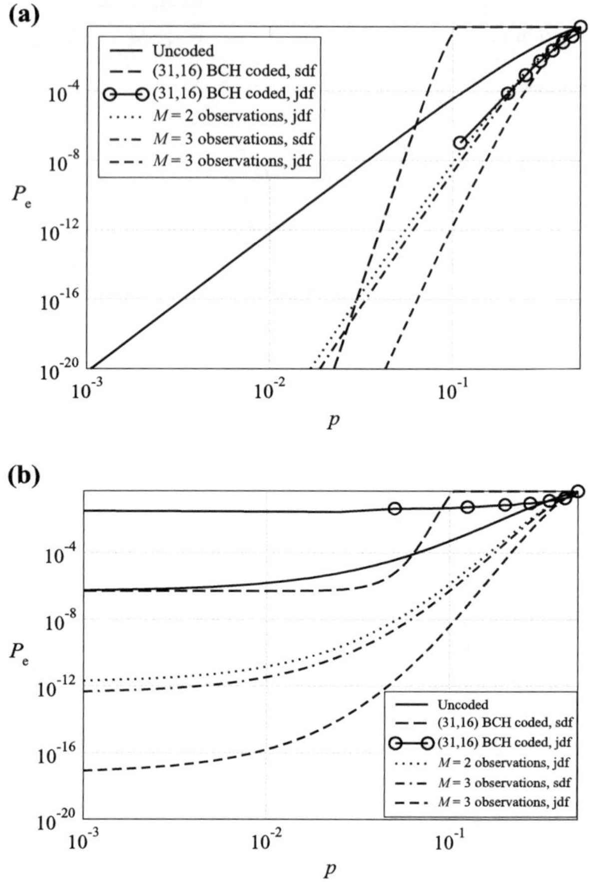


**Fig. 2.31** Probability of decision error, as a function of the BER $p$ at the output of the detector, in a scenario with $n = 16$ sensors and *noisy* phenomenon observations. Two values for the observation SNR are considered: **a** 20 dB and **b** 10 dB. Various sensor network architectures are considered

account the observation noise. The modified channel LLRs can be expressed as follows:

$$\mathcal{L}^{(i)}_{\text{ch-enhanced}} = \mathcal{L}_{\text{ch}} + \mathcal{L}^{(i)}_{\text{a-priori}} \quad i = 1, \ldots, n_{\text{cod}}$$

where

$$\mathcal{L}_{\text{ch}} \triangleq \ln \frac{p(r_i | c_i = 0)}{p(r_i | c_i = 1)} = \frac{r_i}{N_0}$$

and

$$\mathcal{L}^{(i)}_{\text{a-priori}} \triangleq \ln \frac{\mathbb{P}\{c_i = 0\}}{\mathbb{P}\{c_i = 1\}} = \begin{cases} \ln \frac{\Upsilon(0, -r_i, 1)}{\Upsilon(1, -r_i, 1)} & \text{if } i = 1, \ldots, n \\ 0 & \text{if } i = n+1, \ldots, n_{\text{cod}} \end{cases}$$

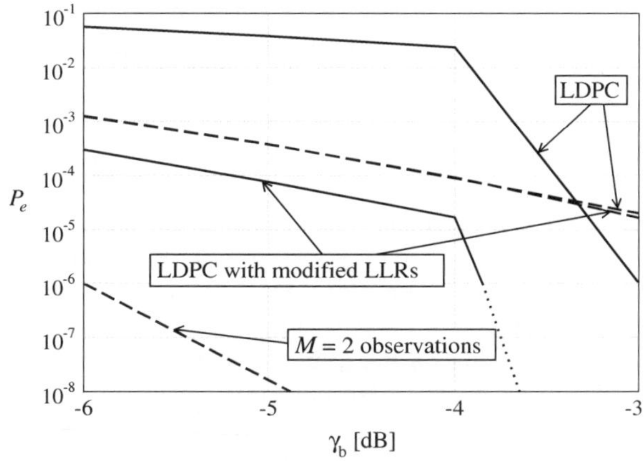


**Fig. 2.32** Probability of decision error, as a function of the SNR at the AP, in a scenario with $n = 100$ sensors, AWGN communication links, and *noisy* phenomenon observations. Two values for the observation SNR are considered: (i) 10 dB (*dashed lines*) and (ii) 20 dB (*solid lines*). Various sensor network architectures are considered

where $\Upsilon$ has been defined in Sect. 2.4.3.2. From the results in Fig. 2.32, one can observe that the use of multiple observations is still the winning strategy also in a large-scale sensor network.[13] However, the enhanced LPDC coded scheme (with modified channel LLRs) outperforms the LDPC coded scheme at large observation SNRs, since a statistical knowledge of the observation noise helps the decoding process. In fact, when the communication noise level is too high, a communication error might compensate an error in the phenomenon estimation at the sensors (due to a too high observation noise level). On the other hand, when the communication links to the AP are reliable (i.e., the communication noise is sufficiently small) an error in the phenomenon estimation might not be compensated and, therefore, the AP might not be able to correctly reconstruct the phenomenon status. Finally, note that in the standard LDPC coded case the performance with an observation SNR equal to 10 dB is better than that associated with an observation SNR equal to 20 dB when the SNR at the AP is sufficiently low. This is due to the fact that for small values of the observation SNR a larger number of codewords is actually used by the sensor network and, consequently, the error correction capabilities of the LDPC code are better exploited. However, when the SNR at the AP increases, the "beneficial" impact of the observation noise is reduced by the presence of reliable communication links.

## 2.5 Concluding Remarks

In this chapter, we have characterized the performance of sensor networks where a spatially constant phenomenon is under observation. First, we have characterized

[13] Note that in Fig. 2.32 only the curve with observation SNR equal to 10 dB is shown in the case with multiple observations. The curve associated with an observation SNR equal to 20 dB and multiple observations is even lower.

the behavior of clustered sensor networks with distributed detection in the presence of multi-level majority-like information fusion. Upon the derivation of a communication-theoretic analytical framework, we have shown that, in the considered scenarios, uniform clustering, i.e., balanced tree network architectures, leads to a lower probability of decision error than non-uniform clustering, i.e., unbalanced tree network architectures. In the former case, the probability of decision error depends *only* on the number of decision levels and *not* on the specific clustering configuration. An information-theoretic perspective has also been presented. Then, the impact of noisy communication links has been investigated. Our results show that the presence of noise in the communication links has a strong bearing on the ultimate achievable performance.

Then, an analytical framework to compute the *network lifetime* of clustered sensor networks subject to a physical layer-oriented QoS condition has been derived. In the presence of ideal reclustering, the network lifetime is the longest possible. On the other hand, in the presence of a fixed clustered configuration, our results show that the number of clusters has a strong impact on the network lifetime. More precisely, the network lifetime is maximized if there are a *few large clusters* (at most four). In all cases, the QoS condition has a strong impact on the network lifetime: the more stringent this condition is, the shorter the network lifetime is. We have also evaluated the cost associated with the reclustering procedure, from both *time delay* and *energy consumption* perspectives. Our results show that reclustering is not useful when phenomenon observations are *rare*, since the network spends more time in transferring control messages than useful data. The impact of noisy communication links, modeled as BSCs, on the network lifetime has also been investigated, showing that the higher the noise level, the shorter the network lifetime. However, also in this scenario reclustering can prolong the network lifetime.

Although the previous analysis was based on the assumption of constant sensor SNR across the sensors, we have proposed an analytical framework to take into account different observation SNRs not known at the AP. In order to model this scenario, four possible sensor SNR profiles (linear, quadratic, cubic, and hyperbolic) have been introduced and we have characterized them by using a *slope coefficient* and the *maximum sensor SNR*. For increasing steepness of the (ordered) sensor SNR profile, i.e., for an increasingly irregular realistic sensor SNR profile, the best performance is obtained by selecting a lower number of sensors (those with highest SNRs). In a scenario with common *average* sensor SNR, the profile which guarantees the best performance is the *cubic*. This is due to the fact that it corresponds to the profile with the largest (in relative terms) number of sensors with SNR higher than the average value. Therefore, a general conclusion is that, for a given *average sensor SNR*, the best performance is obtained when the variance of the sensor SNR is large, i.e., the sensor SNR profile is irregular. The presence of noisy communication links has also been considered. In this case, we have shown that the more irregular the sensor SNR profile, the milder the impact of the noise level in the communication links.

The analytical framework has been enriched with simulation and experimental results (in terms of probability of decision error, throughput, and delay) relative to IEEE 802.15.4-based clustered sensor networks with information fusion. The obtained results confirm the validity of our analytical framework in realistic networking scenarios. Moreover, it has been possible to characterize realistic SNR profiles.

Finally, we have studied how to combine detection, decoding, and fusion at the AP in sensor networks for distributed detection of a spatially constant binary phenomenon. To this end, we have embedded simple distributed channel codes (either block or repetition) into sensor network architectures. The performance of the proposed schemes has been analyzed in scenarios with noisy observations and communications. In all cases, the use of *multiple observations* (i.e., repetition coding) guarantees the best performance, with respect to simple systematic block coding strategies, for practical values of the probability of decision error. This leaves the design of powerful distributed channel codes as an open problem. Considering scenarios with distributed LDPC coding, our results show that knowledge, at the AP, of the observation noise can significantly improve the decoding process, i.e., it can help in reducing the negative effects of the communication noise.

## 2.6 Further Readings

Recent years have witnessed an increasing interest for the use of distributed detection techniques in sensor networks [40], especially for civilian applications [41], e.g., environmental monitoring [42]. The application of distributed detection techniques in the military field has, on the other hand, a long history. In all cases, the goal of a sensor network with distributed detection is to identify the status of a phenomenon of interest through a collaborative action of the sensors [43]. The increasing interest for sensor networks has, therefore, spurred a significant activity on the design of efficient distributed detection techniques, in order to obtain fault-tolerant networks with the longest possible lifetime [44].

Several communication-theoretic-oriented approaches have been proposed to study decentralized detection [45–52]. In [1], the authors follow a Bayesian approach for the minimization of the probability of decision error at the AP and study optimal fusion rules. Most of the proposed approaches are based on the assumption of *ideal* communication links between the sensors and the AP. However, in a realistic communication scenario, these links are likely to be *noisy* [53]. The impact of noisy communication links on the design of optimal fusion rules is evaluated in [3, 54–57]. A practical and widely used model for the noisy communication links is the BSC [3, 49, 54–56]. In [3], a few techniques are proposed to make the system more robust against the noise. In [51], the author considers Minimum Mean Square Error (MMSE) parameter estimation in sensor networks. Use of censoring algorithms at the sensors has also been studied for the design of decentralized detection schemes [58]. In [59], the authors analyze aspects related to compression of observed data (using distributed source coding) and data transmission.

Information-theoretic approaches have also been proposed for the study of sensor networks with decentralized detection. In [50], the authors propose a framework to characterize a sensor network in terms of its entropy and false alarm/missed detection probabilities. Information theory has also been used to tackle the problem of optimally placing sensors over a given surface to meet the chosen design criterion. In [60], the mutual information is evaluated in a scenario with censoring sensors which transmit their local likelihood ratios, by maximizing the probability of correct decision [61]. In order to optimally place the sensors over a given surface, system entropy and mutual information are considered in [62, 63], respectively. In [64], an information-theoretic approach is proposed to solve, with limited complexity, the problem of sensor selection and placement for target localization and tracking. Decentralized detection algorithms, based on the evaluation of the sensor network mutual information, have also been proposed to design intelligent systems that recognize, in a robust manner, a target in a scene which rapidly changes [65].

The impact of communication constraints, e.g., limited bandwidth and presence of noise, is considered in [66], where a randomization paradigm for decentralized detection is proposed to overcome the communication bottle-neck. In [7], the authors consider the problem of decentralized detection in *wireless* sensor networks where communication links are affected by fading. In the latter scenario, the optimal distributed detection strategy is first derived, on the basis of the integration of the communication and fusion phases, and then suboptimal (requiring a limited a priori knowledge of the channel state) strategies are developed. This approach is further extended in [54], where the authors optimize the local decision strategy in sensor networks with fading, and in [67], where the authors propose a decentralized detection strategy based on censoring sensors, which transmit only when their local likelihood ratios are sufficiently large.

For what concerns the issue of energy efficiency, motivated by recent theoretical results in the area of network coding [68–73], significant research activity has been devoted to the development of specific channel coding strategies. Although preliminary works focus on scenarios with ideal communication channels, the impact of communication noise has also been investigated [74]. Moreover, distributed network coding strategies for the multi-access relay channel, i.e., a channel where source nodes can send their information to the destination through a common relay node, have been investigated [75, 76].

The problem of extending the sensor network lifetime is a direct consequence of the energy efficiency in scenarios with battery-powered nodes. In particular, the derivation of upper bounds for the sensor network lifetime has been exploited. In [77–85], various analyses are carried out according to the particular sensor network architecture and the definition of sensor network lifetime. In [86], a simple formula, independent of these parameters, is provided for the computation of the sensor network lifetime and a Medium Access Control (MAC)

protocol is proposed to maximize the sensor network lifetime. In [87], a distributed MAC protocol is designed in order to maximize the network lifetime. In [88], network lifetime maximization is considered as the main criterion for the design of sensor networks with data gathering. In [89], the authors consider a realistic sensor network with nodes equipped with TinyOS, an event-based operating system for networked sensor motes. In this scenario, the network lifetime is evaluated as a function of the average distance of the sensors from the central data collector. In [90], an analytical framework, based on the Chen-Stein method of Poisson approximation, is proposed in order to find the critical time at which isolated nodes, i.e., nodes without neighbors in the network, begin to appear, due to the deaths of other nodes. Although this method is derived for generic networks where nodes are randomly deployed and can die in a random manner, this can also be applied to sensor networks. Finally, an important area of application of wireless sensor networking is the medical field. In [91], an analysis of network lifetime using IEEE 802.15.4 sensor networks is derived for this kind of applications.

## References


1. W. Shi, T.W. Sun, R.D. Wesel, Quasi-convexity and optimal binary fusion for distributed detection with identical sensors in generalized gaussian noise. IEEE Trans. Inform. Theory **47**(1), 446–450 (2001)
2. G. Ferrari, R. Pagliari, Decentralized detection in sensor networks with noisy communication links. In: F. Davoli, S. Palazzo, S. Zappatore (eds) *Distributed Cooperative Laboratories: Networking, Instrumentation, and Measurements (Signals and Communication Technology)* (Springer, New York, 2006)
3. G. Ferrari, R. Pagliari, Decentralized binary detection with noisy communication links. IEEE Trans. Aerosp. Electron. Syst. **42**(4), 1554–1563 (2006)
4. B. Chen, L. Tong, P.K. Varshney, Channel aware distributed detection in wireless sensor networks. IEEE Signal Process. Mag. **23**(4), 16–26 (2006). Special Issue on Distributed Signal Processing for Sensor Networks
5. Q. Cheng, B. Chen, P.K. Varshney, Detection performance limits for distributed sensor networks in the presence of non-ideal channels. IEEE Trans. Wirel. Commun. **5**(11), 3034–3038 (2006)
6. R. Niu, B. Chen, P.K. Varshney, Fusion of decisions transmitted over Rayleigh fading channels in wireless sensor networks. IEEE Trans. Signal Process. **54**(3), 1018–1027 (2006)
7. B. Chen, R. Jiang, T. Kasetkasem, P.K. Varshney, Channel aware decision fusion in wireless sensor networks. IEEE Trans. Signal Process. **52**(12), 3454–6458 (2004)
8. G. Ferrari, M. Martalò, R. Pagliari, Decentralized detection in clustered sensor networks. IEEE Trans. Aerosp. Electron. Syst., **47**(2), April (2011)
9. A. Papoulis, *Probability, Random Variables and Stochastic Processes* (McGraw-Hill, New York, 1991)
10. R. Meester, R. Roy, *Continuum Percolation* (Cambridge University Press, Cambridge, 1996)
11. G.R. Grimmet, *Percolation* (Springer, New York, 1999)
12. O. Dousse, F. Baccelli, P. Thiran, Impact of interferences on connectivity in ad hoc networks, in *Proceedings of IEEE Conference on Computer Communication (INFOCOM)*, vol. 3, San Francisco, USA, April 2003, pp. 1724–1733

13. L. Booth, J. Brook, M. Franceschetti, R. Meester, Continuum percolation and the geometry of wireless networks. Ann. Appl. Prob. **13**(2), 722–733 (2003)
14. O.K. Tonguz, G. Ferrari, *Ad Hoc Wireless Networks: A Communication-Theoretic Perspective* (Wiley, Chichester, 2006)
15. T.M. Cover, J.A. Thomas, *Elements of Information Theory* (Wiley, New York, 1991)
16. G. Ferrari, M. Martalò, R. Pagliari, On multi-level decentralized detection in sensor networks, in *Proceedings of International Conference on Intelligent Systems and Computing: Theory and Applications (ISYC)*, Ayia Napa, Cyprus, July 2006
17. Opnet Website, http://www.opnet.com
18. N. I. of Standards and T. N. Website, http://www.nist.gov
19. G. Ferrari, P. Medagliani, M. Martalò, A. Muzzini, Zigbee wireless sensor networks with data fusion, in *Proceedings of International Symposium on Communications, Control and Signal Processing (ISCCSP)*, St. Julians, Malta, March 2008, pp. 472–477
20. Cross-Bow, Wireless sensor networks, http://www.xbow.org
21. Atmel Corporation, Atmel microcontroller, http://www.atmel.com/atmel/acrobat/doc2467.pdf
22. Texas Instruments, RF/IF and ZigBee solutions, http://www.ti.com/
23. R.E. Ziemer, *Elements of Engineering Probability & Statistics* (Prentice-Hall, Upper Saddle River, 1997)
24. G. Ferrari, M. Martalò, Extending the lifetime of sensor networks through adaptive reclustering, EURASIP J. Wirel. Commun. Netw. **2007**, 20 pages, Special Issue on "Novel Techniques for Analysis & Design of Cross-Layer Optimized Wireless Sensor Networks", article ID 31809. doi:10.1155/2007/31809
25. J.H. Conway, R.K. Guy, *The Book of Numbers* (Springer, New York, 1996)
26. IEEE 802.15.4 Std, *Wireless Medium Access Control (MAC) and Physical Layer (PHY) Specifications for Low-Rate Wireless Personal Area Networks (LR-WPANs)* (IEEE Computer Society Press, 2003), pp. 1–679, ISBN: 0-7381-3677-5
27. G. Ferrari, P. Medagliani, S. Di Piazza, M. Martalò, Wireless sensor networks: performance analysis in indoor scenarios. EURASIP J. Wirel. Commun. Netw. **2007**, 14 pages (2007), Special Issue on "MobileMAN (Mobile Multi-hop Ad Hoc Networks): From Theory to Reality, article ID 81864. doi:10.1155/2007/81864
28. J. Ma, M. Gao, Q. Zhang, L.M. Ni, W. Zhu, Localized low-power topology control algorithms in IEEE 802.15.4-based sensor networks, in *Proceedings of IEEE International Conference on Distributed Computing Systems*, Columbus, OH, USA, June 2005, pp. 27–36
29. G. Ferrari, P. Medagliani, M. Martalò, Performance analysis of Zigbee wireless sensor networks with relaying, in *Proceedings of International Workshop on Distributed Cooperative Laboratories (Ingrid)*, Santa Margherita Ligure, Italy, April 2007
30. G. Ferrari, R. Pagliari, M. Martalò, Decentralized binary detection with non-constant SNR profile at the sensors. Int. J. Sens. Netw. **4**(1), 23–36 (2008). Special Issue on Energy-Efficient Algorithm and Protocol Design in Sensor Networks
31. J. Tsitsiklis, Decentralized detection by a large number of sensor. Math. Control Signals Syst. (MCSS) **1**(2), 167–182 (1988)
32. S. Alhakeem, P.K. Varshney, A unified approach to the design of decentralized detection systems. IEEE Trans. Aerosp. Electron. Syst. **31**(1), 9–20 (1995)
33. P. Willett, B. Tober, P. Swaszek, Fully-connected non-hierarchical decentralized detection networks, in *Proceedings of IEEE Conference on Control Applications*, Dayton, OH, USA, September 1992, pp. 404–409
34. A.B. Carlson, P.B. Crilly, J.C. Rutledge, *Communication Systems, An Introduction to Signals and Noise in Electrical Communication*, 4th edn (McGraw-Hill, New York, 2002)
35. A.F. Molisch, *Wireless Communications* (Wiley, Chichester, 2005)
36. J.G. Proakis, *Digital Communications*, 4th edn (McGraw-Hill, New York, 2001)
37. X. Hu, E. Eleftheriou, D. Arnold, Regular and irregular progressive edge-growth tanner graphs. IEEE Trans. Inform. Theory **51**(1), 386–398 (2005)
38. R.G. Gallager, *Low Density Parity Check Codes* (MIT Press, Cambridge, 1963)

39. F.R. Kschischang, B.J. Frey, H.A. Loeliger, Factor graphs and the sum-product algorithm. IEEE Trans. Inform. Theory **47**(2), 498–519 (2001)
40. I. Akyildiz, W. Su, Y. Sankarasubramaniam, E. Cayirci, A survey on sensor networks. IEEE Commun. Mag. **40**(8), 102–114 (2002)
41. C.Y. Chong, S.P. Kumar, Sensor networks: evolution, challenges, and opportunities. Proc. IEEE **91**(8), 1247–1256 (2003)
42. S.N. Simic, S. Sastry, Distributed environmental monitoring using random sensor networks, in *Proceedings of International Work. on Inform. Processing in Sensor Networks (IPSN)*, Palo Alto, CA, USA, April 2003, pp. 582–592
43. J.N. Tsitsiklis, Decentralized detection. In: H.V. Poor, J.B. Thomas (eds) *Advances in Statistical Signal Process*, vol. 2 (JAI Press, Greenwich, 1993), pp. 297–344
44. R. Verdone, D. Dardari, G. Mazzini, A. Conti, *Wireless Sensor and Actuator Networks: Technologies, Analysis and Design* (Elsevier, London, 2008)
45. R.R. Tenney, N.R. Sandell, Detection with distributed sensors. IEEE Trans. Aerosp. Electron. Syst. **17**(4), 501–510 (1981)
46. A. Reibman, L. Nolte, Design and performance comparison of distributed detection networks. IEEE Trans. Aerosp. Electron. Syst. **23**(6), 789–737 (1987)
47. R. Viswanathan, P.K. Varshney, Distributed detection with multiple sensors—Part I: fundamentals. Proc. IEEE **85**(1), 54–63 (1997)
48. J.F. Chamberland, V.V. Veeravalli, Decentralized detection in sensor networks. IEEE Trans. Signal Process. **51**(2), 407–416 (2003)
49. H. Gharavi, K. Ban, Multihop sensor network design for wide-band communications. Proc. IEEE **91**(8), 1221–1234 (2003)
50. I.Y. Hoballah, P.K. Varshney, An information theoretic approach to the distributed detection problem. IEEE Trans. Inform. Theory **35**(5), 988–994 (1989)
51. Z.Q. Luo, An isotropic universal decentralized estimation scheme for a bandwidth constrained ad hoc sensor network. IEEE J. Select. Areas Commun. **23**(4), 735–744 (2005)
52. R. Blum, A. Kassam, H. Poor, Distributed detection with multiple sensors: Part II. Proc. IEEE **85**(1), 64–79 (1997)
53. T.S. Rappaport, *Wireless Communications, Principles & Pratice*, 2nd edn (Prentice-Hall, Uper Saddle River, 2002)
54. B. Chen, P.K. Willett, On the optimality of the likelihood-ratio test for local sensor decision rules in the presence of nonideal channels. IEEE Trans. Inform. Theory **51**(2), 693–699 (2005)
55. M. Madishetty, V. Kanchumarthy, C.H. Gowda, R. Viswanathan, Distributed detection with channel errors, in *Proceedings of Southeastern Symposium on System Theory (SSST)*, Tuskegee University, AL, USA, March 2005, pp. 302–306
56. A.R. Reibman, L.W. Nolte, Optimal design and performance of distributed signal detection systems with faults. IEEE Trans. Acoust., Speech Signal Process. **38**(10), 1771–1782 (1990)
57. S.C.A. Thomopoulos, L. Zhang, Distributed decision fusion in the presence of networking delays and channel errors. Inform. Sci. **66**(1–2), 91–118 (1992)
58. N. Patwari, A.O. Hero, Hierarchical censoring for distributed detection in wireless sensor networks, in *Proceedings of International Conference on Acoustics, Speech and Signal Processing (ICASSP)*, vol 4. Hong Kong, April 2003, pp. 848–851
59. M. Gastpar, M. Vetterli, P.L. Dragotti, Sensing reality and communicating bits: a dangerous liaison. IEEE Signal Process. Mag. **23**(4), 70–83 (2006)
60. K. Yamasaki, T. Ohtsuki, Design of energy-efficient wireless sensor networks with censoring, on–off, and censoring and on–off sensors based on mutual information, in *Proceedings of IEEE Vehicular Technology Conference (VTC)*, vol 2, Stockholm, Sweden, May 2005, pp. 1312–1316
61. Y. Lin, B. Chen, P.K. Varshney, Decision fusion rules in multi-hop wireless sensor networks. IEEE Trans. Aerosp. Electron. Syst. **41**(2), 475–488 (2005)
62. N.A.C. Cressie, *Statistics for Spatial Data* (Wiley, New York, 1993)

63. C. Guestrin, A. Krause, A.P. Singh, Near-optimal sensor placement in Gaussian processes, in *Proceedings of International Conference on Machine Learning*, vol. 119, Bonn, Germany, August 2005, pp. 265–272
64. H. Wang, K. Yao, D. Estrin, Information-theoretic approaches for sensor selection and placement in sensor networks for target localization and tracking. J. Commun. Netw. **7**(4), 438–449 (2005)
65. T. Ikeda, H. Ishiguro, M. Asada, Adaptive fusion of sensor signals based on mutual information maximization, in *Proceedings of International Conference on Robotics and Automation (ICRA)*, vol 3, Taipei, Taiwan, September 2003, pp. 4398–4402
66. F. Gini, F. Lombardini, L. Verrazzani, Decentralised detection strategies under communication constraints. IEE Proc. Radar Sonar Navig. **145**(4), 199–208 (1998)
67. R. Jiang, B. Chen, Fusion of censored decisions in wireless sensor networks. IEEE Trans. Wirel. Commun. **4**(6), 2668–2673 (2005)
68. R. Ahlswede, N. Cai, S.Y.R. Li, R.W. Yeung, Network information flow. IEEE Trans. Inform. Theory **46**(4), 1204–1216 (2000)
69. S.Y.R. Li, R.W. Yeung, N. Cai, Linear network coding. IEEE Trans. Inform. Theory **49**(2), 371–381 (2003)
70. R. Koetter, M. Medard, An algebraic approach to network coding. IEEE/ACM Trans. Netw. **11**(5), 782–795 (2003)
71. X. Bao, J. Li, Matching code-on-graph with network-on-graph: adaptive network coding for wireless relay networks, in *Proceedings of Allerton Conference on Communication, Control and Computing (ALLERTON)*, Urbana Champaign, IL, USA, September 2005
72. C. Fragouli, E. Soljanin, Information flow decomposition for network coding. IEEE Trans. Inform. Theory **52**(3), 829–848 (2006)
73. D. Tuninetti, C. Fragouli, On the throughput improvement due to limited complexity processing at relay nodes, in *Proceedings of IEEE Symposium on Information Theory (ISIT)*, Adelaide, Australia, September 2005, pp. 1081–1085
74. M. Xiao, T. Aulin, A physical layer aspect of network coding with statistically independent noisy channels, in *Proceedings of IEEE International Conference on Communication (ICC)*, vol 9, Instanbul, Turkey, June 2006, pp. 3996–4001
75. C. Hausl, P. Dupraz, Joint network-channel coding for the multiple-access relay channel, in *Proceedings of IEEE Conference on Sensor and Ad Hoc Communications and Networks (SECON)*, vol 3, New York, NY, USA, June 2006, pp. 817–822
76. C.-C. Chang, H.-N. Lee, Space-time mesh codes for the multiple-access relay network: space vs. time diversity benefits, in *Proceedings of Workshop on Network Coding, Theory, and Applications (NetCod)*, San Diego, CA, USA, January 2007
77. A. Kansal, A. Ramamoorthy, M. Srivastava, G. Pottie, On sensor network lifetime and data distortion, in *Proceedings of IEEE Symposium on Information Theory (ISIT)*, Adelaide, Australia, September 2005, pp. 6–10
78. S. Arnon, Deriving an upper bound on the average operation time of a wireless sensor network. IEEE Commun. Lett. **9**(2), 154–156 (2005)
79. F. Ordonez, B. Krishnamachari, Optimal information extraction in energy-limited wireless sensor networks. IEEE J. Select. Areas Commun. **22**(6), 1121–1129 (2004)
80. H. Zhang, J. Hou, On deriving the upper bound of lifetime for large sensor networks, in *Proceedings of ACM International Symposium on Mobile Ad Hoc Networking and Computing (MOBIHOC)*, Tokyo, Japan, May 2004, pp. 121–132
81. Z. Hu, B. Li, On the fundamental capacity and lifetime limits of energy-constrained wireless sensor networks, in *Proceedings of IEEE Real-Time and Embedded Technology and Applications Symp. (RTAS)*, Toronto, Canada, May 2004, pp. 2–9
82. D.M. Blough, P. Santi, Investigating upper bounds on network lifetime extension for cell-based energy conservation techniques in stationary ad-hoc networks, in *Proceedings of ACM International Conference on Mobile Computing and Networking (MOBICOM)*, Atlanta, GA, USA, September 2002, pp. 183–192

83. M. Bhardwaj, T. Garnett, A.P. Chandrakasan, Upper bounds on the lifetime of sensor networks, in *Proceedings of IEEE International Conference on Communication (ICC)*, vol 119, Helsinki, Finland, June 2001, pp. 785–790
84. M. Bhardwaj, A.P. Chandrakasan, Bounding the lifetime of sensor networks via optimal role assignments, in *Proceedings of IEEE Conference on Computer Communications (INFOCOM)*, vol 3, New York, NY, USA, June 2002, pp. 1587–1596
85. V. Rai, R.N. Mahapatra, Lifetime modeling of a sensor network, in *Proceedings of Design, Automation and Test in Europe (DATE)*, vol 1, Messe Munich, Germany, March 2005, pp. 202–203
86. Y. Chen, Q. Zhao, On the lifetime of wireless sensor networks. IEEE Commun. Lett. **9**(11), 976–978 (2005)
87. Q. Zhao, A. Swami, L. Tong, The interplay between signal processing and networking in sensor networks. IEEE Signal Process. Mag. **23**(4), 84–93 (2006)
88. K. Kalpakis, K. Dasgupta, P. Namjoshi, Maximum lifetime data gathering and aggregation in wireless sensor networks, University of Maryland, Baltimore, Tech. Rep., 2002, http://www.csee.umbc.edu/~kalpakis/
89. M.E.S. Coleri, T.J. Koo, Lifetime analysis of a sensor network with hybrid automata modelling, in *Proceedings of International Workshop on Wireless Sensor Networks and Applications (WSNA)*, Atlanta, USA, September 2002, pp 98–104
90. M. Franceschetti, R. Meester, Critical node lifetime in random networks via the Chen-Stein method. IEEE Trans. Inform. Theory **52**(6), 2831–2837 (2006)
91. N.F. Timmons, W.G. Scanlon, Analysis of the performance of IEEE 802.15.4 for medical sensor body area networking, in *Proceedings of IEEE Conference on Sensor and Ad Hoc Communications and Networks (SECON)*, Santa Clara, CA, USA, October 2004, pp. 16–24

# Chapter 3
# Distributed Detection of Spatially Non-constant Phenomena

In this chapter, we study sensor networks with distributed detection of a *spatially non-constant* phenomenon. In particular, we consider *binary* phenomena characterized by a generic number of status changes (from state "0" to state "1" or vice-versa) across the sensors. We first derive the Mean Square Error (MMSE) fusion algorithm at the Access Point (AP). Then, we propose simplified (sub-optimum) fusion algorithms at the AP, with a lower computational complexity. While we first consider a scenario with *ideal* communication links between the sensors and the AP, we then extend our framework to scenarios with *noisy* communication links.

The structure of this chapter is the following. In Sect. 3.1, we derive MMSE and simplified fusion rules at the AP in a scenario with ideal communication links and both single and multi-boundary phenomena. In Sect. 3.2, we extend the previous fusion rules by taking into account the noise in the communication links between the sensors and the AP. In Sect. 3.3, numerical results on the performance of the proposed fusion algorithms are presented. In Sect. 3.4, the computational complexity of the proposed algorithms, in terms of the number of required operations, is presented. Finally, concluding remarks are given in Sect. 3.5 and a brief review of the literature is presented in Sect. 3.6.

## 3.1 Ideal Communication Links

We assume that the status of the phenomenon under observation is characterized by a *generic* number $n_{\rm bs}$ of boundaries. For the ease of simplicity, the status of the phenomenon will be supposed independent from sensor to sensor. The existence of *correlation* between sensors would require an extension of the derived algorithms. This extension is investigated in [1]. Moreover, the proposed simplified algorithms do not require the knowledge of a possible correlation among the sensors. In particular, we preliminary investigate the performance when the communication

C. Buratti et al., *Sensor Networks with IEEE 802.15.4 Systems*,
Signals and Communication Technology, DOI: 10.1007/978-3-642-17490-2_3,

links between the sensors and the AP are *ideal*, i.e., no noise is introduced during data transmission.

Denote the overall phenomenon status as $\boldsymbol{H} = [H_1, \ldots, H_n]$ with $H_i = 0$ or $H_i = 1$ $(i = 1, \ldots, n)$.[1] As in Chap. 2, the signal observed at the $i$th sensor can be expressed, according to the observable model in (2.1), as

$$r_i = c_{\mathrm{E},i} + n_i \quad i = 1, \ldots, n$$

and the common Signal-to-Noise Ratio (SNR) at the sensors can be defined as follows:

$$\mathrm{SNR}_{\mathrm{sensor}} = \frac{s^2}{\sigma^2}.$$

Each sensor processes (through proper quantization) the observed signal and the output value by the $i$th sensor is denoted as $d_i \triangleq f_{\mathrm{quant}}(r_i)$, where the function $f_{\mathrm{quant}}(\cdot)$ depends on the specific quantization strategy. In the following, we consider (1) binary quantization and (2) absence of quantization. In [2], the proposed analytical framework is extended to scenarios with multi-level quantization at the sensors. Upon the reception of the messages sent by the sensors, the goal of the AP is to reconstruct, through an MMSE or simplified fusion strategy, the status of the distributed binary phenomenon $\boldsymbol{H}$. More precisely, in the considered setting the AP needs to estimate correctly the position of the boundary.

### 3.1.1 MMSE Fusion Rule

The following assumptions are expedient to simplify the derivation of the MMSE detection strategy:

- changes of the phenomenon status are not admitted in correspondence to the first and last sensors: the number of boundaries must then be such that $1 \leq n_{\mathrm{bs}} \leq n - 2$ (in particular, $H_n = H_{n-1}$);
- the phenomenon status is perfectly known at the first sensor. In particular, we assume $H_1 = 0$.

According to the above assumptions, the $n_{\mathrm{bs}}$ boundaries $\{\alpha_1, \ldots, \alpha_{n_{\mathrm{bs}}}\}$ have to simultaneously satisfy the following inequalities:

$$2 \leq \alpha_1 < \alpha_2 < \cdots < \alpha_{k-1} < \alpha_k < \cdots < \alpha_{n_{\mathrm{bs}}} \leq n - 1. \tag{3.1}$$

[1] Note that, while in Chap. 2 we consider $H_0 = 0$ and $H_1 = 1$, now the subscript of the phenomenon $H$ designates the specific sensor. The change of notation is due to the sake of analytical simplicity.

Therefore: between 1 and $\alpha_1 - 1$ the phenomenon status is "0"; between $\alpha_1$ and $\alpha_2 - 1$ the phenomenon status is "1"; and so on. In order for the boundary distribution to be realistic, the following conditions must necessarily hold:

$$\alpha_{k-1} < \alpha_k \leq (n-1) - (n_{\text{bs}} - k) = n - n_{\text{bs}} + k - 1 \quad k = 2, \ldots, n_{\text{bs}}. \tag{3.2}$$

For each value of $k$, condition (3.2) formalizes the intuitive idea that the $k$th boundary cannot fall beyond the $(n - 1 - n_{\text{bs}} + k)$th position, in order for the successive (remaining) $n_{\text{bs}} - k$ boundaries to have admissible positions.

#### 3.1.1.1 Binary Quantization

In this scenario, the $i$th sensor makes a decision comparing its observation $r_i$ with a threshold value $\tau_i$, and computes a local binary decision $d_i \in \{0, 1\}$, i.e., $f_{\text{quant}}(r_i) = U(r_i - \tau_i)$, where $U(\cdot)$ is the unit step function. To optimize the system performance, the thresholds $\{\tau_i\}$ need to be properly selected. As in Chap. 2 (scenarios with detection of spatially constant phenomena), a common value $\tau$ at all sensors is considered. This choice is intuitively motivated by the fact that the sensor SNR is constant across the sensors. In the presence of a spatially non-constant phenomenon, the threshold $\tau$ needs to be optimized in order to minimize the distance between the true phenomenon and its estimate at the AP. In [2], it is shown that the optimized value of $\tau$ corresponds to $s/2$, for every value of the number of sensors $n$.

Denoting as $\boldsymbol{\alpha}$ the sequence of boundaries $(\alpha_1, \ldots, \alpha_{n_{\text{bs}}})$, the MMSE estimation strategy leads to $\widehat{\boldsymbol{\alpha}} = \mathbb{E}[\boldsymbol{\alpha}|\boldsymbol{d}]$ [3]. Using the assumptions introduced at the beginning of this subsection, the generic component of the vector $\widehat{\boldsymbol{\alpha}}$ can be written as[2]

$$\widehat{\alpha}_k = \mathbb{E}[\alpha_k|\boldsymbol{d}] = \sum_{\alpha_k=1}^{n} \mathbb{P}\{\alpha_k|\boldsymbol{d}\} = \sum_{\alpha_k=k+1}^{n-n_{\text{bs}}+k-1} \alpha_k \mathbb{P}\{\alpha_k|\boldsymbol{d}\} \quad k = 1, \ldots, n_{\text{bs}} \tag{3.3}$$

where the upper and lower bounds of the sum in the last term are properly modified in order to take into account the constraint (3.2). The computation of (3.3) can be carried out by applying the following approach. The probability $\mathbb{P}\{\alpha_k|\boldsymbol{d}\} (k = 1, \ldots, n_{\text{bs}})$ can be obtained by marginalizing the joint probabilities of proper boundaries' sequences. By applying the Bayes formula and the total probability theorem [4], after a few manipulations the conditional probability mass function (p.m.f.) of $\boldsymbol{\alpha}$ can be expressed as

$$\mathbb{P}\{\boldsymbol{\alpha}|\boldsymbol{d}\} = \mathbb{P}\{\boldsymbol{d}|\boldsymbol{\alpha}\}\mathbb{P}\{\boldsymbol{\alpha}\} \left[ \sum_{\alpha_1=2}^{n-n_{\text{bs}}} \cdots \sum_{\alpha_k=k+1}^{n-n_{\text{bs}}+k-1} \cdots \sum_{\alpha_{n_{\text{bs}}}=n_{\text{bs}}+1}^{n-1} \mathbb{P}\{\boldsymbol{d}|\boldsymbol{\alpha}\}\mathbb{P}\{\boldsymbol{\alpha}\} \right]^{-1}. \tag{3.4}$$

[2] For ease of notational simplicity, in (3.3) we use the same symbol $\alpha_k$ to denote both the random variable (in the second term) and its realization (in the third and fourth terms). The same simplified notational approach will be considered in the remainder of Sect. 3.1. The context should eliminate any ambiguity.

The first multiplicative term at the right-hand side of (3.4) can be written as

$$\mathbb{P}\{\boldsymbol{d}|\boldsymbol{\alpha}\} = \prod_{i=1}^{n} \mathbb{P}\{d_i|\boldsymbol{\alpha}\} = \prod_{i=1}^{\alpha_1-1} \underbrace{\mathbb{P}\{d_i|\boldsymbol{\alpha}\}}_{H_i=0} \prod_{j=\alpha_1}^{\alpha_2-1} \underbrace{\mathbb{P}\{d_j|\boldsymbol{\alpha}\}}_{H_j=1} \cdots \prod_{q=\alpha_{n_{\text{bs}}}}^{n} \underbrace{\mathbb{P}\{d_q|\boldsymbol{\alpha}\}}_{H_q=0 \text{ or } 1} \tag{3.5}$$

where we have used the fact that the sensors' decisions are conditionally independent. Note that, in the last $n - \alpha_{n_{\text{bs}}} + 1$ terms, $H_i = 0$ if $n_{\text{bs}}$ is *even*, whereas $H_i = 1$ if $n_{\text{bs}}$ is *odd*. The component conditional probabilities at the right-hand side of (3.5) can be written as

$$\mathbb{P}\{d_i|\boldsymbol{\alpha}\} = \begin{cases} \mathbb{P}\left\{ n_i \underset{d_i=1}{\overset{d_i=0}{\lessgtr}} \tau \right\} & \text{if } i \in \mathscr{I}_0(\boldsymbol{\alpha}) \\ \mathbb{P}\left\{ n_i \underset{d_i=1}{\overset{d_i=0}{\lessgtr}} \tau - s \right\} & \text{if } i \in \mathscr{I}_0(\boldsymbol{\alpha}) \end{cases}$$
$$= \begin{cases} (1-d_i)\left[1 - \Phi\left(\frac{\tau}{\sigma}\right)\right] + d_i \Phi\left(\frac{\tau}{\sigma}\right) & \text{if } i \in \mathscr{I}_0(\boldsymbol{\alpha}) \\ (1-d_i)\left[1 - \Phi\left(\frac{\tau-s}{\sigma}\right)\right] + d_i \Phi\left(\frac{\tau-s}{\sigma}\right) & \text{if } i \in \mathscr{I}_1(\boldsymbol{\alpha}) \end{cases}$$

where

$$\mathscr{I}_\ell(\boldsymbol{\alpha}) \triangleq \{\text{indexes } i \text{ such that } H_i = \ell|\boldsymbol{\alpha}\} \quad \ell = 0, 1$$

and $\Phi(x)$ has been introduced in (2.53).

The second multiplicative term at the right-hand side of (3.4) can be written, using the chain rule [4], as

$$\mathbb{P}\{\boldsymbol{\alpha}\} = \prod_{i=1}^{n_{\text{bs}}} \mathbb{P}\{\alpha_i|\alpha_{i-1}, \ldots, \alpha_1\} = \prod_{i=2}^{n_{\text{bs}}} \mathbb{P}\{\alpha_i|\alpha_{i-1}\}\mathbb{P}\{\alpha_1\} \tag{3.6}$$

where we have used the fact that the position of the $i$th boundary depends only on the position of the (previous) $(i-1)$th boundary. The multiplicative terms at the right-hand side of (3.6) can be evaluated by observing that each boundary is spatially distributed according to the constraints in (3.2). In particular, by using combinatorics, one obtains

$$\mathbb{P}\{\alpha_1\} = \frac{1}{n - n_{\text{bs}} + 1}$$
$$\mathbb{P}\{\alpha_k|\alpha_{k-1}\} = \frac{1}{n - n_{\text{bs}} + k - \alpha_{k-1}} \quad k = 2, \ldots, n_{\text{bs}}.$$

The last term at the right-hand side of (3.4) (i.e., the denominator) can be easily computed by observing that it is composed by terms similar to those evaluated in (3.5) and (3.6).

Finally, the a posteriori probabilities of the boundaries' positions in (3.3) can be obtained by proper marginalization of the joint conditional p.m.f. in (3.4):

$$\mathbb{P}\{\alpha_k|\boldsymbol{d}\} = \sum_{\sim\{\alpha_k\}} \mathbb{P}\{\alpha_1,\ldots,\alpha_{n_{\mathrm{bs}}}|\boldsymbol{d}\} \quad k=1,\ldots,n_{\mathrm{bs}}$$

where $\sum_{\sim\{y_i\}} f(y_1,y_2,\ldots,y_n)$ is a short-hand notation for $\sum_{y_1}\cdots\sum_{y_{i-1}}\sum_{y_{i+1}}\cdots\sum_{y_n} f(y_1,y_2,\ldots,y_n)$ [5].

#### 3.1.1.2 Absence of Quantization

In this case, the observations at the sensors are not quantized and a local likelihood value, such as the conditional probability density function (p.d.f.) of the observable, is transmitted from each sensor to the AP. Obviously, this is not a realistic scenario, since an infinite bandwidth would be required to transmit a p.d.f. value. However, investigating this case allows to derive useful information about the limiting performance of the considered distributed detection schemes, since transmission of the p.d.f. of the observables does not entail any information loss at the sensors.

The estimated boundaries can be written, according to the assumptions outlined at the beginning of Sect. 3.1, as

$$\widehat{\alpha}_k = \mathbb{E}[\alpha_k|\boldsymbol{r}] = \sum_{\alpha_k=k+1}^{n-n_{\mathrm{bs}}+k-1} \alpha_k \mathbb{P}\{\alpha_k|\boldsymbol{r}\} \quad k=1,\ldots,n_{\mathrm{bs}}. \tag{3.7}$$

The probabilities at the right-hand side of (3.7) can be obtained, as in Sect. 3.1.1.1, through proper marginalization of joint conditional probabilities of the following type:

$$\mathbb{P}\{\boldsymbol{\alpha}|\boldsymbol{r}\} = p(\boldsymbol{r}|\boldsymbol{\alpha})\mathbb{P}\{\boldsymbol{\alpha}\} \cdot \left[\sum_{\alpha_1=2}^{n-n_{\mathrm{bs}}} \cdots \sum_{\alpha_i+1}^{n-n_{\mathrm{bs}}+i-1} \cdots \sum_{\alpha_{n_{\mathrm{bs}}}=\alpha_{n_{\mathrm{bs}}-1}+1}^{n-1} p(\boldsymbol{r}|\boldsymbol{\alpha})\mathbb{P}\{\boldsymbol{\alpha}\}\right]^{-1}.$$

Under the assumption of independent sensors' observations, it holds that

$$p(\boldsymbol{r}|\boldsymbol{\alpha}) = \prod_{i=1}^{n} p(r_i|\boldsymbol{\alpha})$$

where

$$p(r_i|\boldsymbol{\alpha}) = \begin{cases} p_{\mathcal{N}}(r_i) & \text{if } i \in \mathcal{I}_0(\boldsymbol{\alpha}) \\ p_{\mathcal{N}}(r_i - s) & \text{if } i \in \mathcal{I}_1(\boldsymbol{\alpha}) \end{cases} \tag{3.8}$$

and $p_{\mathcal{N}}(u) \triangleq \frac{1}{\sqrt{2\pi\sigma^2}} \exp(-\frac{u^2}{2\sigma^2})$.

#### 3.1.1.3 Remarks

We would like to remark that the MMSE strategy outlined above is based, regardless of the quantization strategy, on the assumption of knowledge of the number of boundaries $n_{\text{bs}}$ at the AP. However, in several scenarios this knowledge may not be a priori available and $n_{\text{bs}}$ should be properly estimated. The performance of the MMSE algorithm with knowledge of $n_{\text{bs}}$ at the AP can then be used as a benchmark for the performance of the simplified (and feasible) detection algorithms introduced in Sect. 3.1.3.

### *3.1.2 Simplified Fusion Rule with a Single Boundary*

Since the computational complexity of the MMSE fusion strategy rapidly increases with the number of sensors [6], in this subsection we derive, under the assumption of single boundary phenomena, a simplified low-complexity fusion algorithm. The key idea of this simplified algorithm consists in approximating the MMSE boundary estimate $\widehat{\alpha} = \mathbb{E}[\alpha|\boldsymbol{d}]$, which involves a *statistical* average, with a simpler *deterministic* expression. Note that the proposed approach relies on the fact that our goal is to estimate a *single* boundary.

#### 3.1.2.1 Binary Quantization

In this case, the boundary position is estimated as follows:

$$\widehat{\alpha} \simeq \underset{1 \le j \le n}{\operatorname{argmin}} \left\{ \sum_{i=1}^{j-1} |d_i|^2 + \sum_{i=j}^{n} |d_i - 1|^2 \right\}. \tag{3.9}$$

The intuition behind the estimation strategy in (3.9) is based on the fact that there is a single boundary: the initial sensors' decisions (from the 1st to the $(j-1)$th sensor) are compared with "0," whereas the others (from the $j$th to the $n$th sensor) are compared with "1." The estimated boundary minimizes the simplified cost function $|\boldsymbol{d} - \boldsymbol{d}_j|^2$, where $\boldsymbol{d}_j \triangleq [0, \ldots, 0, \underbrace{1}_{j\text{th position}}, 1, \ldots, 1]$, over all possible values of $j \in \{1, \ldots, n\}$.

#### 3.1.2.2 Absence of Quantization

In this scenario, the a posteriori probabilities of the two hypotheses at each sensor, conditionally on the observables, can be used to derive the proper objective function to be maximized. In this case, one can write[3]

$$\widehat{\alpha} \simeq \underset{1 \leq j \leq n}{\operatorname{argmax}} \left\{ \sum_{i=1}^{j-1} \mathbb{P}\{H_i = 0 | r_i\} + \sum_{i=j}^{n} \mathbb{P}\{H_i = 1 | r_i\} \right\} \tag{3.10}$$

where, using Bayes formula and assuming $\mathbb{P}\{H_i = 0\} = \mathbb{P}\{H_i = 1\} \forall i$,

$$\mathbb{P}\{H_i = \ell | r_i\} = \frac{p(r_i | H_i = \ell)}{p(r_i | H_i = 0) + p(r_i | H_i = 1)} = \frac{p_{\mathcal{N}}(r_i - \ell \cdot s)}{p_{\mathcal{N}}(r_i) + p_{\mathcal{N}}(r_i - s)} \quad \ell = 0, 1.$$

### *3.1.3 Simplified Fusion Rule with Multiple Boundaries*

Obviously, the computational complexity of the MMSE distributed detection strategy in scenarios with an arbitrary number of phenomenon boundaries increases more rapidly than in scenarios with a single phenomenon boundary (see Sect. 3.4 for more details). Therefore, the derivation of simplified fusion algorithms with low complexity (but limited performance loss) is crucial.

A first possible choice is a direct extension of the sub-optimal approach in Sect. 3.1.2 for scenarios with a single phenomenon status change. However, this class of simplified fusion algorithms is not efficient in a scenario with multiple boundaries, since the number of comparisons with all possible sequences of boundaries increases exponentially with the number of sensors. Therefore, we now introduce another class of reduced-complexity fusion algorithms, which do not make use of these comparisons. As before, we distinguish between two possible quantization strategy at the sensors.

#### 3.1.3.1 Binary Quantization

Define the following function:

$$f_{\mathrm{bq}}(k, \boldsymbol{d}_1^k) \triangleq \sum_{i=1}^{k} [\mathbb{P}\{H_i = 0 | d_i\} - \mathbb{P}\{H_i = 1 | d_i\}] \quad k = 1, \ldots, n \tag{3.11}$$

where $\boldsymbol{d}_1^k = (d_1, \ldots, d_k)$. The key idea of our approach is the following. The function $f_{\mathrm{bq}}(k, \boldsymbol{d}_1^k)$ is monotonically increasing (or decreasing), with respect to $k$,

[3] Note that in (3.10) the "argmax" function is used, instead of the "argmin" function used in (3.9), since the objective function needs to be maximized.

while the phenomenon does not change its status. In correspondence to each change of the phenomenon status, the function $f_{\mathrm{bq}}(k, \boldsymbol{d}_1^k)$ changes its monotonic behavior. More precisely, a phenomenon variation from "0" to "1" corresponds to a change, trend-wise, from increasing to decreasing; a phenomenon variation from "1" to "0" corresponds to a change, trend-wise, from decreasing to increasing. Therefore, by detecting the changes of the monotonic behavior of $f_{\mathrm{bq}}$ one can estimate the positions of the boundaries. A graphical description of the behavior of $f_{\mathrm{bq}}$ is shown in Fig. 3.1, where the phenomenon under observation and the function $f_{\mathrm{bq}}$ are shown, together with the estimated boundaries. In this pictorial example, the estimated phenomenon coincides with the observed phenomenon.

Note that the proposed algorithm does not take into account the number of boundaries $n_{\mathrm{bs}}$ in the observed phenomenon. However, as we will highlight in Sect. 2.4.5, our numerical results show that the algorithm estimates accurately the number of boundaries for sufficiently high values of the sensor SNR, i.e., when the quality of the sensors' observations is sufficiently high. Obviously, one may modify the estimation strategy in order to take into account the value of $n_{\mathrm{bs}}$. This will lead to an improvement for small values of the sensor SNR, i.e., a scenario which is not of interest for practical applications. The same considerations on possible refinement of the estimation strategy also hold in the presence of multi-level quantization or in the absence of quantization.

The probability $\mathbb{P}\{H_i = \ell | d_i\} (\ell = 0, 1; i = 1, \ldots, n)$ in (3.11) can be written, by applying the Bayes formula and following an approach similar to that in Sect. 3.1.2.2, as

$$\mathbb{P}\{H_i = \ell | d_i\} = \frac{\mathbb{P}\{d_i | H_i = \ell\}}{\mathbb{P}\{d_i | H_i = 0\} + \mathbb{P}\{d_i | H_i = 1\}}$$

where we have used the fact that $\mathbb{P}\{H_i = 0\} = \mathbb{P}\{H_i = 1\}$ and

$$\begin{aligned}\mathbb{P}\{d_i | H_i = \ell\} &= \begin{cases} \mathbb{P}\{s \cdot \ell + n_i < \tau\} = \mathbb{P}\{n_i < \tau - s \cdot \ell\} & \text{if } d_i = 0 \\ \mathbb{P}\{s \cdot \ell + n_i > \tau\} = \mathbb{P}\{n_i > \tau - s \cdot \ell\} & \text{if } d_i = 1 \end{cases} \\ &= (1 - d_i)\Phi\left(\frac{\tau - s \cdot \ell}{\sigma}\right) + d_i\left[1 - \Phi\left(\frac{\tau - s \cdot \ell}{\sigma}\right)\right]. \end{aligned} \tag{3.12}$$

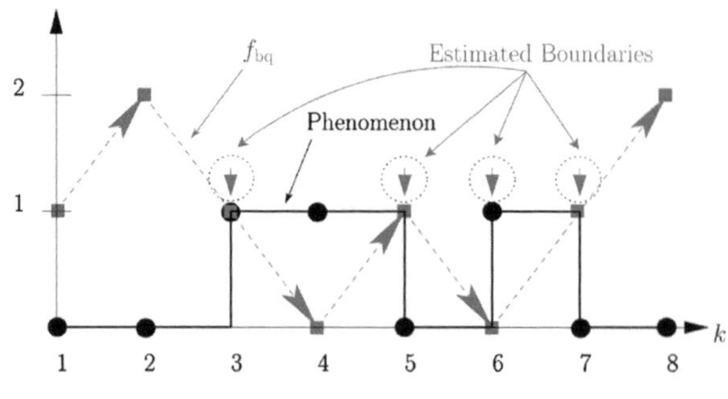


**Fig. 3.1** Illustrative example: the phenomenon under observation (*solid line with circles*) and the corresponding function $f_{\mathrm{bq}}$ in (3.11) (*dashed arrows*). The estimated boundaries are indicated by *vertical arrows*

#### 3.1.3.2 Absence of Quantization

In the absence of quantization at the sensors, one can introduce the following function:

$$f_{\mathrm{nq}}(k, \boldsymbol{r}_1^k) \triangleq \sum_{i=1}^{k} [\mathbb{P}\{H_i = 0|r_i\} - \mathbb{P}\{H_i = 1|r_i\}] \quad k = 1, \ldots, n$$

where $\boldsymbol{r}_1^k = (r_1, \ldots, r_k)$ and

$$\mathbb{P}\{H_i = \ell|r_i\} = \frac{p(r_i|H = \ell)}{p(r_i|H = 0) + p(r_i|H = 1)} \quad l = 0, 1;\ i = 1, \ldots, n$$

with $p(r_i|H = \ell) = p_{\mathcal{N}}(r_i - \ell \cdot s)$. The fusion algorithm at the AP is then identical to that presented in the case with binary quantization, but for the use of $f_{\mathrm{nq}}$ at the place of $f_{\mathrm{bq}}$.

#### 3.1.3.3 Remarks

One should observe that, unlike the MMSE strategy, the simplified detection algorithm (with binary quantization and no quantization, respectively) does *not* require knowledge of the number of boundaries $n_{\mathrm{bs}}$ at the AP. Obviously, we expect that the proposed algorithm will incur a performance degradation with respect to the MMSE algorithm. However, this loss will be limited, as shown with simulation results in Sect. 3.3.

## 3.2 Noisy Communication Links

In this section, we investigate the impact of noisy communication links (between the sensors and the AP) on the structures and performance of the proposed fusion algorithms. In particular, we focus on scenarios with *multi-boundary* phenomena, since the fusion rules for the scenarios with single boundary phenomena and noisy communication links can be easily derived from the equivalent scenarios with ideal communication links.

### *3.2.1 MMSE Fusion Rule*

#### 3.2.1.1 Binary Quantization

In this case, the noisy communication links between the sensors and the AP are modeled as independent Binary Symmetric Channels (BSCs). Here, we denote as $\boldsymbol{d}$

the sequence of binary decisions at the sensors (as in Sect. 3.1.1) and as $\boldsymbol{d}^{\mathrm{AP}}$ the sequence of binary decisions received at the AP. Because of the presence of BSCs, the received decisions $\boldsymbol{d}^{\mathrm{AP}}$ might differ from $\boldsymbol{d}$ (there could be "bit-flipping" in some of the links). In this scenario, the MMSE estimation strategy at the AP becomes:

$$\widehat{\boldsymbol{\alpha}} = \mathbb{E}\left[\boldsymbol{\alpha}|\boldsymbol{d}^{\mathrm{AP}}\right].$$

The analytical framework described in (3.3–3.6) can be applied to this scenario as well, by replacing $\boldsymbol{d}$ with $\boldsymbol{d}^{\mathrm{AP}}$. In particular, the $i$th decision received at the AP can be expressed, using the BSC model, as

$$d_i^{\mathrm{AP}} = \begin{cases} d_i & \text{with probability } (1-p) \\ 1-d_i & \text{with probability } p \end{cases}$$

where $p$ is the cross-over probability of the BSC. After a few manipulations, one obtains:

$$\begin{aligned} \mathbb{P}\{d_i^{\mathrm{AP}}|\boldsymbol{\alpha}\} &= p + (1-2p)\mathbb{P}\left\{ n_i \underset{d_i^{\mathrm{AP}}=1}{\overset{d_i^{\mathrm{AP}}=0}{\lessgtr}} \tau - s\cdot\ell \right\} \\ &= p + (1-2p)\left\{(1-d_i^{\mathrm{AP}})\Phi\left(\frac{\tau - s\cdot\ell}{\sigma}\right) + d_i^{\mathrm{AP}} Q\left(\frac{\tau - s\cdot\ell}{\sigma}\right)\right\} \\ &\quad \text{if } i \in \mathscr{I}_\ell(\boldsymbol{\alpha}) \end{aligned}$$

where $\ell = 0, 1$ and $Q(x)$ has been introduced in Sect. 2.1.2.1.

#### 3.2.1.2 Absence of Quantization

In a scenario with the absence of quantization, i.e., where the sensors transmit real numbers (the likelihood values) to the AP, the BSC model for noisy communication links does not apply. In order to obtain results comparable with those associated with a scenario with binary quantization, we consider Additive White Gaussian Noise (AWGN) communication links. In other words, the $i$th observable at the AP ($i = 1, \ldots, n$), denoted as $r_i^{\mathrm{AP}}$, can be written as

$$r_i^{\mathrm{AP}} = r_i^{\mathrm{sensor}} + n_i^{\mathrm{comm}} \tag{3.13}$$

where $r_i^{\mathrm{sensor}}$ is the observable transmitted by the $i$th sensor and $n_i^{\mathrm{comm}}$ has a Gaussian distribution $\mathcal{N}(0, \sigma_{\mathrm{comm}}^2)$. The value of $\sigma_{\mathrm{comm}}^2$ is set in order to be consistent with the value of the cross-over probability $p$ in the scenario with BSCs.

In particular, in the case with uncoded Binary Phase Shift Keying (BPSK) transmission over AWGN links, the following relation holds [7]:

$$p = Q\left(\sqrt{\frac{E_{\mathrm{b}}}{\sigma^2_{\mathrm{comm}}}}\right). \tag{3.14}$$

Therefore, the value of $\sigma^2_{\mathrm{comm}}$ corresponding to a given value of the cross-over probability $p$ of the equivalent BSC can be obtained from (3.14). This will allow to make a fair performance comparison between the cases with binary quantization and without quantization.

After a few manipulations, one can conclude that the fusion rule described in Sect. 3.1.1.2 still holds, by replacing the conditional p.d.f. in (3.8) with the following:

$$p(r_i|\boldsymbol{\alpha}) = \begin{cases} p_{\mathrm{comm}}(r_i) & \text{if } i \in \mathscr{I}_0(\boldsymbol{\alpha}) \\ p_{\mathrm{comm}}(r_i - s) & \text{if } i \in \mathscr{I}_1(\boldsymbol{\alpha}) \end{cases} \quad i = 1, \ldots, n$$

where $p_{\mathrm{comm}}(r) \triangleq \frac{1}{\sqrt{2\pi(\sigma^2+\sigma^2_{\mathrm{comm}})}} \exp\left[-\frac{r^2}{2(\sigma^2+\sigma^2_{\mathrm{comm}})}\right]$.

## 3.2.2 Simplified Fusion Rule

### 3.2.2.1 Binary Quantization

In order to extend the reduced-complexity fusion algorithm introduced in Sect. 3.1.3.1 for a scenario with ideal communication links to a scenario with BSCs, the objective function in (3.11) must be properly modified. In particular, the following expression for the objective function can be derived:

$$f_{\mathrm{bq}}(k, \boldsymbol{d}_k^{\mathrm{AP}}, p) \triangleq (1-2p) \sum_{i=1}^{k} [\mathbb{P}\{H_i = 0 | d_i^{\mathrm{AP}}\} - \mathbb{P}\{H_i = 1 | d_i^{\mathrm{AP}}\}] \quad k = 1, \ldots, n. \tag{3.15}$$

As one can see, the only difference between (3.11) and (3.15) lies in the term $(1-2p)$. Since $p \in (0, 0.5)$, it follows that the term $(1-2p)$ is always positive. Therefore, this term does not influence the monotonic behavior of the sum at the right-hand side of (3.15) and can be neglected without changing the behavior of $f_{\mathrm{bq}}$. Finally, the conditional probabilities in (3.12) can be extended to a scenario with BSC communication links as follows:

$$\mathbb{P}\{d_i^{\mathrm{AP}}|H_i=\ell\} = p+(1-2p)\mathbb{P}\left\{ n_i \underset{d_i^{\mathrm{AP}}=1}{\overset{d_i^{\mathrm{AP}}=0}{\lesseqgtr}} \tau - s\cdot\ell \right\}$$
$$= (1-d_i^{\mathrm{AP}})\left[p+(1-2p)\Phi\left(\frac{\tau-s\cdot\ell}{\sigma}\right)\right]$$
$$+ d_i^{\mathrm{AP}}\left\{p+(1-2p)Q\left(\frac{\tau-s\cdot\ell}{\sigma}\right)\right\}.$$

As shown in Sect. 3.1.3, the evaluation of these conditional probabilities is sufficient for the implementation of the reduced-complexity fusion algorithm illustrated in Fig. 3.1.

#### 3.2.2.2 Absence of Quantization

As previously stated in Sect. 3.2.1.1, the fusion rule derived for a scenario with ideal communication links in Sect. 3.1.3.2 still applies in the current scenario with noisy communication links, but for the replacement of $P(H_i=\ell|r_i)$ with $P(H_i=\ell|r_i^{\mathrm{AP}})(i=1,\ldots,n;\ell=0,1)$, where $r_i^{\mathrm{AP}}$ is defined in (3.13). After a few simple manipulations, one obtains:

$$\mathbb{P}\{H_i=\ell|r_i^{\mathrm{AP}}\} = \frac{p_{\mathrm{comm}}(r_i-s\cdot\ell)}{p_{\mathrm{comm}}(r_i^{\mathrm{AP}})+p_{\mathrm{comm}}(r_i-s)} \qquad i=1,\ldots,n \quad \ell=0,1.$$

## 3.3 Numerical Results

We now analyze, through Monte Carlo simulations, the performance of the distributed detection schemes previously introduced. We denote as $\varepsilon$ the following quadratic distance between the observed phenomenon $\boldsymbol{H}$ and its estimate $\widehat{\boldsymbol{H}}$ :

$$\varepsilon(\boldsymbol{H},\widehat{\boldsymbol{H}}) \triangleq \left|\langle(\boldsymbol{H}\oplus\widehat{\boldsymbol{H}});(\boldsymbol{H}\oplus\widehat{\boldsymbol{H}})\rangle\right|^2 \tag{3.16}$$

where $\langle\cdot,\cdot\rangle$ is the scalar product, $\oplus$ stands for bit-by-bit ex-or, and $\widehat{\boldsymbol{H}}$ is the estimated phenomenon, directly derived from the estimated boundaries' positions $\widehat{\boldsymbol{\alpha}}$. We will simply refer to $\varepsilon$ as "distance". Note that expression (3.16) for the distance reduces to $\varepsilon(\boldsymbol{H}, \widehat{\boldsymbol{H}}) = |\alpha - \widehat{\alpha}|^2$ in the case of single-boundary phenomena.

The Monte Carlo simulation results are obtained through the following steps:

1. the number of boundaries and their positions are randomly generated according to a uniform distribution[4] (in the case of a single boundary, only its position is randomly generated);
2. the sensors' decisions (or the p.d.f. of the observables, according to the chosen quantization strategy at the sensors) are transmitted to the AP;
3. the AP estimates the boundaries' positions through either the MMSE or the simplified fusion algorithms previously proposed;
4. the distance $\varepsilon$ (between the true phenomenon and its estimate) is evaluated, on the basis of the estimated sequence of boundaries;
5. steps 1÷4 are repeated a sufficiently large number of times, by generating different numbers of boundaries during each simulation run;
6. the average distance $\overline{\varepsilon}$ is finally computed as the arithmetic average of the distances computed at the previous iterations (in step 4 at each iteration).

We point out that the proposed performance analysis leads to the "average" performance over all possible numbers of boundaries. Should one limit the analysis to a *fixed* number of boundaries, it is expected that the performance will either improve (if the fixed number of boundaries is small) or worsen (if the fixed number of boundaries is large).

### 3.3.1 Ideal Communication Links

In Fig. 3.2, the distance $\overline{\varepsilon}$ is shown, as a function of the sensor SNR, in a scenario with *single boundary* phenomena and ideal communication links. Two possible values for the number $n$ of sensors are considered: (i) 16 and (ii) 32. The results with both absence of quantization and binary quantization at the sensors are presented. One can observe that the distance reduces to zero for increasing values of the sensor SNR in all considered scenarios; this is to be expected, since the sensors' observations and, consequently, the data sent to the AP are more and more reliable. For low values of the sensor SNR, instead, the distance increases for increasing values of the number of sensors, since larger values for the estimated boundary are possible and, therefore, the distance may become larger. Note, also, that the performance degradation incurred by the use of quantization, with respect to the unquantized case, increases for increasing numbers of sensors. No result in the case of multi-level quantization is reported here. However, the results in [2]

[4] Obviously, after the position of a boundary is extracted, the following boundary position is randomly chosen among the remaining positions. After all the boundary positions are extracted, they are ordered. This implies that the multiple conditions in (3.1) are satisfied.

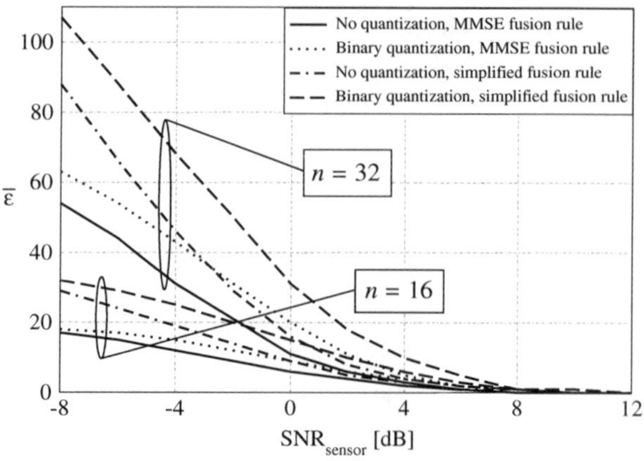


**Fig. 3.2** Distance, as a function of the sensor SNR, in a scenario with *single boundary* phenomena and ideal communication links. Two possible values for the number $n$ of sensors are considered: (i) 16 and (ii) 32. The results with both absence of quantization and binary quantization at the sensors are shown

show that the use of higher-level quantization (e.g., 2 or 3 quantization bits) leads to a minor performance gain. Finally, the scheme with simplified fusion rule at the AP has a performance worse than that of the scheme with the MMSE fusion rule at the AP. However, the performance of the simplified fusion algorithm is close to that of the MMSE fusion rule in the region of interest ($\text{SNR}_{\text{sensor}} \geq 0\,\text{dB}$) and the performance loss reduces to zero for large values of the sensor SNR.

In order to evaluate the loss incurred by the use of the simplified fusion algorithm, it is expedient to introduce the following percentage loss:

$$P \triangleq \sqrt{\underbrace{\frac{\overline{\varepsilon}^{\text{simp}} - \overline{\varepsilon}^{\text{MMSE}}}{\overline{\varepsilon}^{\text{MMSE}}}}_{\text{Term}_1} \cdot \underbrace{\frac{\overline{\varepsilon}^{\text{simp}} - \overline{\varepsilon}^{\text{MMSE}}}{n^2}}_{\text{Term}_2}}. \tag{3.17}$$

The intuition behind the definition of $P$ in (3.17), corresponding to the geometric average of $\text{Term}_1$ and $\text{Term}_2$, is the following. $\text{Term}_1$ represents the relative loss of the simplified fusion rule with respect to the MMSE fusion rule. However, using only this term could be misleading. In fact, for high sensor SNRs, the terms $\overline{\varepsilon}^{\text{simp}}$ and $\overline{\varepsilon}^{\text{MMSE}}$ are much lower than $n^2$ (the maximum possible distance). Therefore, even if $\overline{\varepsilon}^{\text{simp}} > \overline{\varepsilon}^{\text{MMSE}}$ (for example, $\overline{\varepsilon}^{\text{simp}} = 4$ and $\overline{\varepsilon}^{\text{MMSE}} = 1$ with $n = 32$), both algorithms might perform very well. The introduction of $\text{Term}_2$ eliminates this ambiguity, since it represents the relative loss (between MMSE and simplified fusion algorithms) with respect to the maximum (quadratic) distance, i.e., $n^2$. In Fig. 3.3, the behavior of $P$ is shown as a function of the sensor SNR. In the region of interest ($\text{SNR}_{\text{sensor}} \geq 0$ dB), one can observe that $P$ is lower than 15%, i.e., the proposed simplified fusion algorithm is effective. Note that the same considerations can be carried out in a scenario with noisy communication links.

In Fig. 3.4, the distance is shown, as a function of the sensor SNR, in scenarios with multi-boundary phenomena and ideal communication links, with $n = 8$ sensors. No quantization is considered at the sensors and the performance with the simplified fusion algorithm at the AP is compared directly with that obtained using

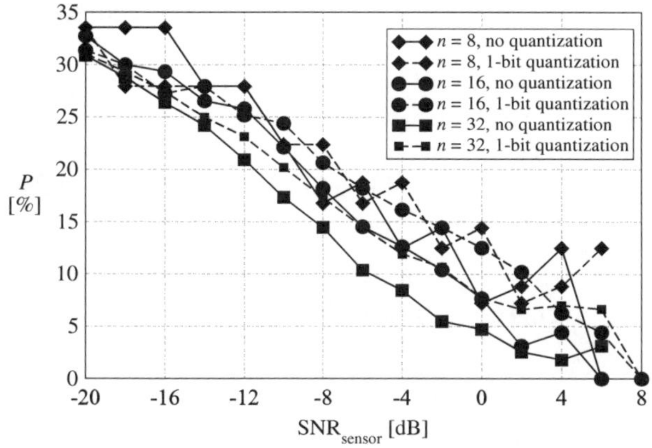


**Fig. 3.3** Percentage loss, as a function of the sensor SNR, in a scenario with a *single boundary* phenomenon and simplified fusion algorithm at the AP. The communication links are ideal. Three different values for the number $n$ of sensors are considered: (i) 8, (ii) 16, and (iii) 32. The performance in the presence of no quantization (*solid lines*) is compared with that using binary quantization at the sensors (*dashed lines*)

**Fig. 3.4** Distance, as a function of the sensor SNR, in a scenario with a *multi-boundary* phenomenon, considering $n = 8$ sensors and absence of quantization (MMSE and simplified fusion algorithms at the AP are considered). The communication links are ideal

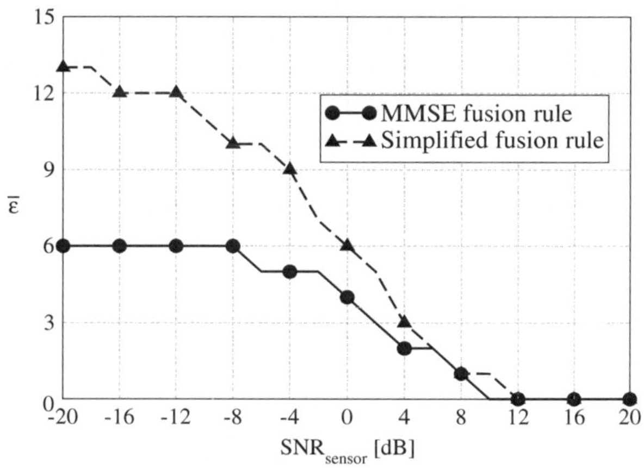


the MMSE fusion rule. As expected, the distance $\overline{\varepsilon}$ reduces to zero for increasing values of the sensor SNR and the performance with the MMSE fusion algorithm is better than that with the simplified fusion algorithm. We recall that the performance with the MMSE fusion rule can be evaluated only in scenarios with a number $n$ of sensors not larger than 8, since the computational complexity becomes unbearable for values of $n$ larger than 8 (the simulations are too lengthy). In order to investigate scenarios with larger numbers of sensors, the use of the reduced-complexity simplified fusion algorithms derived in Sect. 3.1.3 is mandatory. Our results presented in [2] show that the proposed simplified fusion rule is effective for all the considered values of the number of sensors $n$ , i.e., the distance reduces to zero for large values of the sensor SNR. Moreover, the performance does not improve by using more than one quantization bit at the sensors. It remains to be investigated what would be the relative loss of the simplified fusion algorithm, with respect to the MMSE fusion algorithm, in scenarios with multi-boundary phenomena. The fact that the quantization strategy at the sensors has little impact seems to suggest that this relative loss might *not* be negligible.

### 3.3.2 Noisy Communication Links

We finally investigate the impact of noisy communication links on the system performance. In Fig. 3.5, the distance is shown, as a function of the cross-over probability $p$, in a scenario with $n = 8$ sensors and binary quantization—in this case, the communication links are modeled as BSCs. Three values for the sensor SNR are considered: (i) $-10$ dB, (ii) 0 dB, and (iii) 10 dB. Both MMSE and simplified fusion algorithms at the AP are considered. As previously observed in Fig. 3.4, the use of the simplified fusion algorithm at the AP leads to a performance worse than that with the MMSE fusion algorithm. However, the higher is the sensor SNR, the smaller is the difference between the performance of the two algorithms. Moreover, one can observe that the distance might not converge to zero, due to the presence of two *independent* noise components (i.e., observation and communication noises). For a sufficiently high value of the sensor SNR, however, the distance reduces to zero when $p$ tends to zero (as confirmed by the results in Fig. 3.4).

In Fig. 3.6, the distance $\overline{\varepsilon}$ is shown, as a function of the sensor SNR, in a scenario with $n = 8$ sensors, noisy communication links (modeled as BSCs), and binary

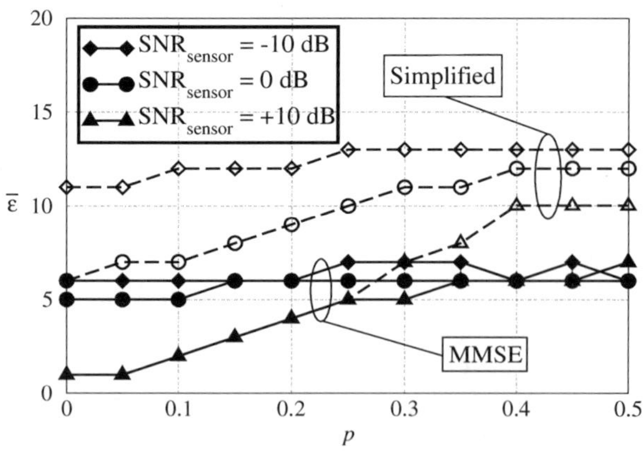


**Fig. 3.5** Distance, as a function of the cross-over probability $p$, in a scenario with $n = 8$ sensors, binary quantization, and noisy communication links (modeled as BSCs). Three values for the sensor SNR are considered: (i) $-10$ dB, (ii) 0 dB, and (iii) 10 dB. Both MMSE and simplified fusion algorithms at the AP are considered

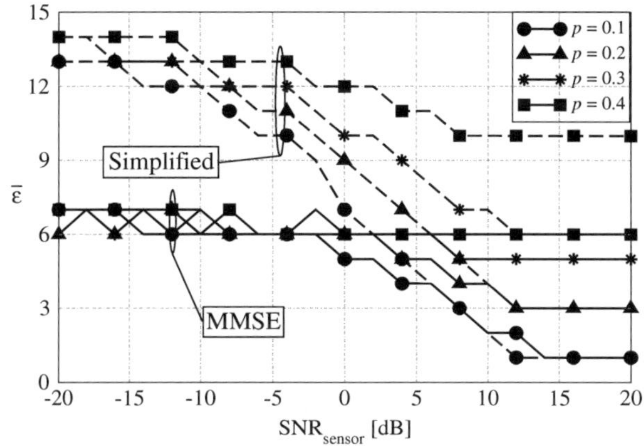


**Fig. 3.6** Distance, as a function of the sensor SNR, in a scenario with $n = 8$ sensors, binary quantization, and noisy communication links (modeled as BSCs). Four different values of the cross-over probability $p$ are considered: (i) 0.1, (ii) 0.2, (iii) 0.3, and (iv) 0.4. Both MMSE and simplified fusion algorithms at the AP are considered

quantization at the sensors. Four different values of the cross-over probability $p$ are considered: (i) 0.1, (ii) 0.2, (iii) 0.3, and (iv) 0.4. The performance with both MMSE and simplified fusion algorithms at the AP is investigated. Unlike the results presented in Sect. 3.1.3 for a scenario with ideal communication links, there appears to be a distance floor (higher than zero) for large values of the sensor SNR. This is to be expected, since the communication noise (independent of the observation noise at the sensors) prevents the AP from correctly recovering the data sent by the sensors. In particular, when the cross-over probability is sufficiently high (e.g., $p = 0.4$), the performance does not depend on the value of the sensor SNR, since the noisy communication links make the data sent by the sensors very unreliable. Finally, one can observe that, for low values of the sensor SNR, the simplified fusion algorithm shows a non-negligible performance loss with respect to the MMSE fusion algorithm. This loss reduces to zero, for increasing values of the sensor SNR, *only* for sufficiently small values of $p$. In other words, if the communication links are not reliable, then increasing the accuracy of the observations at the sensors is useless.

Finally, in Fig. 3.7 the distance $\bar{\varepsilon}$ is shown, as a function of the sensor SNR, in a scenario with $n = 8$ sensors, absence of quantization, and noisy communication links (modeled as AWGN channels). Two different values of the bit error rate $p$ (corresponding to different values of $\sigma^2_{\text{comm}}$ according to (3.14)) are considered: (i) 0.1 and (ii) 0.2. The performance of both MMSE and simplified fusion algorithms at the AP is evaluated. One can observe that, unlike the case with binary quantization at the sensors, the distance reduces to zero when the sensor SNR increases, i.e., no floor appears. Moreover, the distance with the simplified fusion rule at the AP approaches that with the MMSE fusion rule, i.e., it reduces to zero. This means that the proposed simplified fusion algorithm is (asymptotically)

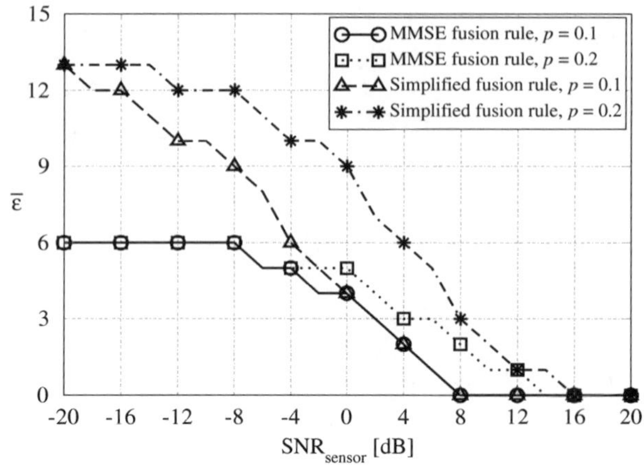


**Fig. 3.7** Distance, as a function of the sensor SNR, in a scenario with $n = 8$ sensors, absence of quantization, and noisy communication links (modeled as AWGN channels). Two different values of the equivalent bit error rate $p$ (corresponding to different values of $\sigma^2_{\mathrm{comm}}$ according to (3.14)) are considered: (i) 0.1 and (ii) 0.2. Both MMSE and simplified fusion algorithms at the AP are considered

effective. Obviously, this is only a theoretical performance limit. In fact, even if the communication links were noisy, the transmission of the "exact" observables (requiring an infinite bandwidth) from the sensors would allow a correct estimation of the true phenomenon. This cannot happen in realistic scenarios with limited transmission bandwidths.

## 3.4 Computational Complexity

It is now of interest to evaluate the improvement, in terms of computational complexity reduction with respect to the MMSE fusion rule, brought by the use of the simplified fusion algorithms. As complexity indicators, we choose the numbers of additions and multiplications (referred to as $n_{\mathrm{s}}$ and $n_{\mathrm{m}}$, respectively) required by the considered fusion algorithms, evaluated as functions of the number of sensors $n$. The following considerations are carried out referring to the formulas relative to the fusion algorithms for the scenario with ideal communication links (i.e., the derivations in Sect. 3.1). However, the same conclusions still hold for scenarios with noisy communication links, since the structures of the proposed fusion algorithms are the same in both scenarios (i.e., only the expressions of the used probabilities and p.d.f. change).

The numbers of operations (in terms of additions and multiplications) required by the MMSE fusion algorithm are $n_{\mathrm{s}}^{\mathrm{MMSE}} = \Theta(n^{2n_{\mathrm{bs}}})$ and $n_{\mathrm{m}}^{\mathrm{MMSE}} = \Theta(n^{2n_{\mathrm{bs}}+1})$, respectively—recall that $n_{\mathrm{bs}}$ is the number of boundaries. The notation

$f(n) = \Theta(g(n))$ means that there exists an $n_0$ such that for $n > n_0$, $\exists c_1 \in (0,1)$, $c_2 > 1$ such that $c_1 g(n) \leq f(n) \leq c_2 g(n)$ [8]. As described at the beginning of Sect. 2.4.5, in the considered simulation set-up the number $n_{\rm bs}$ of boundaries is randomly chosen between 1 and $n-2$. Therefore, one can assume that the phenomenon is characterized, on average, by $(n-2)/2$ boundaries. Under this assumption, the numbers of additions and multiplications required by the MMSE fusion algorithm would be $n_{\rm s}^{\rm MMSE} = \Theta(n^{n-2})$ and $n_{\rm m}^{\rm MMSE} = \Theta(n^{n-1})$. On the other hand, the reduced-complexity fusion algorithm requires only $n$ additions, since no multiplication has to be performed. Therefore, the computational complexity of the proposed simplified fusion algorithm is characterized by $n_{\rm m}^{\rm simp} = 0$ and $n_{\rm s}^{\rm simp} = n$, showing a significant complexity reduction with respect to the MMSE fusion algorithm—this also justifies the non-negligible performance loss at small values of the sensor SNR.

## 3.5 Concluding Remarks

In this chapter, we have analyzed the problem of distributed detection of spatially non-constant binary phenomena, i.e., phenomena with statuses characterized by single or multiple boundaries. We have proposed an analytical framework considering various quantization strategies at the sensors: (i) no quantization at the sensors and (ii) binary quantization. In all cases, the MMSE fusion algorithm at the AP has been derived and the impact of relevant network parameters (e.g., the sensor SNR, the communication noise level, and the number of sensors) has been investigated. Then, low-complexity fusion rules for scenarios with single-boundary and multi-boundary phenomena have been derived. We have shown that the performance penalty introduced by the simplified fusion algorithms is asymptotically (for high sensor SNR and low communication noise level) negligible. Finally, we have quantified the computational complexity reduction brought by the use of the simplified fusion algorithm with respect to the MMSE algorithm. Our results underline that this complexity reduction is pronounced in scenarios with multi-boundary phenomena.

## 3.6 Further Readings

In [6], the authors consider a scenario with a single phenomenon status change (denoted, in the following, as *boundary*) and propose a framework, based on MMSE estimation, to determine the position of this boundary. In [9], under the assumption of proper regularity of the observed boundary, a reduced-complexity MMSE decoder is proposed. In [10], the authors show that an MMSE decoder is unfeasible for large scale sensor networks, due to its computational complexity, and propose a distributed detection strategy based on factor graphs and the sum-product algorithm. Moreover, MMSE-based distributed detection schemes have also been investigated in scenarios with a common binary phenomenon under observation and bandwidth constraints [11]. Finally, in [12, 13] the authors examine the problem of determining boundaries of natural phenomena through proper processing of data collected by sensor networks. In those papers, particular attention is devoted to the estimation accuracy, given in terms of the confidence interval of the results obtained with the proposed framework.

## References


1. A. Abrardo, G. Ferrari, M. Martalò, Non-cooperative wireless orthogonal multiple access schemes with and without relaying, in *IEEE International Symposium on Communications, Control and Signal Processing (ISCCSP)*, St. Julians, Malta, March 2008, pp. 455–460
2. G. Ferrari, M. Martalò, M. Sarti, Reduced-complexity decentralized detection of spatially non-constant phenomena, in *Proceedings of International Workshop on Distributed Cooperative Laboratories (Ingrid)*, Santa Margherita Ligure, Italy, April 2007
3. S.M. Kay, *Fundamentals of Statistical Signal Processing*, vol. 1, Estimation Theory (Prentice-Hall, Upper Saddle River, 1993)
4. A. Papoulis, *Probability, Random Variables and Stochastic Processes* (McGraw-Hill, New York, 1991)
5. F.R. Kschischang, B.J. Frey, H.A. Loeliger, Factor graphs and the sum–product algorithm. IEEE Trans. Inform. Theory **47**(2), 498–519 (2001)
6. R. Nowak, U. Mitra, Boundary estimation in sensor networks: theory and methods, in *Proceedings of International Work. on Information Processing in Sensor Networks (IPSN)*, Palo Alto, CA, USA, April 2003, pp. 80–95
7. J.G. Proakis, *Digital Communications*, 4th edn. (McGraw-Hill, New York, 2001)
8. T.H. Cormen, C.E. Leiserson, R.L. Rivest, C. Stein, *Introduction to Algorithms*, 2nd edn. (MIT Press, Cambridge, 2002)
9. R. Nowak, U. Mitra, R. Willett, Estimating inhomogeneous fields using wireless sensor networks. IEEE J. Sel. Areas Commun. **22**(6), 999–1006 (2004)
10. J. Barros, M. Tückler, Scalable decoding on factor trees: a practical solution for wireless sensor networks. IEEE Trans. Commun. **54**(2), 284–294 (2006)
11. J.-J. Luo, Z.-Q. Luo, Universal decentralized detection in a bandwidth constrained sensor network. IEEE Trans. Signal Process. **53**(8), 2617–2624 (2005)
12. S. Duttagupta, K. Ramamritham, Distributed boundary estimation using sensor networks (Indian Institute of Technology, Mumbai, Tech. Rep., 2006), available at http://www.it.iitb.ac.in/research/techreport/
13. S. Duttagupta, K. Ramamritham, P. Ramanathan, Distributed boundary estimation using sensor networks, in *Proceedings of International Conference on Mobile Ad-hoc and Sensor Systems (MASS)*, Vancouver, Canada, October 2006, pp. 316–325

# Part III
# MAC and Connectivity

# Chapter 4
# Tree-Based Topologies for Multi-Sink Networks

The *connectivity theory* studies networks formed by large numbers of nodes distributed according to some statistics over a limited or unlimited region of $\mathbb{R}^d$, with $d = 1, 2, 3$, and aims at describing the potential set of links that can connect nodes to each other, subject to some constraints from the physical viewpoint (power budget or radio resource limitations). Connectivity depends on the number of nodes per unit area (nodes' density) and on the transmit power. The choice of an appropriate transmit power level is an important aspect of network design as it affects network connectivity. In fact, with a high transmit power a large number of nodes are expected to be reached via direct links. On the other hand, a low transmit power would increase the possibility that a given node cannot reach any other node, that is, it is *isolated*.

In ad hoc networks, the best performance is achieved when data generated by a node can flow along the network and reach any possible endpoint. Therefore, the goal of connectivity is to make it possible for any node to reach any other node, perhaps in a multi-hop fashion. Although Wireless Sensor Networks (WSNs) are sometimes thought of as a special case of ad hoc networks, they present a substantial difference, that is, nodes are at least of two different types: sensor and sink nodes. The purpose of this kind of networks is to process data originated by sensors, and sinks are in charge of collecting such data. Thus, the goal of connectivity is somewhat different here because it is sufficient for any sensor node to be able to reach at least one sink node, either directly or through other sensor nodes. That provided, the network is said to be *fully-connected*.

This chapter proposes statistical models to characterize connectivity in WSNs, providing useful general insights on network parameters' design rules, such as node density and transmission power.

More precisely, a multi-sink WSN, where sensors transmit data to one sink selected among many through multi-hop communication, is considered. Sensors are organised in trees, rooted at the sinks. The optimal design of these trees, assuming that sensors and sinks are uniformly and randomly distributed over an

C. Buratti et al., *Sensor Networks with IEEE 802.15.4 Systems*,
Signals and Communication Technology, DOI: 10.1007/978-3-642-17490-2_4,

infinite plane, is treated first. In particular, once the trees height is fixed, the optimum number of children per parent, maximising network connectivity, is derived. This analysis is performed through mathematical approaches and by means of simulations. Then, a mathematical framework is developed to derive some metrics which characterize the network connectivity level. In this case, bounded and unbounded regions are considered.

The chapter is organised as follows. Section 4.1 introduces the aims of the study proposed in this chapter. Sections 4.2 and 4.3 introduce the channel model and some basics and properties on connectivity theory in Poisson Point Process (PPP) field of nodes, used in the rest of the chapter. Section 4.4 defines the scenario studied. In Sect. 4.5 the design of the optimum tree-based topology, showing both mathematical and simulation results, is dealt with. In Section 4.6 the multi-sink multi-hop connectivity model for bounded and unbounded regions is described. Section 4.7 gives some concluding remarks, whereas further readings on these topics are resumed in Sect. 4.8.

## 4.1 Aims of the Study

A multi-sink WSN, collecting data from the environment through the sampling of some physical entities and sending them to some external user, through multiple sinks, is considered. The reference application is spatial/temporal process estimation [1] and the environment is observed through queries/respond mechanisms: queries are periodically generated by the sinks, and sensor nodes respond by sampling and sending data. Through a simple polling model, sinks periodically issue queries, making all sensors perform sensing and communicate their measurement results back to the sinks they are associated with. The user, by collecting samples taken from different locations, and observing their temporal variations, can estimate the realisation of the observed process. Good estimates require sufficient data taken from the environment.

Often, the data must be sampled from a specific portion of space, even if the sensor nodes are distributed over a larger area. Therefore, only a location-driven subset of sensor nodes must respond to queries. The aim of the query/response mechanism is then to acquire the largest possible number of samples from the area.

The data taken from the area where sensors are distributed are transmitted to a centralised unit by means of wireless links connecting sensors to sinks, which collect the samples and forward them to the unit through a proper network. If few sensor nodes are deployed and the target area is small, a single sink can be used. When the number of sensors or the target area is large, nodes are often organised in clusters: one sink per cluster forwards the queries to sensors, and collects the responses.

Sinks are sometimes specifically deployed in optimised and planned locations with respect to sensors. However, opportunistic exploitation of the presence of sinks, connected to the centralised unit through a mobile radio interface, is an option in some cases. Under these circumstances, many sinks can be present in the monitored space, but their positions are unknown and unplanned; therefore, achievement of a sufficient amount of samples is not guaranteed, because sensor nodes might not reach any sink (and thus be isolated) due to the limited transmission range.

According to the type of enabling technology used (e.g., Bluetooth or IEEE 802.15.4), different network topologies might be conveniently created such as, for instance, trees, rings, or cluster-based topologies [2, 3]. For WSNs, where the set of destination nodes (i.e., the sinks) is separated by those of sources (namely, sensor nodes), tree-based topologies seem to be more efficient than the others. In fact, routing is much simpler, and also distributed data aggregation mechanisms are more efficient. Moreover, as stated in Chap. 1, this topology is one of the topologies defined by the Zigbee Alliance [4, 5], therefore suitable for IEEE 802.15.4 networks. When dealing with a multiple sink scenario, formally a forest of (disjoint) trees is formed. In this scenario, being an uncoordinated environment, network connectivity is a relevant issue and is basically dominated by the randomness of the radio channel and the density of sinks.

In this chapter, connectivity issues in tree-based multi-sink WSNs, by considering two separate studies (with different aims), are dealt with.

The first study focuses on properly designing the tree-based topology on the basis of connectivity requirements. The objective of this study is to maximise the number of samples reported to the sink(s), that is, network coverage, whereas the tree height should be set keeping energy consumption under control. The study has been carried out through simulations and mathematical analysis. In particular, we study: (i) a multiple level tree topology using a deterministic Medium Access Control (MAC), based on Bluetooth or the Contention Free Period (CFP) of the IEEE 802.15.4 superframe (allocation of Guaranteed Time Slots (GTSs) to nodes) and (ii) a three-level topology using both the Contention Access Period (CAP) and CFP of the IEEE 802.15.4 superframe. The latter case is studied through simulation [6], while the former can be mathematically handled through a statistical approach. The mathematical model is derived, assuming that both sensors and sinks are uniformly distributed over an infinite area. It is shown that in both cases (i) and (ii), once the tree height is fixed, network coverage is maximised by a proper choice of the average number of nodes at each level (and, therefore, of the average number of children per parent). However, the choice of the tree height has a relevant impact on such optimisation.

In the second study, instead, a bounded scenario where, once again, sinks and sensors are uniformly and randomly distributed, is accounted for. In this scenario, the probability that sensor nodes are connected to at least one sink is mathematically derived. Starting from such a result, the probability that all nodes, or a subset of them, are connected, is computed. The work is based on previous papers published in the literature that provided results in the case of an infinite plane [7, 8]. Our approach differs from the previous ones, since it takes into consideration bounded scenarios, a situation which of course is way more realistic and requires suitable consideration of the border effects. The analysis is first performed in the case of single-hop communication (i.e., every sensor transmits the sensed data directly to a sink). Then, the multi-hop case (i.e., sensors may also act as routers) is considered.

## 4.2 Channel and Link Models

Many works in the WSNs scientific literature assume deterministic distance-dependent and threshold-based packet capture models. In other words, all nodes within a circle centered at the transmitter, with given radius, can receive a packet sent by the transmitting one; if a receiver is outside the circle, reception is impossible [9–11]. While the threshold-based capture model, which assumes that a packet is captured if the signal-to-noise ratio (in the absence of interference) is above a given threshold, is a good approximation of real capture effects, the deterministic channel model does not represent realistic situations in most cases. The use of realistic channel models is therefore of paramount importance in wireless systems. In this book, a narrow-band channel, accounting for the power loss due to propagation effects, including a distance-dependent path loss and slow/fast channel fluctuations, is considered. In [1], results of experiments made with nodes using the IEEE 802.15.4 standard at 2.4 GHz Industrial Scientific Medical (ISM) band, deployed in different environments (grass, asphalt, indoor, etc.), are reported. The measurements provide inputs for understanding the basic aspects of narrow-band propagation in typical WSNs scenarios at 2.4 GHz. In particular, suitable comparison between the measurements performed and some simple analytical expressions has been conducted for different environments [1]. Considering the received power in logarithmic scale, it was found that, in general, a Gaussian model can approximate the measurements fairly well, with different values of the standard deviation. Also some papers in the literature report results achieved in similar environments, and the Gaussian model seems to be accredited. Note that, since the scenario is stationary, the assumption of a (slow-varying) shadowing environment is acceptable, as a log-normal distribution models the randomness of the geometry (presence of obstacles, etc.). This is also done in other papers in the literature on WSNs (see, e.g., [12]). Channel reciprocity is also assumed.

It is assumed that the ratio between the transmit power, $P_{\mathrm{T}}$, and the received power, $P_{\mathrm{R}}$, is given by $k \cdot d^{\beta} \cdot S$, where $k$ is the propagation coefficient, $d$ is the distance between the transmitter and the receiver, $\beta$ is the attenuation coefficient which commonly ranges from 2 to 5, and $S$ is the long-term (shadowing) fading component. We define $L = k \cdot d^{\beta} \cdot S$ as the average (with respect to fast fading) loss (in linear scale). By introducing the logarithmic scale, we obtain

$$L[\mathrm{dB}] = k_0 + k_1 \ln d + s[\mathrm{dB}], \tag{4.1}$$

where $k_0 = 10\log_{10} k$, $k_1 = \beta\frac{10}{\ln 10}$, and $s$ [dB] is a Gaussian random variable (r.v.), with zero mean and variance $\sigma_s^2$. Note that in (4.1) the dependence on the distance is through a natural logarithmic function, instead of a more typical base 10 log function. However, the transformation is quite simple, and this notation is the same used in works taken from the literature whose results are used as starting point in this section [13]. This channel model was also adopted by Orriss and Barton [7] and other authors [14, 15]. For instance, by suitably setting $k_1$, it is possible to accommodate an inverse square law relationship between power and distance ($k_1 = 8.69$) or an inverse fourth-power law ($k_1 = 17.37$).

For what concerns the link model, a radio link between two nodes is said to exist, which means that the two nodes are *connected* or *audible* from each other,[1] if $L < L_{\text{th}}$, where $L_{\text{th}}$ represents the maximum loss tolerable by the communication system. The threshold $L_{\text{th}}$ depends on the transmit power and the receiver sensitivity. By solving (4.1) for the distance $d$ with $L = L_{\text{th}}$, the transmission range can be defined as follows:

$$TR = e^{\frac{L_{\text{th}} - k_0 - s}{k_1}}, \tag{4.2}$$

which corresponds to the maximum distance between two nodes at which communication can still take place. Such range defines the connectivity region of the sensor. Note that by characterizing the values of $s$, in different links, as independent r.v.'s, we have different values of $TR$ for every nodes' pair. This means that any sensor observes a different realization of the r.v. $TR$ depending on the direction of the potential interlocutor, thus acquiring a jaggy wireless footprint. In other words, circles to predict sensor connectivity are not used here. However, by setting $\sigma_s = 0$, channel fluctuations are neglected and an ideal reference transmission range can be defined as

$$TR_{\text{i}} = e^{\frac{L_{\text{th}} - k_0}{k_1}}, \tag{4.3}$$

which corresponds to the radius of the circular deterministic footprint.

According to this channel model, we can also define the probability that two nodes are audible, denoted as $C(d)$, as the probability that $L < L_{\text{th}}$, i.e.,

$$C(d) = \mathbb{P}\{L < L_{\text{th}}\} = 1 - 0.5\,\text{erfc}\left(\frac{L_{\text{th}} - k_0 - k_1 \ln(d)}{\sqrt{2}\sigma_s}\right) \tag{4.4}$$

where $\mathbb{P}\{\mathcal{E}\}$ denotes the probability of the event $\mathcal{E}$ and erfc() is the complementary error function. The disk model is obtained by considering $\sigma_s \to 0$, i.e., $s \equiv 0$, thus leading to the following expression for $C(d)$:

$$C(d) = \begin{cases} 1 & \text{for} \quad d \le TR \\ 0 & \text{for} \quad d > TR. \end{cases} \tag{4.5}$$

[1] Link reciprocity is assumed.

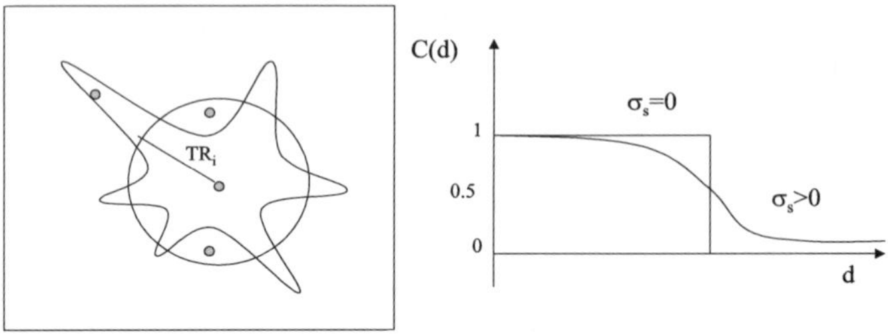


**Fig. 4.1** Link connectivity with and without shadowing effects

As we can see in Fig. 4.1, taking into account a specific transmitting node, the effect of the shadowing is to make audible some nodes that are not reachable when adopting the disk model ($\sigma_s = 0$) because outside of the circle with radius *TR*. On the other hand, shadowing also makes some nodes inside the circle non audible.

## 4.3 Connectivity Properties in PPP Fields

Let us consider a number of nodes randomly distributed over a field. It is worth noting that, due to the random position of nodes and channel fluctuation effects, the number of nodes which are connected[2] to whatever a node in the field is not deterministic. This is true regardless of the connectivity model we are considering. Therefore, the number of nodes connected to a give node, $n$, is a r.v. whose statistical properties depend on the connectivity models we are using and on the spatial distribution of nodes. In particular, when the position of nodes is distributed according to a PPP, we can apply the following theorem:

**Theorem** *Assume a Poisson distribution of nodes in a m-dimensional space and consider a reference node, denoted by $R_N$, located somewhere in the scenario. Let $d, C(d)$, and $n$ be the euclidean distance between a generic node and $R_N$, the probability that a generic node is connected with $R_N$, and the number of nodes which are connected with $R_N$, respectively. Then, $n$ is a Poisson r.v.*

*Proof*. The proof is a consequence of the *Marking Theorem* for Poisson processes [1]. □

As a result of the previous property, the probability mass function (p.m.f.) of $n$ is

$$\mathbb{P}\{n = n_1\} \triangleq P(n_1, \mu) = \frac{\mu^{n_1}}{n_1!} e^{-\mu}, \tag{4.6}$$

where $\mu = \mathbb{E}\{n\}$, being $\mathbb{E}\{\cdot\}$ the expectation. $\mu$ depends on the connectivity model chosen, and on the area in which nodes are distributed. In particular, when the

2 In the sense that the two nodes can reliably communicate to each other.

channel model of Eq. 4.1 is used, the mean number of nodes audible within a range of distances $r_1$ and $r$, to a generic node ($r \geq r_1$), is denoted as $\mu_{r_1,r}$ and can be written as [7, 13]

$$\mu_{r_1,r} = \pi\rho[\Psi\{a_1, b_1; r\} - \Psi\{a_1, b_1; r\}], \tag{4.7}$$

where $\rho$ is the initial nodes' density,

$$\Psi(a_1, b_1; r) = r^2\Phi(a_1 - b_1 \ln r) - e^{\frac{2a_1}{b_1}+\frac{2}{b_1^2}}\Phi(a_1 - b_1 \ln r + 2/b_1), \tag{4.8}$$

and $a_1 = (L_{\text{th}} - k_0)/\sigma_s, b_1 = k_1/\sigma_s$ and $\Phi(x) = \int\limits_{-\infty}^{x} (1/\sqrt{2\pi})e^{-u^2/2}\,\mathrm{d}u$. By letting $r_1 = 0$ and $r \to \infty$ (which correspond to consider an infinite area where nodes are distributed), $\Psi(a_1, b_1; r)$ vanishes and we can write

$$\begin{aligned}\mu_{0,\infty} &= -\pi\rho\Psi(a_1, b_1; r_1)\\ &= \pi\rho \exp[(2a_1/b_1) + (2/b_1^2)]\\ &= \pi\rho \exp[(2(L_{\text{th}} - k_0)/k_1) + (2\sigma_s^2/k_1^2)].\end{aligned} \tag{4.9}$$

Note that the mean value of $\mu = \mathbb{E}\{n\}$, is equal to $\mu_{0,\infty}$ also in the case that nodes are distributed over a finite plane, but border effects are negligible, which means that the exponential $\Psi(a_1, b_1; r)$ is close to zero.

## 4.4 Reference Scenario

We assume that sensors and sinks are uniformly and randomly distributed over the bi-dimensional plane with densities $\rho_s$ and $\rho_0$, respectively, with the latter much smaller than the former. We denote as $\Delta$ the ratio between these two densities, that is $\Delta = \rho_s/\rho_0$.

The sensor nodes deployed in the monitored area (that could be bounded or unbounded) need to communicate the sensed data to one sink, responsible for collection of information from the area. Communication can take place through multi-hop paths. Sensors are assumed to be split into $T$ groups (that we call *levels*) obtained through a random procedure which lets nodes belonging to each level be all uniformly distributed in the area (bounded or unbounded). The nodes are then connected through a hierarchical architecture, where nodes at a given level need to connect to nodes at a lower level to reach a sink (sinks belonging to the lowest level, in our formalism, see Fig. 4.2). As an example, it takes three hops to a node belonging to level 3 to reach the sink: two nodes (one belonging to level 2 and the other belonging to level 1) will act as relays. This assumption, that we denote as a-priori level partitioning, accounts for networks where a node belongs to one out of $T$ categories of devices, each one having different physical features.

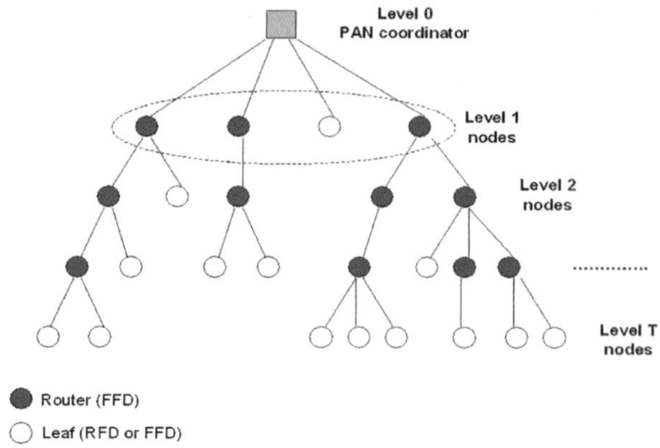


**Fig. 4.2** ZigBee-compliant tree network topology

The expression a-priori stems from the fact that the partitioning procedure occurs independently from the nodes' positions. Just to give a practical example, in IEEE 802.15.4 [16] networks, devices (such as the 13192 Evaluation Boards by Freescale [17] operating on a peer-to-peer topology, can be either Full Function Device (FFD) or Reduced Function Device (RFD): hence, since an RFD device may only talk to an FFD one, if the latter belongs to level $i$, the former will necessarily belong to level $i-1$. We emphasize that the nodes are then grouped with a-priori fixed densities: in fact, regardless of whether we are dealing with two diverse boards or with the same board running two different pieces of software, both the hardware (in the first case) and the software (in the second) remain the same for the entire operational time of the network (e.g., the software may not be re-compiled on-the-fly). Hence, although it is not the optimal situation from a connectivity perspective (not all possible paths to the sinks are exploitable), the a-priori partitioning assumption is noteworthy because it is widely adopted in practice. Moreover, connectivity models for two-dimensional $T$-hop networks under more general conditions are still being studied [18].

We denote as $\rho_i$ the node density at level $i$, with $i \in [1, \ldots, T]$, and we assume that a node belongs to level $i$ with a probability $p_i$ (equal for all nodes), fixed a-priori as stated above; therefore, $\rho_i = p_i \rho_s$. Whatever the strategy used, the density of nodes at all levels must satisfy the following constraint:

$$\sum_{i=1}^{T} \rho_i = \rho_s. \tag{4.10}$$

## 4.5 On the Design of Optimum Tree-Based Topologies

The aim of this section is to optimally design the tree topology, accounting for connectivity issues. We assume that the air interface imposes a maximum number of nodes that can be connected to a given node. As an example, if Bluetooth is used [19], a maximum number of seven slaves can be connected to the master of the piconet. In particular, we will denote as $c_i$ the capacity of level $i-1$ nodes, that is the maximum number of $i$th level nodes that can be serviced by an $(i-1)$th level node. When the capacity does not depend on $i$, we denote it as $c$ and also refer to the maximum number of children per parent in the tree.

Two different scenarios are addressed in this section: (i) a multiple level tree topology using a deterministic MAC protocol, that could be based on Bluetooth or IEEE 802.15.4 if a maximum number of seven children per parent is imposed, that is $c \leq 7$ (in this case, in fact, all nodes can use GTSs); and (ii) a three-level tree topology using both CAP and CFP of IEEE 802.15.4, where the capacity constraint could be imposed or not, and the contention-based MAC protocol is accounted for. The first scenario is studied through a mathematical analysis, whereas the second one is implemented through simulations. In Sects. 4.5.1 and 4.5.2, the multi-level tree scenario is dealt with; whereas the three-level tree is described in Sects. 4.5.3 and 4.5.4.

### *4.5.1 The Multi-Level Tree: Mathematical Analysis*

Being sensors and sinks Poisson distributed over the infinite bi-dimensional plane, the number of samples reported to a generic sink through the tree is, once again, a r.v., denoted as $n$, having a probability distribution $f(n)$. When a capacity constraint is imposed, $n$ is upper-bounded by $n_{\max} = \sum_{i=1}^{T} c^i$; if, more generally, no constraint is imposed, then $n$ is unlimited. The probability that the number of samples received by a given sink is above (or equal to) a fixed fraction $x$ of the mean $\Delta$ is given by:

$$R = \mathbb{P}\{n \geq x\Delta\} = \sum_{x\Delta}^{\infty} f(n), \tag{4.11}$$

assuming $x\Delta$ is an integer. If $x\Delta$ is not an integer, the extension is straightforward.

Once $f(n)$ is known (this distribution is derived in the following), the only degree of freedom, in order to properly design the trees, is the set of values $\rho_i$ $(i = 1, \ldots, T)$, that need to be designed according to the constraint (4.10).

For the sake of simplicity, here it is assumed that the ratio between the node density at a given level and the one at the next higher level is set to a common value $\eta$, but for the $T$th level that will include the remaining nodes. Formally,

$$\rho_i/\rho_{i-1} = \eta \quad i = 1, \ldots, T-1 \text{ and } \rho_T/\rho_{T-1} \leq \eta. \tag{4.12}$$

Therefore, $\eta$ is the mean number of children per parent, and the probability of blocking (i.e., the transmission of the samples collected to the higher level is not possible because of capacity limits, or collisions) will be the same at all levels from $T-1$ to 1. It is worth noting that this choice is not necessarily optimized, as the optimum choice should reflect a compromise between the cost of blocking the transmission at higher levels (where a node needs to report the many samples collected by its children) and the overall network energy efficiency.

As a result, $T$ and $\eta$ should be fixed according to the constraint (4.10) and expressions given in (4.12). Clearly, when $\eta$ increases, the minimum value $T$ needed to satisfy the constraint (4.10) will decrease. In particular, using equations (4.12) and (4.10) the minimum value of $T$ is found using the following formula:

$$\sum_{i=1}^{T-1} \eta^i + \eta^{T-1} \cdot \rho_T / \rho_{T-1} = \Delta. \tag{4.13}$$

On the other hand, if $\eta$ significantly exceeds the air interface capacity, the probability of blocking will increase. Thus, the objective of our analysis is to derive the value of $\eta$ which optimises $R$.

According to the channel model described above, a node can hear a transmitting one if $L \leq L_{\text{th}}$; thus, the number of level $i$ sensors audible at a random point on the plane has a Poisson distribution with the following mean:

$$\mu_i = \pi \rho_i e^{\frac{2(L_{\text{th}} - k_0)}{k_1} + \frac{2\sigma_s^2}{k_1^2}}, \tag{4.14}$$

which corresponds to Eq. 4.9 by replacing $\rho$ with $\rho_i$. As all sensors at all levels are randomly distributed, this applies to the number of level $i$ sensors audible from every other sensor. This result is derived, in a different context, in [7].

Assuming that every sensor will seek service at the loudest sensor at the next level, in [20] it is shown that the number of level $i$ sensors seeking service at a given level $i-1$ node has a Poisson distribution with a given mean. In our case, the mean number of level 1 sensors seeking service at a given sink is $\left[1 - e^{\mu_0}\right] \frac{\mu_1}{\mu_0}$. The ratio $\frac{\mu_1}{\mu_0}$ is the mean number of level 1 sensors per level 0 one, and the factor $\left[1 - e^{\mu_0}\right]$ eliminates those which cannot hear at least one level 0.

To deal with the hierarchical case we define a probability generating function $\Pi_i(s)$ for the number of level $i$ sensors being serviced by a given level $i-1$ sensor. Then the probability generating function for the number of level $i+1$ sensors being serviced by a given level $i-1$ sensor through level $i$ sensors (a three-level hierarchy) is

$$\Pi_i(s\Pi_{i+1}(s)). \tag{4.15}$$

Here, within the bracket, the term $\Pi_{i+1}(s)$ "counts" the level $i+1$ sensors reporting to a given level $i$ one, and the additional $s$ adds the latter before the report is sent up to the next level.

The extension to higher level hierarchies is immediate. Thus the probability generating function for the number of level $i+2$ sensors being serviced by a given level $i-1$ sensor through level $i$ and level $i+1$ sensors (a four-level hierarchy) is

$$\Pi_i(s\Pi_{i+1}(s\Pi_{i+2}(s))), \tag{4.16}$$

and similarly for yet higher levels.

In these circumstances, denoting as $\mu_{(i)}$ the mean number of level $i$ sensors being serviced by a given level $i-1$ sensor, it follows that the mean number of level $i+1$ sensors being serviced by a given level $i-1$ sensor through level $i$ sensors is

$$\mu_{(i)}(\mu_{(i+1)}+1), \tag{4.17}$$

while for the four-level hierarchy this becomes $\mu_{(i)}\mu_{(i+1)}\mu_{(i+2)} + \mu_{(i)}\mu_{(i+1)} + \mu_{(i)}$. With no capacity limitation, this probability generating function is that of the Poisson distribution of the number of sensors seeking service described above. With capacity limitation, we start with that Poisson distribution (whose mean we take as $\mu$), but cumulate all probabilities from the term in $s^{c_i}$ onwards. The probability generating function therefore becomes

$$\Pi_i(s) = \sum_{u=0}^{c_i-1} \frac{\mu^u s^u}{u!} e^{-\mu} + s^{c_i} \sum_{u=c_i}^{\infty} \frac{\mu^u}{u!} e^{-\mu}. \tag{4.18}$$

The number of levels in the hierarchy depends on $\eta$ and $\Delta$. At one extreme, if $\eta \geq \Delta$, then all sensors are at level 1, and we have a two-level hierarchy. If $\eta < \Delta$, then the density of level 1 sensors is $\rho_0\eta$, leaving a density $\rho_0(\Delta-\eta)$ of sensors to allocate to lower levels: these will all remain at level 2 if $\eta^2 \geq \Delta - \eta$. Otherwise, the density of level 2 sensors will be $\rho_0\eta^2$, leaving a density of $\rho_0(\Delta-\eta-\eta^2)$ for level 3 or lower. Repeating this process for as long as necessary, one finds that, in general, the hierarchy will be of level $T+1$ if $0 < \Delta - \eta - \eta^2 - \cdots - \eta^{T-1} \leq \eta^T$.

### 4.5.2 Mathematical Analysis Results

The following parameters are set: $k_0 = 40\,\text{dB}$, $k_1 = 15$, $\sigma_s = 4\,\text{dB}$, $L_{\text{th}} = 110\,\text{dB}$, $\rho_0 = 10^{-4}\,\text{m}^{-2}$ and $\Delta = 100$. The default requirement is to have at least 90 samples received at each sink (therefore, $x = 0.9$). Capacity limit is $c = 7$.

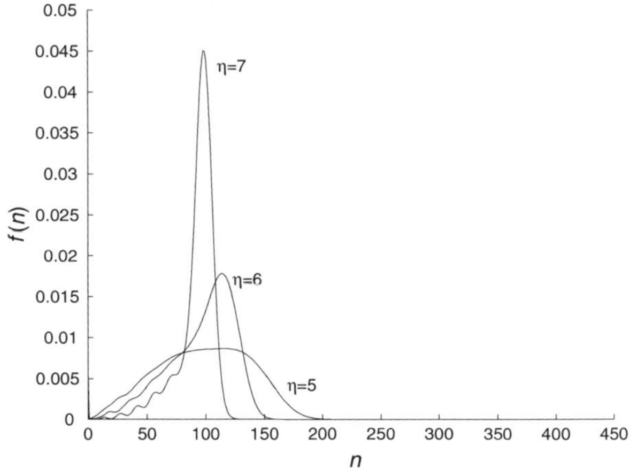


**Fig. 4.3** Probability distribution $f(n)$ of the number of nodes serviced as a function of $n$, for various values of $\eta$. In all cases, $T = 3$.

Let us consider $T = 3$. According to the constraints (4.10) and (4.12), the values of $\eta$ that should be considered approximately range from 4.3 to 9.5. Figure 4.3 shows the probability $f(n)$ as a function of $n$ for $\eta = 5, 6, 7$. As expected, the means tend to converge to $\Delta$ when $\eta$ increases, as all trees will find a sufficient number of nodes to fill all levels. Is it worth noting that as $\eta$ gets larger, the variance of these statistics decreases.

Figure 4.4 shows the effect of $x$ on the probability $R$, shown on the vertical axis as a function of $\eta$, for $T = 3$, again. According to the relevant variations on the standard deviation of $f(n)$, the curves vary significantly depending on $x$. In all cases, an optimum value of $\eta$ can be determined by these curves, depending on the requirement set. Note that the optimum value is close to seven (i.e., the capacity limit), or a bit larger. The sudden decrease of $R$ after the maximum is determined by the increase in the blocking probability. Also note that, as expected, when $x$ is larger, the probability $R$ gets smaller.

Figure 4.5 shows $R$ as a function of $\eta$ for $\Delta = 50, c = 7$, having set $k_0 = 40\,\text{dB}$, $k_1 = 13.03, \sigma_s = 3.5\,\text{dB}$, $L_{\text{th}} = 95\,\text{dB}$, and $\rho_0 = 4 \times 10^{-4}\,\text{m}^{-2}$. According to the constraints (4.10) and (4.12), the values of $\eta$ that should be considered depend on $T$. Here we consider the options $T = 4, 3$, and 2: $\eta$ can approximately range from 2.4 to 3.2, from 3.2 to 6.5, and from 6.5 to 10, respectively. Note that the cases with $T = 2$ and 3 should converge for $\eta = 6.5$, where we have a four-level tree with the lowest level empty, or a three-level tree with the lowest level having node density which is $\eta$ times larger than that at the higher level. The same holds for $T = 4$ and 3 at $\eta = 3.2$. Note that the larger $x$ (i.e., a more stringent requirement is set), the smaller the probability $R$, as expected. However, the most important aspect stands in the maximum value of $R$; depending on $x$, optimum performance is achieved for $T = 2$ or 3. In other words, the optimum tree height depends on the coverage requirement. From Fig. 4.5 it can be seen that for $x = 0.7$ the optimum topology requires $T = 2, \eta \approx 7.7$, while for $x = 0.9$ it is given by the

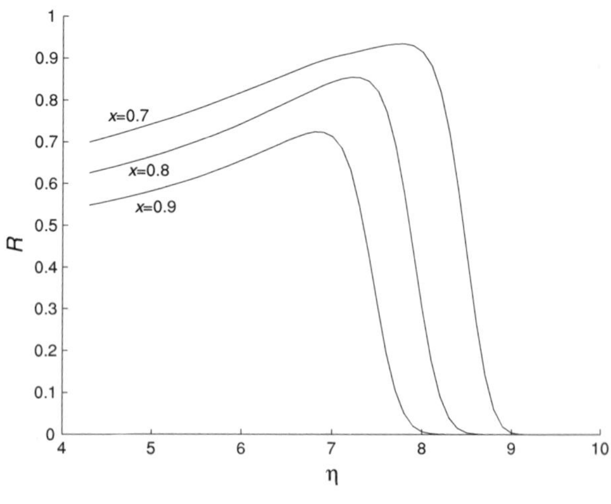


**Fig. 4.4** $R$, as a function of $\eta$, for various values of $x$. In all cases, $T = 3$

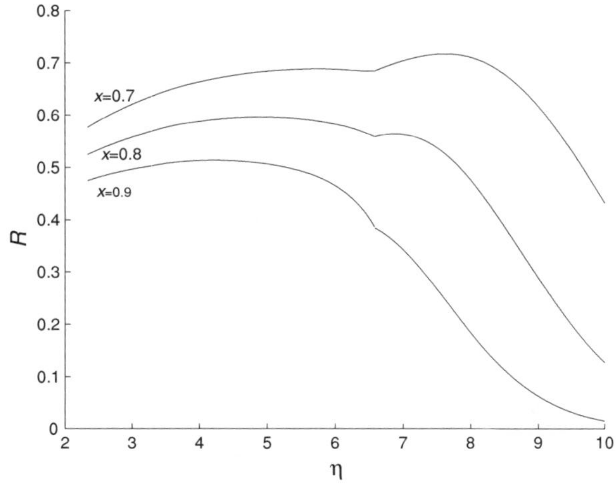


**Fig. 4.5** $R$ as a function of $\eta$. $\Delta = 50, c = 7$

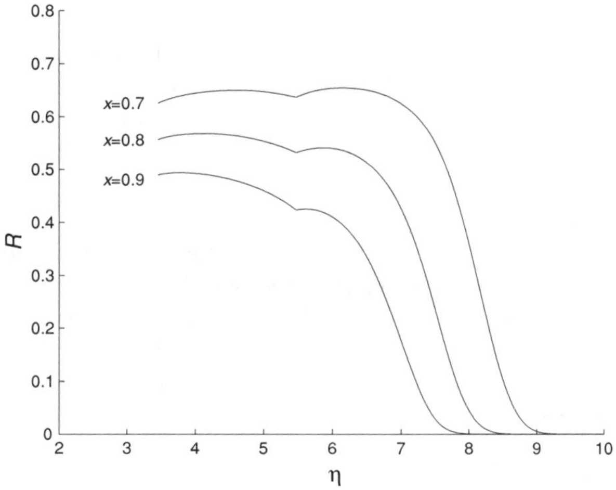


**Fig. 4.6** $R$, as a function of $\eta$, in a scenario with $\Delta = 200$ and $c = 7$

pair $T = 3, \eta \approx 4.5$. In [6], this effect is more thoroughly discussed and it is shown that this depends on the shape of the number of nodes reporting to a given sink distribution.

In Fig. 4.6, we set $\Delta = 200$ and we leave the values of the other parameters as in Fig. 4.5. The two cases $T = 4$ ($\eta$ ranging in this case from 3.5 to 5.5) and 3 (from 5.5 to 9.5) are considered. Similar considerations to the case of Fig. 4.5 can be done. However, given the larger average number of nodes per tree, with respect to Fig. 4.5, the optimum topology requires $T = 3$ or 4.

Finally, in Fig. 4.7 we show the behavior of $R$, for different capacities, having fixed $\Delta = 50, T = 2$, and $x = 0.7$. In particular, the capacity limit for sinks is $c_1 = 13$ while $c_2$ ranges from 3 to 13. The graph shows that reducing $c_2$ affects only the left part of the curves, at least if the value is not too low. This can be

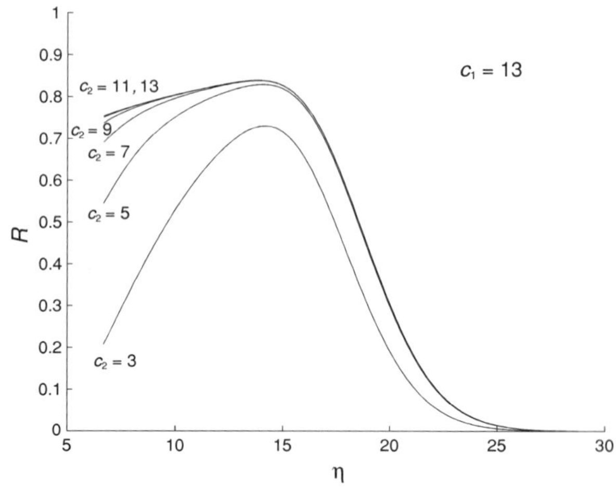


**Fig. 4.7** $R$, as a function of $\eta$, in a scenario with $\Delta = 50, c_1 = 13, T = 2$, and $x = 0.7$

motivated by the fact that for lower values of $\eta$, the lowest level in the tree is the one hosting the majority of nodes, and a capacity limitation at the penultimate level strongly affects the possibility to collect information from the field. On the opposite, for large values of $\eta$, the lowest level tends to become empty, and such capacity limitation does not affect significantly the probability $R$.

By comparing the curve for $x = 0.7$ in Fig. 4.5 to those of Fig. 4.7, it can be seen that the capacity increase from 7 to 13 clearly shows an improvement on network coverage. However, $R$ does not reach unity. Indeed, it was found that with $c$ tending to infinity, $R$ monotonically increases with $\eta$, and the maximum is reached for $\eta = 50$ where $R$ becomes approximately 0.98. The difference between this value and unity is due to the statistical behaviour of the number of nodes per tree: even if there are no capacity limitations and network connectivity is assured, the probability of any given numbers of nodes being connected to a sink does not reach unity because there is non-zero probability of trees with very few nodes (even zero, with low probability).

Finally, results having fixed $\rho_0 = 5 \times 10^{-4}\,\mathrm{m}^{-2}, \Delta = 10, k_0 = 40\,\mathrm{dB},\ k_1 = 13.03, \sigma_s = 3.5\,\mathrm{dB}$, and $L_{\mathrm{th}} = 95.6\,\mathrm{dB}$, are shown.

Figure 4.8 shows $R$, as a function of $\eta$, for $x = 0.7$, 0.8, and 0.9. As we can see, by increasing $x, R$ decreases, as expected. According to the constraints (4.10) and (4.12), the values of $\eta$ that should be considered depend on $T$. Here we consider $T = 2$, 3, and 4, which correspond to $\eta$ ranging in [2.71, 10], [1.74, 2.71], and [1.4, 1.74], respectively. Note that the cases with $T = 2$ and 3 converge for $\eta = 2.71$, because in this point we have a four-level tree with the lowest level empty, or a three-level tree with the lowest level having nodes density which is $\eta$ times larger than that at the higher level. The same holds for $T = 3$ and 4, at $\eta = 1.74$. In all cases, the maximum value of $R$ is reached for $T = 2$, whereas the optimum value of $\eta$ decreases by increasing $x$. As one can see, even in the case $x = 0.7, R$ assumes a maximum value equal to 0.4, that is quite low: the reason is that performance has

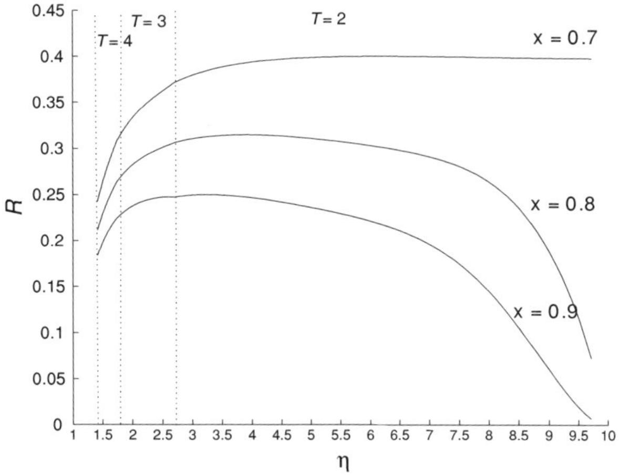


**Fig. 4.8** $R$, as a function of $\eta$, for $\rho_0 = 5 \times 10^{-5}\,\mathrm{m}^{-2}$ and $x = 0.7, 0.8$, and $0.9$

been evaluated for a network with low density, which has connectivity problems. Thus, in Fig. 4.9, we show $R$, as a function of $\eta$, for a network having $\rho_0 = 5 \times 10^{-4}\,\mathrm{m}^{-2}$ and $\Delta = 10$, for $T = 2$, 3, and 4 and $x = 0.7$, 0.8, and 0.9. As we can see, this network is more connected than the one considered in Fig. 4.8; in fact, $R$ reaches the values of 0.85 for $x = 0.7$. The optimum value of $R$ is reached for $T = 2$ and for $\eta = 10$, 7, and 6.3 in the three cases $x = 0.7$, 0.8, and 0.9, respectively. Moreover, we can note that for $x = 0.8$ and 0.9, when $\eta$ assumes value larger than 7, $R$ decreases: this is due to the capacity limit imposed ($c = 7$) which affects $R$, for large values of $\eta$.

### *4.5.3 The Three-Level Tree: Simulation Environment*

Simulation results are generated through a C language simulation tool specifically developed to model the environment and protocols described in the following. The reference scenario considered consists of a number of nodes randomly and uniformly distributed over a square area (having side $B$ meters, so that $\rho_s = 1/B^2$), which is Poisson distributed with given mean, $N_s$. Both single-sink and multi-sink scenarios are simulated. In the first case only one sink is located in a given position of the area and we fix $N_s$ such that $\Delta = N_s$. In the multi-sink scenario, instead, a number of sinks are Poisson distributed in the square, with given mean, $I$. As stated in Sect. 4.1, in the simulation environment IEEE 802.15.4 devices are considered. Therefore, the sinks are the Personal Area Network (PAN) coordinators, managing a PAN, composed of a given number of sensors and formed according to a procedure described in the following. Nodes work in beacon-enabled mode, so that sinks periodically send beacon packets. The network must be able to provide the information detected by nodes to the sinks within the

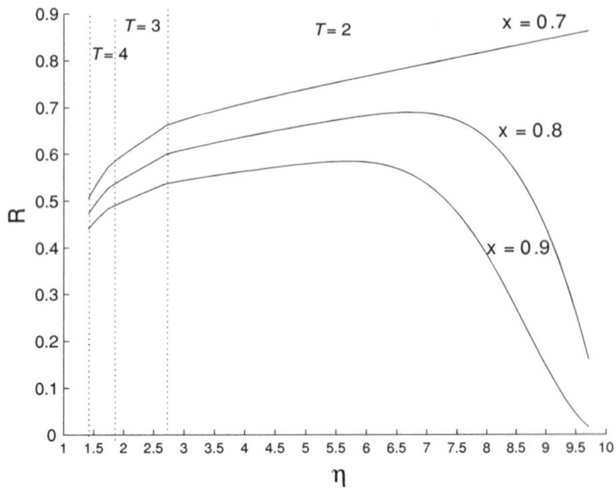


**Fig. 4.9** $R$, as a function of $\eta$, for $\rho_0 = 5 \times 10^{-4}\,\mathrm{m}^{-2}$ and $x = 0.7, 0.8$, and $0.9$

superframe starting with the transmission of the beacon from sinks. We denote as *round* the period of time between two successive beacon packets sent by the sinks (i.e., the beacon interval). It is also assumed that all sinks are time synchronised, i.e., they transmit the beacon packets at the same time.

Note that here we do not consider the Zigbee tree-based topology: tree formation and the access to the channel is managed through a different communication protocol, described in the following.

Finally, we impose a capacity constraint, thus we fix a maximum number of children per parent.

#### 4.5.3.1 The Tree-Based Topology

The network is organised in a three-level hierarchical topology: the sink is at level zero, level one is constituted by nodes denoted as Cluster Heads (CHs), and level two is constituted by non CH nodes. Nodes elect themselves CHs with probability $p_1$. Therefore, we have $\rho_1 = p_1 \cdot \rho_s$ and consequently $p_1 = \eta/\Delta = \eta/I$. Recall that $\eta$ is the mean number of children per parent. Having fixed $I$, and by varying $p_1$, this observation allows to draw curves of $R$, that can be easily derived through simulation, as a function of $\eta$.

The tree is formed according to the following steps.

1. PANs formation—each sink transmits a beacon packet and nodes select the PAN to belong to on the basis of the received power: each node selects the sink from which it receives the highest power.
2. Clusters formation—in each PAN a certain number of nodes elect themselves CHs, with probability $p_1$. Each CH broadcasts a packet informing of its role and those nodes that did not elected themselves as CHs (non CHs) select their CHs

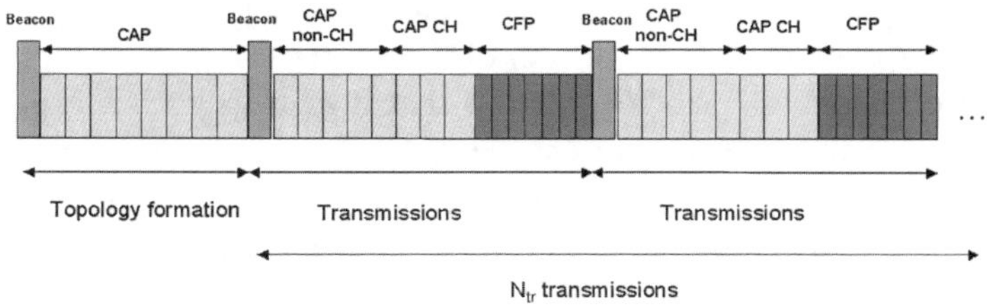


**Fig. 4.10** The IEEE 802.15.4 superframe used in the communication protocol

to transmit to, on the basis of the power received by each CH. In particular, once again, each node selects the loudest CH.

3. Transmissions—each non CH node transmits its packet to the respective CH, which, on its turn, sends all packets received, plus the one it generated, to the selected remote sink via a direct link. If a non CH node does not receive correctly any broadcast packets coming from CHs, or there are not elected CHs in a PAN, its packet is lost (transmissions from level two to level zero are not allowed).

As will be clarified in the following, two superframes are needed for exploiting the protocol: a superframe is used for PANs and clusters formation and another one is devoted to transmissions. In particular, a superframe for the tree formation, is followed by $N_{tr}$ superframes where sample transmissions take place. Therefore, trees are formed every $N_{tr}$ rounds (see Fig. 4.10). This is reasonable, under the assumption that the channel has a coherence time equal to $N_{tr}$ rounds.

#### 4.5.3.2 MAC Layer Protocol

The beacon-enabled mode, with acknowledgement (ACK) transmission, is considered.

Three kinds of packet can be transmitted in the network: the beacon, having a size of 62 bytes (i.e., it is transmitted in $124\,T_s$, since a bit rate of 250 kbit/s is used); the ACK packet sent to notify the correct reception of a data packet, having a size of 5 bytes, and data packets, containing the measurement result and having a size of 25 bytes (with a payload of 10 bytes).

At the end of the topology formation phase, it is assumed that the sinks are aware of the topology. This is possible because of CH broadcasts and the ACK packets sent by non CHs to notify broadcast reception are received by the final sink too (assuming reciprocal links, if a node correctly receives the beacon, it can reach the sink). Since CHs have to transmit to the final sink all packets received inside their clusters plus the one generated by themselves, the loss of a CH packet implies the loss of a large number of samples. For this reason we decide to assign GTSs to CHs. Then, when the sink transmits the beacon which starts the sample transmission phase, it assigns the GTSs to the CHs whose clusters are larger. In other words, we introduce a priority for those CHs which have the largest cluster sizes. Moreover the sink assigns a specific channel to each cluster, that is each non

CH belonging to a given cluster uses the same channel: in this way, collisions between clusters are avoided, while non CHs compete during the CAP on a given channel (in case of more than 16 clusters a spatial frequency reuse is performed). Thus, both CAP and CFP are present: the CAP duration ranges from $T_{\mathrm{CAP}_{\max}} = 960 \cdot 2^{SO} \cdot T_s - 124\,T_s$, when there are no CHs, and GTSs are not allocated, to $T_{\mathrm{CAP}_{\min}} = 60 \cdot 2^{SO} \cdot T_s \cdot (16 - 7) - 124\,T_s$, when there are seven or more than seven CHs and all GTSs are allocated (see Fig. 4.10). A large value of *SO* is set so that the packets (and the inter-frame space) could be contained in the minimum duration of a GTS ($60 \cdot 2^{SO}\,T_s$) and seven GTSs could be allocated (see Chap. 1). When the number of CHs is lower than seven, all the CAP is used by non CHs that have to transmit to their CHs, through CSMA/CA. When, instead, the number of CHs is larger than seven, the CFP is used and the CAP duration is $T_{\mathrm{CAP}_{\min}}$. In this case, the CAP is subdivided into two parts: the first part, $T_{\mathrm{CAP}_{\mathrm{nonCH}}}$, set to the $C\%$ of $T_{\mathrm{CAP}_{\min}}$, is devoted to non CHs transmissions, whereas the second part, $T_{\mathrm{CAP}_{\mathrm{CH}}}$, set to $(100 - C)\%$ of $T_{\mathrm{CAP}_{\min}}$, is devoted to transmissions of the CHs that do not have a GTS assigned. These nodes use the default frequency to transmit to the final sink, thus they could collide. In the following, we show curves for different values of *C*.

No mechanism to handle hidden terminals is performed and, therefore, collisions occur and some packets are lost. To realistically account for collisions, capture effect is taken into consideration: we assume a packet is captured by the receiver, even in the case of packet collisions (i.e., simultaneous transmissions of packets by different nodes), if

$$\frac{P_{\mathrm{R}0}}{\sum_{i=1}^{N_c} P_{\mathrm{R}i}} > \alpha \tag{4.19}$$

where $P_{\mathrm{R}0}$ is the power received from the useful signal; $P_{\mathrm{R}i}$ is the *i*th interferer power; $N_c$ is the number of colliding packets; and $\alpha$ is the capture threshold (i.e., the protection ratio), set to 4 dB. When condition (4.19) is not fulfilled, the packet is lost and the receiving node does not transmit the ACK packet.

### *4.5.4 Simulation Results*

In this subsection, we report the numerical results obtained through simulations, in the mono- and multi-sink scenarios. 1000 rounds are simulated and then, ten different and uncorrelated realisations of node locations are considered. At each round the packet error rate, obtained dividing the number of samples lost by the number of nodes in the network, is computed and, at the end, *R* is evaluated.

The packet losses are caused by the following events:

- a node is isolated: it does not receive the beacon packet or it does not receive any CH broadcast packet and it cannot reach directly the final sink; this event has very low probability with the system parameters considered in this chapter;

- a node tries to access the channel for more than $NB_{max}$ consecutive times without success (see Chap. 1);
- a node does not succeed in correctly transmitting its packet by the end of the superframe portion devoted to it;
- when a capacity constraint is imposed it may happen that some nodes (CHs or non CHs) cannot transmit their packets to their parent (the selected sink, or CH).

As in the mathematical analysis, the objective is to maximise the probability $R$ that the number of samples received by the final sink is above (or equal to) a fixed fraction $x$ of the mean $\Delta$. To do this, we study the behaviour of $R$ by varying $\eta$ (and thus $p_1$): results show that there exists an optimum value of $\eta$, maximising $R$. This optimum number can be easily motivated by the need to compromise between the load within clusters, which depends on their size and is controlled by increasing the number of CHs, and the probability of collisions among CH packets, that can be minimised by decreasing the number of CHs.

The first results are obtained in the single-sink scenario, by setting $N_s = \Delta = 100, B = 100\,\text{m}$ (i.e., $\rho_0 = 10^{-4}\,\text{m}^{-2}$), $SO = 10, k_0 = 40\,\text{dB},\ k_1 = 15, \sigma_s = 4\,\text{dB}$, and $L_{th} = 110\,\text{dB}$.

In Fig. 4.11, $R$ is shown as a function of $\eta$, by varying the threshold $x$. In all cases, the percentage $C$ is set to 70. No capacity constraints are imposed here. By increasing $x, R$ decreases, as expected. As we can see, the curves show a maximum, that is there exists an optimum value of $\eta$, denoted as $\eta_{opt}$, for which $R$ assumes a maximum value. In fact, when $\eta$ is too low, GTSs are not exploited and the number of non CHs which compete for the channel is large and collisions inside clusters have higher probabilities. On the other hand, when $\eta$ increases, the number of CHs using the CAP becomes large and the collision probability increases in the superframe portion devoted to CHs. We note that, by varying $x$, the value of $\eta_{opt}$ remains approximately the same.

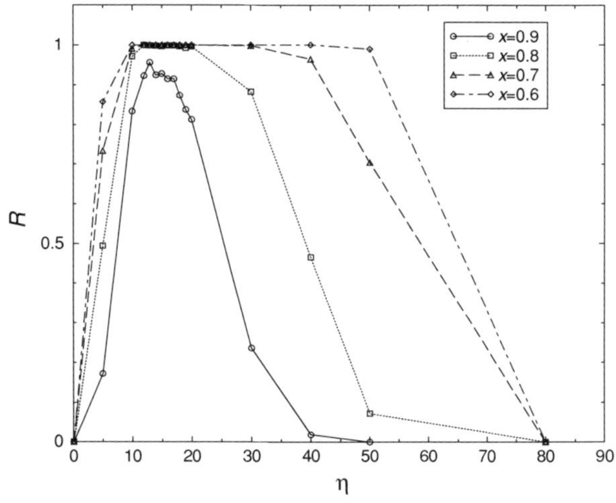


**Fig. 4.11** $R$, as a function of $\eta$, in the case $C = 70$

In Fig. 4.12, instead, the case with $C = 100$ is investigated. In this case, the maximum number of CHs that can be connected to the final sink is seven; therefore, it is equivalent to impose the constraint $c_1 = 7$, whereas no constraint is imposed on $c_2$. In this way, if the number of CHs is larger than seven the samples gathered by those having smaller clusters are lost. As one can see in Fig. 4.12, the curves reach a maximum in correspondence to an optimum value $\eta_{\mathrm{opt}}$; once again, by increasing $x, R$ decreases and $\eta_{\mathrm{opt}}$ is approximately the same for the different values of $x$.

If we compare Figs. 4.11 and 4.12, we can observe that when $T_{\mathrm{CAP_{CH}}}$ is set to zero, performance worsens, because when the number of CHs is larger than seven all their packets are lost. Consequently, the $\eta_{\mathrm{opt}}$ values in this case are lower than the ones obtained with $T_{\mathrm{CAP_{CH}}}$ set to 70% of $T_{\mathrm{CAP_{min}}}$, because the network works better when the number of CHs is lower.

In the following figures, a comparison between the results obtained through simulations and the mathematical model for the single-sink scenario, while setting $T = 2$ (three-level tree), is provided. Owing to the different strategies to access the channel, contention-based in simulations and contention-free in the mathematical framework, the comparison does not aim at validating the model, but at showing how the use of different MAC protocols impacts the performance.

The following values for the parameters are set: $N_s = \Delta = 50, B = 50$ m (i.e., $\rho_0 = 4 \times 10^{-4}\,\mathrm{m}^{-2}$), $SO = 10, k_0 = 40\,\mathrm{dB}$, $k_1 = 13.03, \sigma_s = 3.5\,\mathrm{dB}$, and $L_{\mathrm{th}} = 92\,\mathrm{dB}$.

Figure 4.13 shows $R$, as a function of $\eta$, for $\Delta = 50, c = 13$, with $x$ taking values 0.7, 0.8, 0.9. Simulation and mathematical results are reported. With such large value of node capacity, for large $\eta$ the number of collisions during the CAP can be large. In fact, this scenario is characterised by soft capacity constraints. As a result, the optimum value of $\eta$ is smaller than in the case of deterministic access, considered in the mathematical model. Simulations report better performance for

**Fig. 4.12** $R$, as a function of $\eta$, in the case $C = 100$

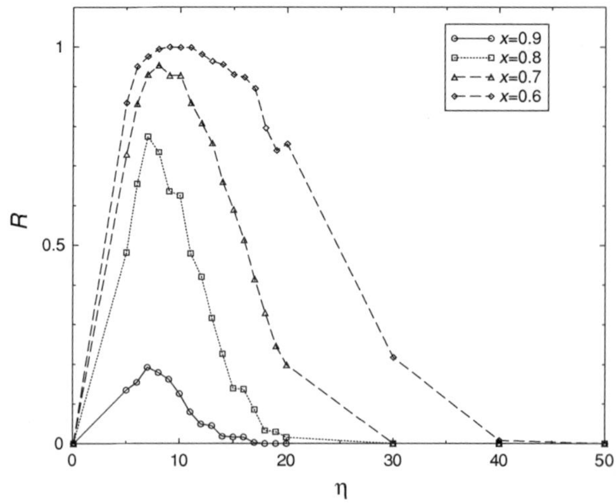

the optimum values of $\eta$, because of the border effects introduced by the limited area considered in the simulated scenario.

With smaller capacity values (see Fig. 4.14 with $c$ set to 7), leading to situations where coverage is limited by hard capacity constraints, we found that simulation results give smaller values of $R$ than the mathematical analysis. However, in this case (where collisions play a minor role) the optimum value of $\eta$ found with the mathematical and simulation approaches coincide, confirming the motivation given above to the different optimum values of $\eta$.

Finally, Fig. 4.15 reports simulation outcomes achieved for the same set of parameters as in Fig. 4.13. The trends for the various values of $c_2$ are very similar, and the differences are motivated by the effects mentioned in the previous subsections.

In Fig. 4.16, we compare results obtained in the single-sink and in the multi-sink scenarios. For a fair comparison, we set a common value of sink's density, $\rho_0 = 4 \times 10^{-4}\,\mathrm{m}^{-2}$, and $\Delta = 50$. As a consequence, the square side varies with the number of sinks: if the latter is equal to one, $B = 50\,\mathrm{m}$, in case $I = 9, B = 150\,\mathrm{m}$, and so on. The other parameters are set as follows: $SO = 10, k_0 = 40\,\mathrm{dB},\ k_1 = 13.03, \sigma_s = 3.5\,\mathrm{dB}$, and $L_{\mathrm{th}} = 92\,\mathrm{dB}$. The figure reports $R$ as a function of $\eta$ (equal to $p_1 \cdot \Delta$) in three different situations: (i) the single-sink deterministic scenario (with sink located in the centre of the area); (ii) the single-sink random scenario (where the sink is located in a random position); and (iii) the multi-sink case. We set $SO = 10, x = 0.7$, and $c_1 = c_2 = 13$. This makes $R$ decreasing rapidly when $\eta$ takes values larger than 13 (many nodes belong to level 1 but they are blocked, and a few level 2 nodes connect to the level 1 nodes that are accepted by the sink). In the deterministic single-sink case, according to the propagation parameters used, and the side of the area, the majority of nodes can hear the sink and there are no isolated nodes; thus, losses are due to MAC failures and capacity constraints. In the multi-sink case, instead, $R$ assumes smaller values, because of a larger

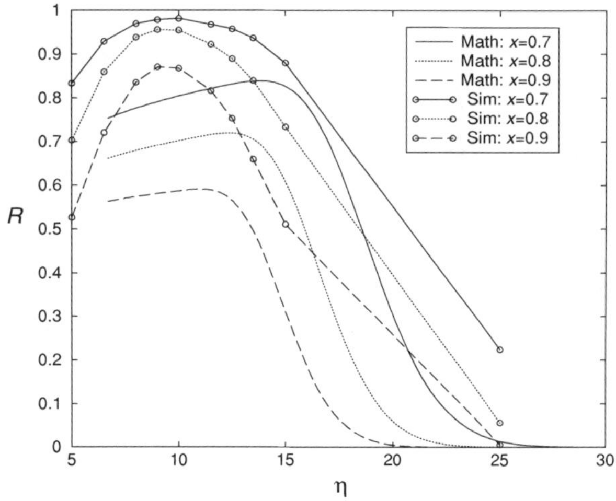


**Fig. 4.13** $R$, as a function of $\eta$, in a scenario with $\Delta = 50$ and $c = 13$. Mathematical and simulation results are reported

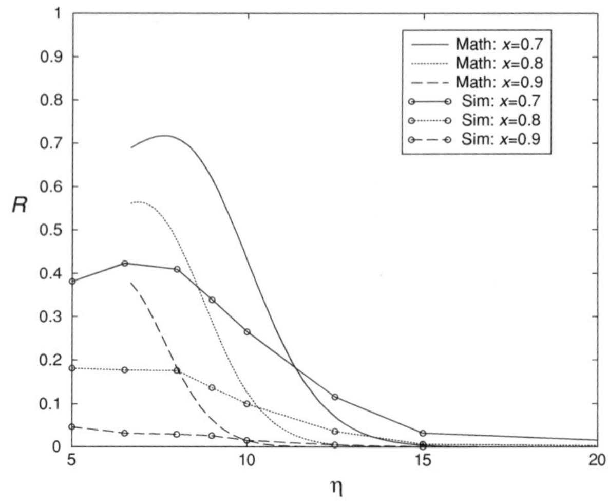


**Fig. 4.14** $R$, as a function of $\eta$, in a scenario with $\Delta = 50$ and $c = 7$. Mathematical and simulation results are reported

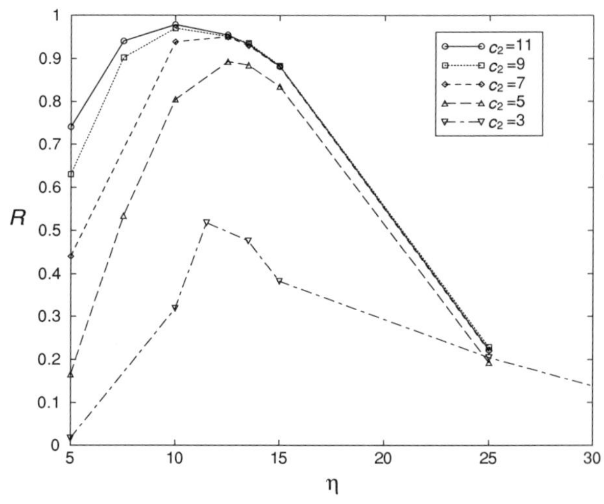


**Fig. 4.15** $R$, as a function of $\eta$, in a scenario with $\Delta = 50$, $c_1 = 13$, and $x = 0.7$. Only simulation results are presented

probability to have isolated nodes that can hear no sinks. As $I$ increases, the dispersion in the number of nodes that join a sink decreases, and the distribution of the PAN sizes has smaller variance. As a result, $R$ increases.

The distributions of the PAN sizes are reported in Fig. 4.17 for $I = 100$ and 200. They are compared to a Poisson distribution having proper mean: according to [21], in an infinite plane the PAN sizes should be Poisson distributed with mean that can be calculated starting from node and sink densities, and propagation and physical layer parameters. The figure shows that the limited area brings to larger variances in such distributions with respect to the infinite plane case. As a result of such discussion, one can conclude that the multi-sink scenario gives smaller values of $R$ when $\eta$ is given, owing to the larger variances of the PAN size distributions.

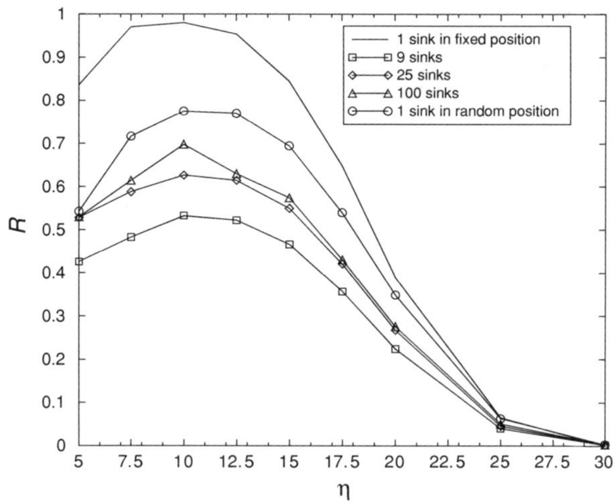


**Fig. 4.16** $R$, as a function of $\eta$, in the single-sink and multi-sink scenarios, with different values of $I$, having fixed $\Delta = 50$ and $c = 13$

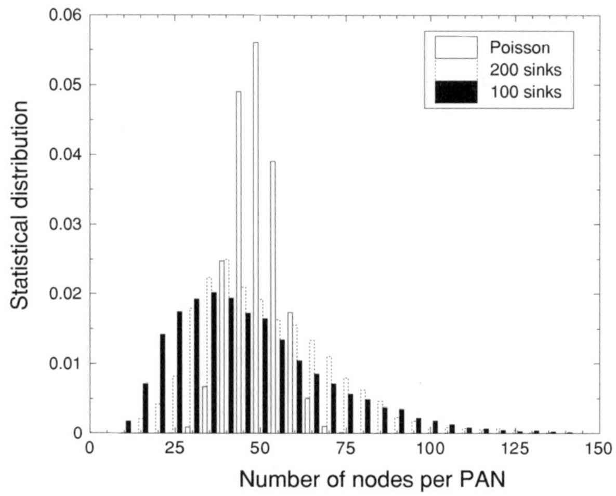


**Fig. 4.17** Statistical distribution of the number of nodes per PAN in the 100 and 200 sink scenarios

In Fig. 4.18, we show $R$, as a function of $\eta$, for a network having side $B = 150$ m (thus $I = 9$), having fixed $x = 0.7$ and $c_1 = 13$, for $c_2$ ranging from 3 to 13. The curves behaviour is the same observed in Fig. 4.7, obtained through the mathematical model. The values of $R$, obtained through simulations, are lower than the corresponding values obtained through the mathematical model. This is due to MAC failures and to the fact that we consider a network with $I = 9$, which is affected by border effects (here, in fact, different packets are lost for connectivity issues), whereas by increasing $I$ (as shown in Fig. 4.16), $R$ increases (losses due to connectivity issues decrease) and for high values of $I$ we could reach the value obtained in the mathematical model (we cannot show here results obtained for larger values of $I$, owing to too long simulation time needed). The other difference is that the maximum value of $R$ is obtained for different values of $\eta$; this is due to

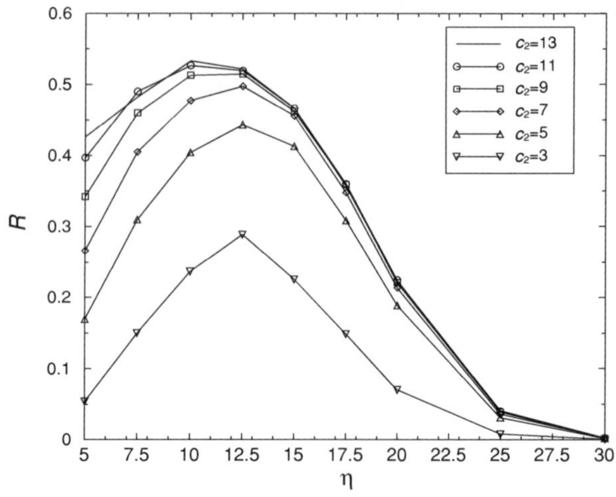


**Fig. 4.18** Multi-sink scenario ($I = 9$). $R$, as a function of $\eta$, for different values of $c_2$, having fixed $c_1 = 13$

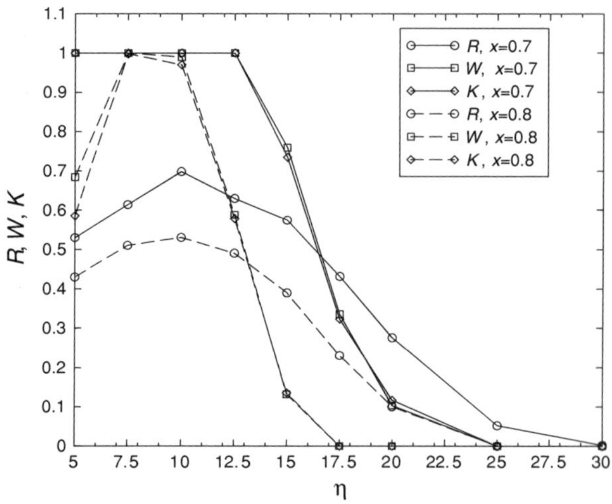


**Fig. 4.19** Multi-sink scenario ($I = 9$). $R, W$, and $K$, as functions of $\eta$, for $x = 0.7$ and $x = 0.8$, having fixed $c_1 = c_2 = 13$

the fact that the mathematical model requires an a priori definition of which level each node belongs to, whereas in the simulation environment a real topology formation algorithm is considered.

Finally, in Fig. 4.19 two new performance metrics (functions of $x$), denoted as $W$ and $K$, are introduced. $W$ is defined as the probability that the number of packets correctly received in the network, considering all sinks, is larger than a percentage $x$ of the real number of nodes in the network; $K$ is defined as the probability that the number of packets correctly received in the network, considering all sinks, is larger than a percentage $x$ of the mean number of nodes in the network. Thus, in Fig. 4.19 we show $R, W$, and $K$, as functions of $\eta$, for

$x = 0.7$ and $x = 0.8$ for a network distributed in a square area, having side $B = 250\,\mathrm{m}$ ($I = 100$), $\rho_s = 0.2\,\mathrm{m}^{-2}$, $\rho_0 = 0.002\,\mathrm{m}^{-2}$ and $c = 13$. As we can see, the $W$ and $K$ curves are overlapped; on the other hand, $W$ and $K$ show different values with respect to $R$. When $\eta$ is smaller than the capacity limit, $W$ and $K$ tend to one because the distribution of the total number of nodes attached to any sink shows a smaller variance with respect to the distribution of the number $n$ of nodes attached to a generic sink; when capacity limits become significant (for larger $\eta$), the different clusters tend to be equally limited in size and this reduces the probability of large values of the sum of all cluster sizes, thus reducing $W$ and $K$.

## 4.6 Connectivity of Multi-Sink Multi-Hop WSNs in Bounded Regions

In this section, we mathematically derive the probability that sensor nodes uniformly distributed over the monitored area are connected to at least one sink, where multiple sinks are also uniformly distributed over the same region. Starting from such a result, we also derive the probability that all nodes, or a subset of them, are connected. This derivation is performed assuming a link power loss which takes both dependance on the distance and random channel fluctuations into account (the channel model of Eq. 4.1) and considering border effects due to finiteness of the deployment region. The latter is assumed to be a square, as it often happens (see, e.g., [22]), because of its simplicity. Nonetheless, rectangular networks exhibit very similar connectivity properties unless one side is much greater than the other [23].

The work is based on previous works [7, 23] devoted to single hop networks. Here, bounded scenarios are accounted for, and this requires suitable consideration of the border effects. It is also shown that this model converges to the ones applied in the case of infinite plane, when the bounded region has an area which is sufficiently large.

The analysis is first carried out in the case of single-hop communication (i.e., every sensor transmits the sensed data directly to a sink). Then, the multi-hop case (i.e., sensors may also act as routers) is considered assuming tree-based topologies of various heights and widths. Finally, the mean energy consumed by the network is evaluated, and the tradeoff between connectivity and energy consumption is shown.

In the remainder of this section, the connectivity model for infinite networks, which represents the starting point of the analysis, is introduced. Then, in Sect. 4.6.2 the bounded region is introduced and the full and partial connectivity probabilities are derived for the single-hop case. In Sect. 4.6.3, the multi-hop case is considered. In Sect. 4.6.4, the mean energy consumption is examined and numerical results are shown in Sect. 4.6.5.

### 4.6.1 Connectivity in Unbounded Single-Hop Networks

The first scenario consists of an infinite bi-dimensional plane with sensors and sinks distributed according to a homogeneous PPP, with densities $\rho_s$ and $\rho_0$, respectively. Since the channel model described by Eq. 4.1 is used, the number of audible sinks within a range of distances $r_1$ and $r$ from a generic sensor node ($r \geq r_1$), denoted as $n_{r_1,r}$, is Poisson distributed with mean $\mu_{r_1,r}$, given by Eq. 4.7 by simply substituting $\rho$ with $\rho_0$. Then by letting $r_1 = 0$ and $r \to \infty$, we obtain

$$\mu_{0,\infty} = \pi\rho_0 \exp[(2(L_{\text{th}} - k_0)/k_1) + (2\sigma_s^2/k_1^2)]. \tag{4.20}$$

Equation 4.20 represents the mean value of the total number, $n_{0,\infty}$, of audible sinks for a generic sensor, obtained considering an infinite plane [7]. Its non isolation probability is simply the probability that the number of audible sinks is larger than zero and is given by

$$q_\infty = 1 - e^{-\mu_{0,\infty}}. \tag{4.21}$$

### 4.6.2 Connectivity in Bounded Single-Hop Networks

When moving to networks of nodes located in bounded domains, two important changes happen. First, even with $\rho_0$ unchanged, the number of sinks that are audible from a generic sensor will be lower due to geometric constraints (a finite area contains (on average) a lower number of audible sinks than an infinite plane). Second, the mean number of audible sinks will depend on the position $(x, y)$ in which the sensor node is located in the region that we consider. The reason for this is that sensors which are at a distance $d$ from the border, with $d \sim TR_\text{i}$, have smaller connectivity regions and thus the average number of audible sinks is smaller. These effects, known in literature as *border effects* [10], are accounted for in our model.

The result (4.7) can be easily adjusted to show that the number of audible sinks within a sector of an annulus having radii $r_1$ and $r$ and subtending an angle $2\theta$, is once again Poisson distributed with mean

$$\mu_{r_1,r;\theta} = \theta\rho_0[\Psi(a_1, b_1; r) - \Psi(a_1, b_1; r_1)] \quad 0 \leq \theta \leq \pi. \tag{4.22}$$

If the annulus extends from $r$ to $r + \delta r$, and $\theta = \theta(r)$, this mean value becomes

$$\mu_{r,r+\delta r;\theta} = \theta(r)\rho_0 \frac{\delta\Psi(a_1, b_1; r)}{\delta r}\delta r \quad 0 \leq \theta \leq \pi. \tag{4.23}$$

Consider now a polar coordinate system whose origin coincides with a sensor node. As a consequence of (4.23), if a region is located within the two radii $r_1$ and $r_2$ and its points at a distance $r$ from the origin are defined by a $\theta(r)$ law (see [23,

Fig. 1]), then the number of audible sinks in such a region is again Poisson distributed with mean $\mu_{r_1,r_2;\theta(r)} = \int_{r_1}^{r_2} \theta(r)\rho_0 \frac{d\Psi(a_1,b_1;r)}{\mathrm{d}r}\,\mathrm{d}r$, that is, from (4.8) and after some algebra,

$$\mu_{r_1,r_2;\theta(r)} = \int_{r_1}^{r_2} 2\theta(r)\rho_0 r\Phi(a_1 - b_1 \ln r)\,\mathrm{d}r. \tag{4.24}$$

Now consider a square *SA* of side $B$ (dimension: [m]) and area $A = B^2$, where sensors and sinks are uniformly distributed with densities $\rho_s$ and $\rho_0$, respectively. Equation 4.24 is suitable for expressing the mean number of audible sinks from an arbitrary point $(x, y)$ of *SA*, provided that such point is considered as a new origin and that the boundary of *SA* is expressed with respect to the new origin as a function of $r_1, r_2$, and $\theta(r)$. In order to apply equation (4.24) to this scenario and obtain the average number, denoted as $\mu(x, y)$, of audible sinks from the point $(x, y)$, the origin of a reference system has to be set in $(x, y)$, partition *SA* in eight subregions $(S_{r,1} \ldots S_{r,8})$ by means of circles whose centers lie in $(x, y)$ (see [23, Fig. 2]). Owing to the properties of Poisson r.v.'s, the contribution of each region can be summed and we obtain an exact expression for

$$\mu(x, y) = \sum_{i=1}^{8} \int_{r_{1,i}}^{r_{2,i}} 2\theta_i(r) \cdot \rho_0 \cdot r \cdot \Phi(a_1 - b_1 \ln r)\,\mathrm{d}r, \tag{4.25}$$

which is the mean number of sinks in *SA* that are audible from $(x, y)$, where $r_{1,i}, r_{2,i}, \theta_i(r)$ are reported in [23, Tables 1–2]. If we assume a single-hop network, a sensor potentially located in $(x, y)$ is isolated (i.e., there are no audible sinks from its position) with probability $p(x, y) = e^{-\mu(x,y)}$ and it is non isolated with probability

$$q(x, y) = 1 - e^{-\mu(x,y)}. \tag{4.26}$$

Having assumed that sensor nodes are uniformly and randomly distributed in *SA*, if we now want to predict the probability that a randomly chosen sensor node is not isolated, we need to average $q(x, y)$ on *SA*. In fact, the probability that a randomly chosen sensor node is not isolated (which is an ensemble measure) and the average non isolation probability over a single realization coincide due to the ergodicity of stationary Poisson processes (see [24, p. 104]). This was also verified by simulation.

Recalling that we have considered the lower half of the first quadrant, which is one eighth of the totality, we have

$$\overline{q} = \frac{8}{A} \int_0^{B/2} \int_0^{x} q(x, y)\,\mathrm{d}y\,\mathrm{d}x. \tag{4.27}$$

For the sake of simplicity, we define the function $F_{\text{con}}(\cdot,\cdot)$ to be equal to the right side of (4.27), so that

$$\overline{q} = F_{\text{con}}(\rho_0, B). \tag{4.28}$$

Several results may be derived from (4.28). First, we compute the *probability that the network is fully connected*, Z, (i.e., every sensor can directly reach at least one sink). Assume that we have $k$ sensors in *SA* with positions $(x_1, y_1), (x_2, y_2), \ldots,$ $(x_k, y_k)$. By indicating with $F$ the event of full connectivity and with $n_s$ the number of sensors in a scenario, we have

$$\mathbb{P}\{F|n_s = k; (x_1, y_1), \ldots, (x_k, y_k)\} = \prod_{i=1}^{k} q(x_i, y_i), \tag{4.29}$$

where we assumed that sensors connect to the sink independently from each others, which is a realistic assumption in networks that are not capacity-limited and $\mathbb{P}\{\mathscr{E}\}$ denotes the probability of the event $\mathscr{E}$.

Now, by deconditioning with respect to the nodes positions, we have

$$\begin{aligned}
\mathbb{P}\{F|n_s = k\} &= \underbrace{\int \cdots \int}_{2k} \prod_{i=1}^{k} q(x_i, y_i) f_{X_1,Y_1}(x_1, y_1) \cdots \\
&\quad \cdots f_{X_k,Y_k}(x_k, y_k)\, \mathrm{d}x_1\, \mathrm{d}y_1 \cdots \mathrm{d}x_k\, \mathrm{d}y_k \\
&= \left[\int\int q(x_1, y_1) f_{X_1,Y_1}(x_1, y_1)\, \mathrm{d}x_1\, \mathrm{d}y_1\right] \cdots \\
&\quad \cdots \left[\int\int q(x_k, y_k) f_{X_k,Y_k}(x_k, y_k)\, \mathrm{d}x_k\, \mathrm{d}y_k\right],
\end{aligned}$$

where

$$f_{X_i,Y_i}(x_i, y_i) = \begin{cases} 1/A, & (x_i, y_i) \in SA \\ 0, & \text{otherwise} \end{cases}$$

is the probability density function p.d.f. of the position of the $i$th node. Note now that since the same assumption (i.e., uniform distribution) holds for all nodes, thus we have

$$\begin{aligned}
\mathbb{P}\{F|n_s = k\} &= \left[\int\int q(x, y) f_{X,Y}(x, y)\, \mathrm{d}x\, \mathrm{d}y\right]^k \\
&= \left[\frac{1}{A} \int_{-B/2}^{B/2} \int_{-B/2}^{B/2} q(x, y)\, \mathrm{d}x\, \mathrm{d}y\right]^k \\
&= \left[\frac{8}{A} \int_{0}^{B/2} \int_{0}^{x} q(x, y)\, \mathrm{d}y\, \mathrm{d}x\right]^k = \overline{q}^k.
\end{aligned} \tag{4.30}$$

Since $n_s$ is Poisson distributed with mean $\rho_s A$, we can decondition (4.30) with respect to $n_s$ and obtain

$$Z = \mathbb{P}\{F\} = \sum_{k=1}^{+\infty} \mathbb{P}\{F|n_s = k\} \cdot \mathbb{P}\{n_s = k\} = \sum_{k=1}^{+\infty} \bar{q}^k \cdot \frac{e^{-\rho_s A}}{k!} (\rho_s A)^k. \qquad (4.31)$$

Equation 4.31 represents the probability that a sensor network performs at best (*full connectivity*), but the event $F$ turns out to be a strict requirement for most of them. In other words, for many applications it is sufficient to guarantee that a certain amount of sensors can transmit their data to the sinks. For this reason, it is of interest to compute the probability of the event, $C_j$, of having at least a number, $j$, of connected sensor (*partial connectivity*). We first consider the event $C_j^*$ of having exactly $j$ connected sensors. When $n_s = k$, the probability of having $j$ connected sensors is

$$\mathbb{P}\{C_j^*|k\} = \binom{k}{j} \bar{q}^j (1 - \bar{q})^{k-j} \quad j \leq k, \qquad (4.32)$$

where the binomial coefficient $\binom{k}{j} = \frac{k!}{j!(k-j)!}$ accounts for all the possible ways to group $j$ sensors out of $k$. Note that for the events $C_j$ and $C_j^*$ the following holds:

$$C_j = \{C_j^* \cup C_{j+1}^* \cup \cdots \cup C_k^*\}. \qquad (4.33)$$

Thus, if we consider the event $C_j$ we need to add contributions similar to (4.32) for all $j$, $j \leq k$, to obtain

$$\mathbb{P}\{C_j|n_s = k\} = \sum_{l=j}^{k} \binom{k}{l} \bar{q}^l (1 - \bar{q})^{k-l} \quad j \leq k. \qquad (4.34)$$

Once again, by deconditioning (4.34) with respect to $n_s$ we have

$$\begin{aligned}\mathbb{P}\{C_j\} &= \sum_{k=j}^{+\infty} \mathbb{P}\{C_j|k\} \cdot \mathbb{P}\{k\} \\ &= \sum_{k=j}^{+\infty} \sum_{l=j}^{k} \binom{k}{l} \bar{q}^l (1 - \bar{q})^{k-l} \cdot \frac{e^{-\rho_s A}}{k!} (\rho_s A)^k. \end{aligned} \qquad (4.35)$$

Note that the outer sum in (4.35) starts at $j$ instead of 1, because when $k < j$ there is no contribution (i.e., the probability of having $j$ connected sensors in a network with less than $j$ sensors is zero). For this reason, we want to highlight the fact that $\mathbb{P}\{C_j\}$ of (4.35) depends also on $\rho_s$: in fact, the probability of having at least $j$ connected sensors is affected, besides $\bar{q}$, also by the total number of sensors in the network (i.e., either connected or not). In order to emphasize this, a new notation, $Z_{\bar{m}}(j)$, is introduced and, after some simple algebra, we have

$$
\begin{aligned}
Z_{\bar{m}}(j) &= \mathbb{P}\{C_j\} \\
&= e^{-\bar{m}} \cdot \sum_{k=j}^{+\infty} \sum_{l=j}^{k} \binom{k}{l} \frac{\bar{m}^k \bar{q}^l (1-\bar{q})^{k-l}}{k!},
\end{aligned}
\tag{4.36}
$$

with $\bar{m} = \rho_s A$ being the average number of sensors in $SA$. Thus, $Z_{\bar{m}}(j)$ of (4.36) has the meaning of probability of having at least $j$ connected sensors in a network with (on average) $\bar{m}$ sensors.

### *4.6.3 Connectivity in Bounded Multi-Hop Networks*

Now we wish to extend our analysis to the case of multi-hop WSNs. Each sensor is allowed to forward its data to another sensor instead of trying to communicate directly with the sinks, with the constraint of a fixed maximum number of hops.

The a-priori partitioning of nodes described in Sect. 4.4 is considered also here. Each node belongs to one out of $T$ levels, meaning that an $i$th level node can send its data only to an $(i-1)$th level node. Therefore, it will take $i$ hops to the former node to communicate with a sink (which is considered a zero level node according to this formalism). This approach is justified by the fact that in some classes of sensor networks each node has a certain probability $p_i$ to be a level $i$ node, with $i \in [0, T]$ ($p_0$ is the probability of being a sink). Thus the parental relations between nodes are in some sense pre-assigned. If $\rho_{\text{tot}}$ is the overall nodes density (i.e., $\rho_{\text{tot}} = \rho_0 + \rho_s$) and $\rho_s$ is the overall sensor nodes density, we have for the generic $i$th level density $\rho_i = \rho_{\text{tot}} \cdot p_i$, $0 \le i \le T$, with $\sum_{i=0}^{T} \rho_i = \rho_{\text{tot}}$ and $\sum_{i=1}^{T} \rho_i = \rho_s$. We also assume that nodes at each level are uniformly distributed in $SA$. We now want to find the probability $\bar{q}_1$ that a randomly chosen sensor is connected and that it is one-hop away from the sink. In terms of the $F_{\text{con}}$ function introduced in (4.28), we can write

$$
\bar{q}_1 = p_1 \cdot F_{\text{con}}(\rho_0, S),
\tag{4.37}
$$

where the two factors account for the events of belonging to the 1st level and being actually connected to a sink, respectively. Note that $\bar{q}_1$ of (4.37) has the same meaning of $\bar{q}$ in (4.27) when $T = 1$. If we consider multi-hop paths, we can define the probability $\bar{q}_i$ that a randomly chosen sensor has a connection to the sink through a path containing at most $i$ hops. In other words, it must be a connected 1st level sensor, or a connected 2nd level sensor, or a connected $i$th level sensor. As an example, the probability $\bar{q}_2$ may be written as

$$
\begin{aligned}
\bar{q}_2 &= p_1 \cdot F_{\text{con}}(\rho_0, B) + p_2 \cdot F_{\text{con}}(\rho_{\text{tot}} \cdot \bar{q}_1, B) \\
&= \bar{q}_1 + p_2 \cdot F_{\text{con}}(\rho_{\text{tot}} \cdot \bar{q}_1, B),
\end{aligned}
\tag{4.38}
$$

where $p_2 \cdot F_{\text{con}}(\rho_{\text{tot}} \cdot \bar{q}_1, B)$ is the probability that the sensor belongs to level 2 and has a connection to any 1st level sensor which is, in turn, connected to a sink. As for $\bar{q}_3$, the chain is one hop longer, thus we need to write

$$\bar{q}_3 = \bar{q}_2 + p_3 \cdot F_{\text{con}}(\rho_{\text{tot}} \cdot p_2 \cdot F_{\text{con}}(\rho_{\text{tot}} \cdot \bar{q}_1, B), B). \tag{4.39}$$

In general, for an $T$-level network we have the recursive expression

$$\bar{q}_T = \bar{q}_{T-1} + p_T \cdot F_{\text{con}}(\rho_{\text{tot}} \cdot p_{T-1} \cdot F_{\text{con}}(\ldots\rho_{\text{tot}} \cdot p_2 \cdot F_{\text{con}}(\rho_{\text{tot}} \cdot \bar{q}_1, B)\ldots, B), B), \tag{4.40}$$

with (4.37) providing expression for $\bar{q}_1$.

We can now introduce the *probability*, denoted as $Z^{(T)}$, *of having all sensors connected in a $T$-level network* by following the same reasoning as in the one-hop case (see Eqs. 4.30–4.31). We recognize that, once the parameters of the network $A$ and $\rho_s$ are fixed, the only difference between the one-hop and the multi-hop case resides in how the non isolation probability is computed, i.e., we have $\bar{q}$ for the one-hop case and $\bar{q}_T$ for the multi-hop case. According to this, we can generalize (4.31) as

$$Z(x) = \sum_{k=1}^{+\infty} x^k \cdot \frac{e^{-\rho_s A}}{k!}(\rho_s A)^k, \tag{4.41}$$

where the structure is fixed and we leave the non isolation probability as variable. Recalling (4.32–4.36), we find that the same holds for (4.36), which yields

$$Z_{\bar{m}}(j; x) = e^{-\bar{m}} \cdot \sum_{k=j}^{+\infty} \sum_{l=j}^{k} \binom{k}{l} \frac{\bar{m}^k x^l (1-x)^{k-l}}{k!}, \tag{4.42}$$

where we set, once again, the non isolation probability as variable. Thus, for $Z^{(T)}$ we can simply use (4.41) with $x = \bar{q}_T$, getting

$$Z^{(T)} = Z(\bar{q}_T) = \sum_{k=1}^{+\infty} \bar{q}_T^k \cdot \frac{e^{-\rho_s A}}{k!}(\rho_s A)^k. \tag{4.43}$$

Similarly, we also compute the probability, denoted as $Z_{\bar{m}}^{(T)}(j)$, of having at least $j$ connected sensors in a $T$-level network with (on average) $m$ sensors by using (4.42) with $x = \bar{q}_T$ and obtain

$$\begin{aligned} Z_{\bar{m}}^{(T)}(j) &= Z_{\bar{m}}(j; \bar{q}_T) \\ &= e^{-\bar{m}} \cdot \sum_{k=j}^{+\infty} \sum_{l=j}^{k} \binom{k}{l} \frac{\bar{m}^k \bar{q}_T^l (1-\bar{q}_T)^{k-l}}{k!}. \end{aligned} \tag{4.44}$$

The way in which the densities $\rho_i$ $(i \geq 1)$ are defined can follow, as an example and without loss of generality, the simple partitioning criterion

$$\rho_{i+1}/\rho_i = \eta, \quad 0 \le i < T, \tag{4.45}$$

where $\eta$ is a constant (i.e., level densities follow an exponential growth, which is kind of a "natural" law in hierarchical networks). Note that (4.45) holds only for $i<T$: in fact, if we fix $\rho_0$ and $\rho_s$, the $T$th level nodes must have density $\rho_T = \rho_s - \sum_{j=1}^{T-1} \rho_j$ in order for the sensor densities to sum up to $\rho_s$. Moreover, by fixing $\rho_0, \rho_s$ and $\eta$ (or equivalently $\rho_0, \Delta = \rho_s/\rho_0$ and $\eta$), there are no longer degrees of freedom and the number $T$ of levels in the network is also consequently assigned.

### 4.6.4 Energy Consumption

We assume that each node consumes energy when transmitting and receiving packets, whereas we neglect the energy spent by the node to stay in idle or sleeping states. We also assume that the sinks do not have energy consumption problems, thus we do not consider the energy spent by them. The mean energy spent in the network for each transmission towards the sink is given by

$$E = \sum_{i=1}^{T} [E_{\rm rx} + E_{\rm tx} \cdot i + E_{\rm rx} \cdot (i-1)] \cdot (\overline{q}_i - \overline{q}_{i-1}), \tag{4.46}$$

where $E_{\rm rx}$ is the energy spent to receive a packet, $E_{\rm tx}$ is the energy spent to transmit a packet, and $\overline{q}_i$ is given by (4.37–4.40). $(\overline{q}_i - \overline{q}_{i-1})$ is the probability that a generic node belongs to level $i$ of a connected tree. The energy spent in the network to deliver a packet from a source node to the final sink, instead, depends on the level at which the source node is located. In particular, if the source node is at level one, the packet can reach the sink through a single transmission; if, instead, the node is at level two its packet must be (i) transmitted by the source node, (ii) received by the level one node, and (iii) transmitted by the latter node to the final sink. Therefore, two transmissions and one reception are needed. We also consider the energy spent by each node to receive the triggering packet coming from its parent in the tree (tree formation). According to the Freescale devices data sheets [17], we set the energy spent to transmit a bit equal to $0.3\,\mu$J/bit and the energy spent to receive a bit equal to $0.33\,\mu$J/bit. Moreover, we set the packet size equal to 20 bytes, therefore $E_{\rm tx} = 48\,\mu$J and $E_{\rm rx} = 52.8\,\mu$J.

### 4.6.5 Numerical Results

Figure 4.20 shows $\overline{q}$ for different sink densities as a function of $L_{\rm th}$, proportional to the transmit power if the receiver sensitivity is fixed. Clearly, as the sink density increases, for a fixed transmit power, it is more likely for a sensor to reach at least

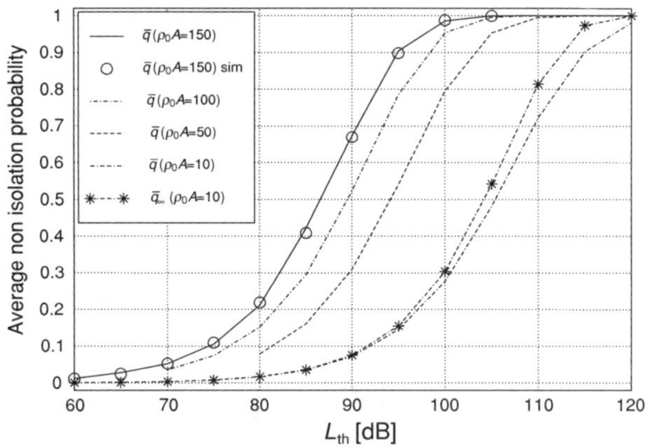


**Fig. 4.20** $\overline{q}$, as a function of $L_{\text{th}}$, for different values of $\rho_0$, with $B = 1000$ m, $k_0 = 40$ dB, $k_1 = 13.03, \sigma_s = 3.5$ dB

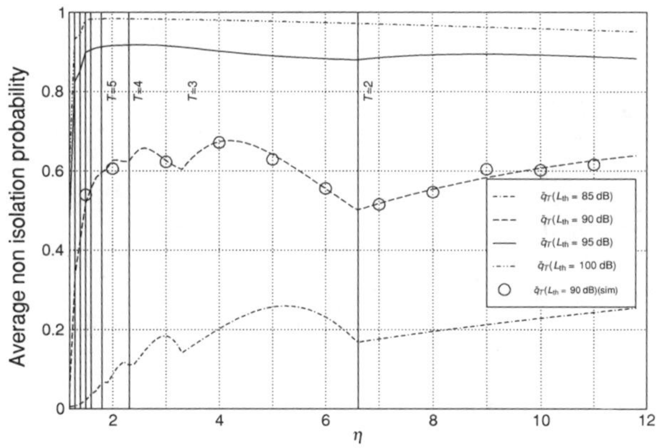


**Fig. 4.21** $\bar{q}_T$, as a function of $\eta$, with $\rho_0 = 50/B^2$, $\rho_s = 2500/B^2$ ($\Delta = 50$), $B = 1000$ m, $k_0 = 40$ dB, $k_1 = 13.03, \sigma_s = 3.5$ dB

a sink and thus $\overline{q}$ also increases. For example, if we want a randomly chosen sensor to be connected with 90% probability, we need $L_{\text{th}} \approx 95$ dB when $\rho_0 A = 150, L_{\text{th}} \approx 98$ dB when $\rho_0 A = 100, L_{\text{th}} \approx 103$ dB when $\rho_0 A = 50$ and $L_{\text{th}} \approx 115$ dB when $\rho_0 A = 10$. Also note the comparison to the curve for $q_\infty$ obtained with no consideration of border effects: the error becomes non negligible for transmission ranges which are of the same size as the side $B$ of the domain (e.g., $TR_{\text{i}}$ ($L_{\text{th}} = 115$ dB) $\approx 316$ m), a typical case for WSNs.

In Figs. 4.21 and 4.22, connectivity results related to multi-hop WSNs are reported. The criterion of a-priori partitioning is used in accordance with (4.45). Observe that, for $T = 5, \eta$ ranges from 1.9 to 2.3. This means that when $\eta = 2.3$ the network has four levels or, equivalently, five levels with the 5th being empty. $\bar{q}_T$ and $Z_{\bar{m}}^{(T)}(j)$ are shown as functions of $\eta$, respectively. They show arches and local optima which depend on the loss threshold $L_{\text{th}}, \eta$, and $T$. In particular, from Fig. 4.21, we conclude that a large value of $T$ is opportune only if $L_{\text{th}}$ (and, consequently, the transmit power) is large enough: in fact, when $T = 5$ ($\eta$ ranging

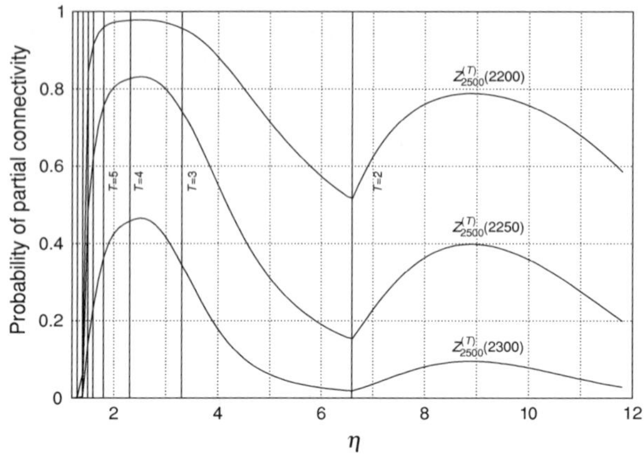


**Fig. 4.22** $Z_{2500}^{(T)}(2200)$, $Z_{2500}^{(T)}(2250)$, and $Z_{2500}^{(T)}(2300)$, as functions of $\eta$, with $L_{\rm th} = 95\,\text{dB}$, $\rho_0 = 50/B^2, \rho_s = 2500/B^2\ (\Delta = 50), B = 1000\,\text{m},\ k_0 = 40\,\text{dB},\ k_1 = 13.03, \sigma_s = 3.5\,\text{dB}$

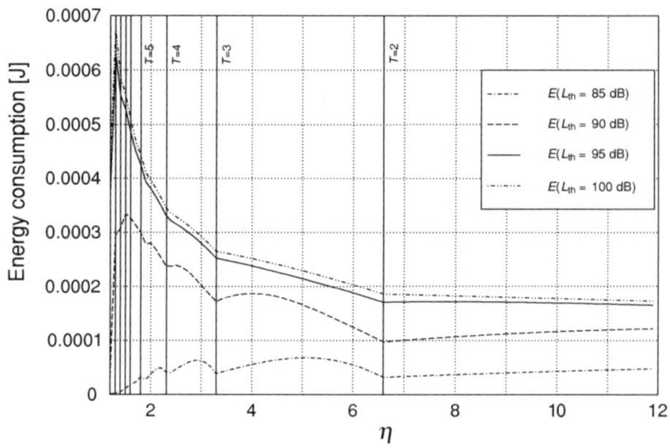


**Fig. 4.23** Average energy consumption $E$ [J], as a function of $\eta$, with $\rho_0 = 50/B^2$, $\rho_s = 2500/B^2$ $(\Delta = 50), B = 1000\,\text{m}$, $k_0 = 40\,\text{dB}, k_1 = 13.03$, $\sigma_s = 3.5\,\text{dB}$

from 1.9 to 2.3) we have global optima for $L_{\rm th} = 95$ dB and $L_{\rm th} = 100$ dB but only local optima for $L_{\rm th} = 85$ dB and $L_{\rm th} = 90$ dB.

Finally, in Fig. 4.23 we show the mean energy spent, denoted as $E$, as a function of $\eta$ and $T$, for different values of $L_{\rm th}$. As one can see, $E$ is an increasing function of $T$, since (on average) more transmissions and receptions are needed to reach the sink. Therefore, for large values of $L_{\rm th}$ a tradeoff between connectivity and energy consumption should be found: in fact, large $T$ improves connectivity but also increases energy consumption. Moreover, the evaluation of the energy consumption behavior is useful to select the optimum values of $\eta$ and $T$, for a desired degree of connectivity. As an example, when we set $L_{\rm th} = 90$ dB, we obtain approximatively the same maximum of $\bar{q}_T$ for $T = 4$ and $T = 3$; however, the consumed energy is notably larger for $T = 4$.

## 4.7 Concluding Remarks

A novel mathematical model for studying the connectivity of multi-sink WSNs over unbounded and bounded regions has been proposed. The practical outcome of this approach is the possibility: (i) to set the proper power level of nodes and their densities, given a requirement in terms of connectivity; (ii) to select the optimum height and average number of children per parent in the tree; (iii) to evaluate the trade-off between connectivity and energy consumption. As an example, results of Fig 4.20 could be useful to fix the sinks density, once the transmit power (i.e., $L_{\text{th}}$) is set: the application requires a minimum average non isolated probability, $\bar{q}$, that must be satisfied and, once $L_{\text{th}}$ is fixed (being defined the technology used), we can obtain the average number of sinks that must be distributed in the network. Similarly, once the sink density is fixed we can obtain the power that must be used for transmissions. Similarly, from Fig. 4.21, as an example, if the application requires $\bar{q}_T \geq 0.6$ and $L_{\text{th}}$is set to 90 dB, $T = 4$ or 3 and $\eta \simeq 4$ should be set. But being the case $T = 3$ less energy expensive, it will be the best choice, satisfying the requirement.

The main limitation of the mathematical models developed in this chapter is that no MAC issues are accounted for. In fact, in Sect. 4.5 a capacity-constrained air interface is assumed, such that resources can be allocated to nodes and no contentions are present. However contention-based protocols are more suitable for WSNs. In this case, a hard capacity constraint, as that introduced here, does not exist, even if a sort of soft constraint could be defined. This constraint is due to the fact that, as will be clear in the following, increasing the number of nodes competing for the channel significantly decreases the success probability, such that not too many nodes should be allowed to try to access the channel simultaneously. To account for a contention-based protocol, in the following chapter the MAC protocol defined by the IEEE 802.15.4 is modeled and in Chap. 6 the model described in Sect. 4.6 of this chapter is integrated with the one presented in Chap. 5, to study WSNs under a new perspective.

## 4.8 Further Readings

Many papers in the literature, based on random graph theory, continuum percolation, and geometric probability [25–29], devoted their attention to connectivity issues of networks. In particular, wireless ad hoc and sensor networks have recently attracted a growing attention [9–11, 15, 30, 31]. A great insight on connectivity of ad hoc wireless networks is provided in [9–11]. Connectivity-related issues of WSNs are addressed in [15, 30]. In [15], while considering channel randomness, the authors restrict the analysis to a single-sink scenario. Single-sink scenarios have attracted more attention so far. Although such scenarios have been more examined, multi-sink scenarios have been increasingly considered. Furthermore, the models based on bounded domains turn out to be of more practical use. As an example, [30] addresses the problem of deploying multiple sinks in a multi-hop limited WSN. However, the work presents a deterministic approach to distribute the sinks on a given region, rather than considering a more general uniform random deployment.

## References

1. R. Verdone, D. Dardari, G. Mazzini, A. Conti, *Wireless Sensor and Actuator Networks: Technologies, Analysis and Design* (Elsevier, London, 2008)
2. P. Santi, *Topology Control in Wireless Ad Hoc and Sensor Networks* (Wiley, Chichester, 2005)
3. C.F. Chiasserini, M.A. Marsan, A distributed self-healing approach to Bluetooth scatternet formation. *IEEE Trans. Wirel. Commun.* **4**(6), 2649–2654 (2005)
4. Zigbee Alliance, http://www.zigbee.org
5. A. Koubaa, M. Alves, E. Tovar, Modeling and worst-case dimensioning of cluster-tree wireless sensor networks, in *Proceedings of IEEE International Real-Time Systems Symposium* (RTSS), Rio de Janeiro, Brazil (2006), pp. 412–421
6. R. Verdone, C. Buratti, J. Orriss, On the design of tree-based topologies for wireless sensor networks, in *Proceedings of IFIP Annual Mediterranean Ad Hoc Networking Workshop* (MedHocNet), Lipari Island, Italy (2006)
7. J. Orriss, S. K. Barton, Probability distributions for the number of radio transceivers which can communicate with one another. *IEEE Trans. Commun.* **51**(4), 676–681 (2003)
8. M. Haenggi, On distances in uniformly random networks. *IEEE Trans. Inf. Theory* **51**(10), 3584–3586 (2005)
9. P. Santi, D.M. Blough, The critical transmitting range for connectivity in sparse wireless ad hoc networks. *IEEE Trans. Mob. Comput.* **2**(1), 25–39 (2003)
10. C. Bettstetter, J. Zangl, How to achieve a connected ad hoc network with homogeneous range assignment: an analytical study with consideration of border effects, in *Proceedings of International Workshop on Mobile and Wireless Communications Network*, Stockholm, Sweden (2002), pp. 125–129
11. C. Bettstetter, On the minimum node degree and connectivity of a wireless multihop network, in *Proceedings of ACM Symposium on Mobile Ad Hoc Networks and Comp.* (Mobihoc), Lausanne, Switzerland (2002), pp. 80–91
12. A. Fanimokun, J. Frolik, Effects of natural propagation environments on wireless sensor network coverage area, in *Proceedings of Southeastern Symposium on System Theory* (SSST), Morgantown, WV, USA (2003), pp. 16–18
13. J. Orriss, A. Phillips, S. Barton, A statistical model for the spatial distribution of mobiles and base stations, in *Proceedings of IEEE Vehicular Technical Conference* (VTC), vol. 1, Los Angeles, CA, USA (1999), pp. 127–130
14. D. Miorandi, E. Altman, Coverage and connectivity of ad hoc networks in presence of channel randomness, in *Proceedings of IEEE Conference on Computer Communication* (INFOCOM), vol. 1, Miami, FL, USA (2005), pp. 491–502
15. E. Salbaroli, A. Zanella, A connectivity model for the analysis of a wireless ad-hoc network of finite area, in *Proceedings of IEEE Conference on Sensor and Ad Hoc Communications and Networks* (SECON), vol. 3, New York, NY, USA (2006), pp. 756–760
16. IEEE 802.15.4 Std, Wireless Medium Access Control (MAC) and Physical Layer (PHY) Specifications for Low-Rate Wireless Personal Area Networks (LR-WPANs). IEEE Computer Society Press, pp. 1–679, October 2003, ISBN: 0-7381-3677-5
17. Freescale, *Freescale Semiconductor's MC13192 Developer's Kit*

18. S. Vural, E. Ekici, Probability distribution of multi-hop-distance in one-dimensional sensor networks. *ACM Comput. Netw. Int. J. Comput. Telecommun. Netw.* **51**(13), 3727–3749 (2007)
19. Bluetooth$^{TM}$, *Specificaton of the Bluetooth System*, vol. 0–3. (IEEE, 2004), http://standards.ieee.org/getieee802/802.15.html
20. J. Orriss, S.K. Barton, R. Verdone, A hierarchical model for a sensor network, in *Proceedings of International Workshop on Wireless, Ad hoc and Sensor Networks* (IWWAN), London, UK (2005)
21. C. Buratti, J. Orriss, R. Verdone, On the design of tree-based topologies for multi-sink wireless sensor networks, in *Proceedings of IEEE NEWCOM/ACORN Workshop*, Vienna, Austria (2006)
22. A. Marcucci, M. Nati, C. Petrioli, A. Vitaletti, Directed diffusion light: low overhead data dissemination in wireless sensor networks, in *Proceedings of IEEE Vehicular Technical Conference* (VTC), vol. 4, Stockholm, Sweden (2005), pp. 2538–2545
23. F. Fabbri, R. Verdone, A statistical model for the connectivity of nodes in a multi-sink wireless sensor network over a bounded region, in *Proceedings of IEEE European Wireless* (EW), Prague, Czech Republic (2008), pp. 1–6
24. D. Stoyan, W.S. Kendall, J. Mecke, *Stochastic Geometry and Its Applications* (Wiley, Chichester, 1995)
25. B. Bollobas, *Random Graphs* (Cambridge University Press, Cambridge, 2001)
26. R. Meester, R. Roy, *Continuum Percolation* (Cambridge University Press, Cambridge, 1996)
27. M.D. Penrose, A. Pistztora, Large deviations for discrete and continious percolation. *Adv. Appl. Probab.* **28**(1), 29–52 (1996)
28. M.D. Penrose, On the spread-out limit for bond and continuum percolation. *Ann Appl Probab* **3**(1), 253–276 (1993)
29. M.D. Penrose, On k-connectivity for a geometric random graph. *Rand Struct Algorithms* **15**(2), 145–164 (1999)
30. Z. Vincze, R. Vid, A. Vidacs, Deploying multiple sinks in multi-hop wireless sensor networks, in *Proceedings of IEEE International Conference on Pervasive Services*, Istanbul, Turkey, 55–63 (2007)
31. H. Pishro-Nik, K. Chan, F. Fekri, On connectivity properties of large-scale sensor networks, in *Proceedings of IEEE Conference on Sensor and Ad Hoc Communications and Networks* (SECON). Santa Clara, CA, USA (2004), pp. 498–507

# Chapter 5
# Performance Analysis of the IEEE 802.15.4 MAC Protocol

As stated in Chap. 1, the IEEE 802.15.4 Medium Access Control (MAC) protocol allows two types of channel access mechanisms: beacon- and non beacon-enabled. The latter case uses unslotted Corrier-Sense Multiple Access with Collision Avoidance (CSMA/CA) channel access mechanism, whereas beacon-enabled networks use both contention-based (a slotted CSMA/CA) and contention-less mechanisms to access the channel. In this chapter, an analytical model for both the modalities is provided. We consider a Personal Area Network (PAN) composed of a number of sensor devices (hereafter denoted as *nodes*), which transmit data to a PAN coordinator (hereafter denoted as *sink*) through direct links or multiple hops. As in Chap. 4, we consider a query-based application: upon reception of a query transmitted by the sink, each node takes one sample of a given phenomenon (e.g., atmospheric pressure or temperature) and tries to transmit its packet using the IEEE 802.15.4 MAC protocol. Once transmission is performed, the node moves to an idle state, till the next query is received. The interval of time between two successive queries is denoted as *round*, and its duration, i.e., the query interval, is denoted as $T_{\mathrm{q}}$. The nodes which do not succeed in accessing the channel by the end of the current round discard the packet; at the next round, a new packet is generated.

Concerning network topologies, both stars and trees are accounted for. As stated in Chap. 1, star topologies are preferable when the PAN area is small: the number of nodes that could be associated to the sink, in fact, should range from three to seven, as it is widely accepted that, in this case, IEEE 802.15.4 does not support larger network sizes [1]. Nevertheless, for the sake of completeness and validation of the model, results for networks with a number of nodes larger than seven are also shown. Trees, instead, are used in large networks. Since trees can be formed only when nodes operate in the beacon-enabled mode [2], this topology is implemented only in this modality.

Given the above scenarios, the aim of the proposed model is to provide an analytical description of the transitions between node states (backoff, sensing,

C. Buratti et al., *Sensor Networks with IEEE 802.15.4 Systems*,
Signals and Communication Technology, DOI: 10.1007/978-3-642-17490-2_5,

transmit, idle) of the CSMA/CA algorithm. The mathematical model developed allows the evaluation of the statistical distribution of the traffic generated by nodes. In particular, the statistical distribution of the delays with which the nodes access to the channel and with which their packets are received by the sink, are provided. The knowledge of the statistics of the traffic generated by the PAN is useful, for example, in those applications where the sink acts as gateway toward an infrastructure-based wireless network, e.g., Universal Mobile Telecommunications System (UMTS). Such knowledge, in fact, is useful to schedule radio resources for the gateway [3].

The model is then finalized to derive (i) the probability that a node succeeds when accessing the channel and in transmitting its packet, (ii) the overall throughput generated by the network, and (iii) the average energy consumption.

In the literature, there are several works devoted to the study of IEEE 802.15.4 networks, some of which are dealing with the mathematical modelling of the MAC protocol (see also Sect. 5.7). Some of these works present models which are not very accurate or not in perfect agreement with simulation results [4–7], others are more accurate [8], but they are all very far from the model presented in this book, owing to the different application scenario at hand. All the previous works, in fact, are based on the Bianchi's model [9] and use a Markov chain to describe node states even if the process representing the backoff time counter is not Markovian, since the value of the backoff counter depends on the past history (i.e., how many times the node has tried to access the channel and found it busy). To use a Markov chain, in fact, Bianchi assumes that at each transmission attempt and regardless of the number of retransmissions suffered (backoff stages in the IEEE 802.15.4 case), each packet collides with constant and independent probability (see [9]). This approximation becomes more accurate as long as the contention window gets longer and the number of nodes accessing the channel gets larger. In particular, this is reasonable for the above mentioned works, being the number of nodes competing for the channel constant in time, but it is not accurate for query-based applications. In the latter case, in fact, the number of nodes accessing the channel decreases by passing time (since nodes have only one packet to be transmitted per superframe upon reception of the query), resulting in a decreasing of the probability that a packet collides. Therefore, the above approximation is not applied and the probabilities (of being in sensing, transmission, and of colliding) are evaluated for the different backoff stages and the different instants of time. As shown by the numerical results, in fact, these probabilities depend on time and on the backoff stage in which the node is.

Sections 5.1 and 5.2 deal with the description of the beacon- and non beacon-enabled MAC protocols and the scenario considered, respectively. In Sects. 5.3 and 5.4, the two models related to the two modalities and the related results are given. In Sect. 5.5, a comparison between the two modalities is provided. Finally, concluding remarks and further readings are given in Sects. 5.6 and 5.7 respectively.

# 5.1 The Non Beacon- and Beacon-Enabled MAC Protocols

The details of the non beacon- and beacon-enabled MAC protocols are reported here even if they have already been introduced in Chap. 1, to facilitate the reading of this chapter.

As stated in Chap. 1, in the non beacon-enabled mode nodes use an unslotted CSMA/CA MAC protocol, implemented using units of time called backoff periods, having a duration denoted as $d_{bo}$, equal to 20 $T_s = 320\,\mu s$, where $T_s$ is the symbol time. Here we consider the 2.45 GHz band, meaning a symbol rate of 62.5 ksymbol/s, which corresponds to $T_s = 16\,\mu s$ [10].

Each node maintains two variables for each transmission attempt: *NB* and *BE*. *NB* is the number of times the CSMA/CA algorithm is required to backoff while attempting the current transmission; this value will be initialized to 0 before each new transmission attempt and cannot assume values larger than $NB_{max}$. *BE* is the backoff exponent related to the maximum number of backoff periods a node waits before attempting to assess the channel. *BE* is initialized to the value of $BE_{min}$, and cannot assume values larger than $BE_{max}$. Figure 5.1 illustrates the steps of the CSMA/CA algorithm, starting from the moment at which the node has data to be transmitted. First, *NB* and *BE* are initialized and then the MAC layer will delay any activity for a random number of backoff periods in the range $\{0, \ldots, 2^{BE} - 1\}$

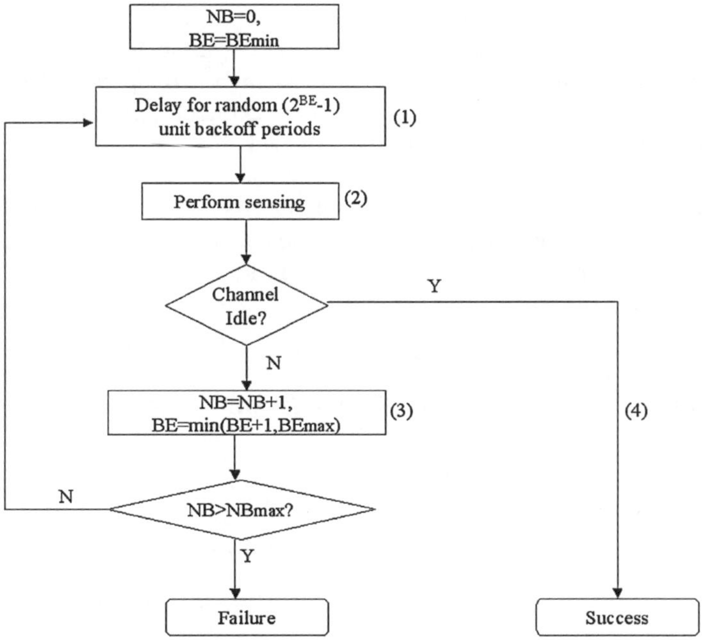


**Fig. 5.1** The IEEE 802.15.4 CSMA/CA algorithm in the non beacon-enabled case

[step (1)]. After this delay, channel sensing is performed for one unit of time [step (2)]. If the channel is assessed to be busy [step (3)], the MAC sublayer will increase both *NB* and *BE* by one, ensuring that *BE* is not larger than $BE_{\max}$. If the value of *NB* is less than or equal to $NB_{\max}$, the CSMA/CA algorithm will return to step (1). If the value of *NB* is larger than $NB_{\max}$, the CSMA/CA algorithm will terminate with a "Failure," meaning that the node does not succeed in accessing the channel. If the channel is assessed to be idle [step (4)], the MAC layer will begin transmission of data immediately ("Success" in accessing the channel).

In the beacon-enabled mode [10], a superframe, starting with the beacon packet (corresponding to the query in the scenario considered here), transmitted by the sink, is established.

The duration of the whole superframe (including active and inactive parts) is *BI*, given by Eq. 1.2. and is equal, in our scenario, to the round duration $T_{\mathrm{q}}$. The duration of the active part of the superframe, containing CAP and CFP, namely the superframe duration, is $SD = 960 \cdot 2^{SO} \cdot T_{\mathrm{s}}$. Note that in the beacon-enabled case, $T_{\mathrm{q}}$ may assume only a finite set of values, whereas in the non beacon-enabled case, $T_{\mathrm{q}}$ may assume any value.

The inactive part (present when $BO > SO$) is used for saving energy (nodes can switch off during this phase) or for exploiting multi-hop. Since in our scenario we assume that nodes enter the idle state after the transmission of the data, we set $SO = BO$ (i.e., $BI = SD = T_{\mathrm{q}}$) in the star topology case (one hop) and $BO > SO$ for trees. A proper setting of the parameters *BO* and *SO* in the latter case is needed: Sect. 5.4.6 tries to provide some guidelines for this setting.

For what concerns the CSMA/CA algorithm used in the CAP portion of the superframe, the only difference with the non beacon-enabled mode is that nodes have to find the channel free for two subsequent backoff periods before transmitting the packet (see Fig. 1.9). To this end, each node maintains another variable, denoted as *CW*, indicating the number of backoff periods that need to be clear of channel activity before the transmission can start. First, *CW* is initialized to 2. Then, once the node senses the channel [step (2)] (see Fig. 1.9), if the channel is found free, *CW* is decremented by 1 and compared with 0: if $CW > 0$, the algorithm returns to step (2) and another sensing phase is implemented; otherwise, a transmission may start.

## 5.2 Reference Scenario and Model Assumptions

We consider a PAN composed of $n$ nodes transmitting packets, having size $z$, equal to $D \cdot 10$ bytes, being $D$ an integer in the range $\{2, \ldots, 13\}$, according to the minimum and maximum possible data packet sizes [10] (however, for the sake of completeness also some results for the case $D = 1$ are shown). The time needed to transmit a packet will be equal to $D \cdot d_{\mathrm{bo}}$, as a bit rate of 250 kbit/s is used; therefore, each packet occupies $D$ backoff periods. We denote the query/beacon size as $z_{\mathrm{B}}$.

Ideal channel conditions are assumed: all nodes can "hear" each other, and, therefore, no hidden terminal problem is accounted for. Similar scenarios and assumptions are considered in many studies in the literature [4–6, 8, 9, 11–15]. Collisions between nodes may occur in case two or more nodes perform channel sensing at the same time, find the channel free and transmit simultaneously their packets. For the sake of energy efficiency, no acknowledgement (ACK) and retransmission mechanism is implemented; therefore, when a packet collides it is definitely lost in that round.

In the model, the resolution time (hereafter denoted as *slot*) is set equal to the backoff period, $d_{bo}$, which corresponds also to the duration of the single sensing phase and to the packet transmission time when $D = 1$.

In the non beacon-enabled mode, it is assumed that all nodes start the backoff algorithm at the same time, when the query transmitted by the sink is received (no propagation delay is present due to short distances), and we fix the origin of the time axis ($t = 0$) at the instant in which all nodes receive the query. Then, the behavior of the network from $t = 0$ to the instant in which all possible transmissions have taken place is modelled (Fig. 5.2).

In the beacon-enabled case, instead, the origin of the time axis is fixed at the beginning of the superframe ($t = 0$), so that nodes will receive the beacon and will

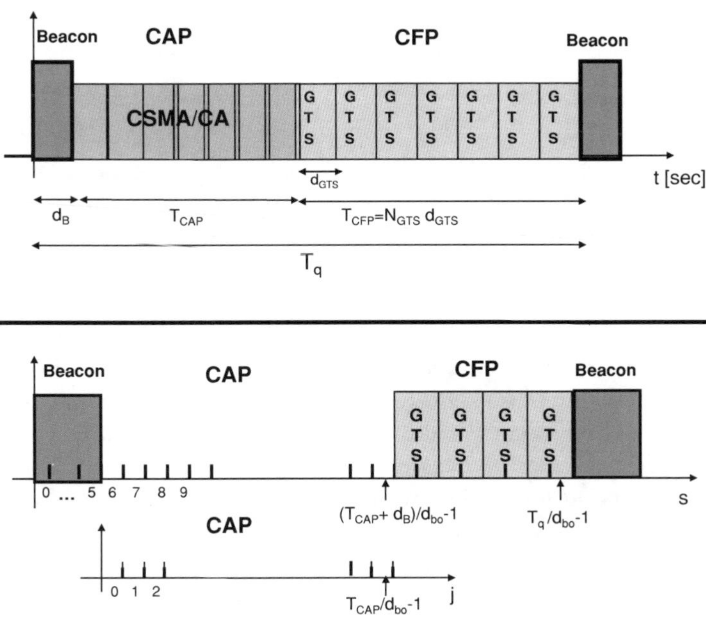


**Fig. 5.2** The IEEE 802.15.4 superframe, considering the time axis, $t$, (*above part*), and the number of slots, $s$, (*below part*)

start the CSMA/CA algorithm in the instant $t = d_{\mathrm{B}}$ (see Fig. 5.2, upper part). Since alignment between the first backoff period of each node and the beginning of the beacon transmission is required, $d_{\mathrm{B}}$ will be equal to the beacon transmission time only in case it is multiple of $d_{\mathrm{bo}}$: otherwise, it will be larger. In this way, the alignment between the first backoff period of each node and the beginning of the beacon transmission is ensured.

Finally, we assume that a packet is lost if it is not correctly received by the end of the round (in the non beacon-enabled case) or the superframe (in the beacon-enabled mode).

## 5.3 The Non Beacon-Enabled Model

In this section, the non beacon-enabled model is presented, starting from the modelling of node states, passing through the description of the finite-state transition diagram developed to model all the possible states in which a node could be and the transitions between the states, ending with the derivation of performance metrics. Only star topologies are considered, since the tree-based Zigbee topology must be implemented in beacon-enabled mode.

### 5.3.1 Node States

Generally speaking, a node accessing the channel during a round can be in one of four states: backoff, sensing, transmission, or idle. However, if after sensing the channel is free, transmission immediately starts, followed by a sequence of idle states till the end of the round. Thus, given the objectives of this model, only the backoff and sensing states must be modeled.

The node state is modelled as a bidimensional process $Q(\hat{t}) = \{BO_{\mathrm{c}}(\hat{t}), BO_{\mathrm{s}}(\hat{t})\}$, where $\hat{t}$ is an integer, representing the time slot and, more precisely, the $j$th slot (from $j \cdot d_{\mathrm{bo}}$ to $(j+1) \cdot d_{\mathrm{bo}}$) is denoted by $\hat{t} = j.BO_{\mathrm{c}}(\hat{t})$ and $BO_{\mathrm{s}}(\hat{t})$ represent the backoff time counter and the backoff stage at time $\hat{t}$, respectively. Both are time-discrete stochastic processes assuming discrete values. Therefore, the process is a chain, but not a Markovian chain [16] because $BO_{\mathrm{c}}(\hat{t})$ is not a memoryless process as its value depends on its history (its value depends on how many times the node has tried to access the channel without success).

The initial value of backoff time counter, $BO_{\mathrm{c}}(0)$, is uniformly distributed in the range $\{0,\ldots, W_{NB}-1\}$, where $W_{NB} = 2^{BE}$ is the dimension of the contention window and $NB \in \{0,\ldots, NB_{\max}\}$. The value of $BE$ depends on the second process characterizing the state: $BO_{\mathrm{s}}(\hat{t})$. We can identify $NB_{\max}+1$ different backoff stages obtained by considering the different possible combinations of the pair

**Table 5.1** The backoff stages.

| $BO_s$ | $NB$ | $BE$ | $W_{NB} = 2^{BE}$ |
|---|---|---|---|
| 0 | 0 | $BE_{\min}$ | $W_0 = 2^{BE_{\min}}$ |
| 1 | 1 | $BE_{\min} + 1$ | $W_1 = 2^{BE_{\min}+1}$ |
| … | … | … | … |
| $NB_{\max}$ | $NB_{\max}$ | $BE_{\max}$ | $W_{NB_{\max}} = 2^{BE_{\max}}$ |

$(NB, BE)$. In Table 5.1, the different backoff stages with the correspondent $W_{NB}$ values (denoted as $W_0, \ldots, W_{NB_{\max}}$) are shown.

The IEEE 802.15.4 MAC protocol states that at the beginning of the backoff algorithm, each node sets $NB = 0$ and $BE = BE_{\min}$. Then, each time the channel is sensed busy, $NB$ and $BE$ are increased by 1. When $BE$ reaches its maximum value, there is no more increase. The case $BO_s = NB_{\max}$ is the last case, because here $NB$ reaches its maximum value, and it cannot be further increased.

Because there exists a maximum value for $NB$, there will also be a maximum delay affecting the transmission of a packet. This maximum is reached in case a node extracts at every backoff stage the higher backoff time counter and at the end of each backoff stage it always finds the channel busy. In this case, the node is in backoff state for $\sum_{k=0}^{NB_{\max}} (W_k - 1)$ slots and in sensing for $NB_{\max} + 1$ slots. Therefore, the last slot in which a transmission can start is

$$\hat{t}_{\max} = \sum_{k=0}^{NB_{\max}} W_k, \tag{5.1}$$

and the last slot in which a transmission can finish is $(\hat{t}_{\max} + D - 1)$. Sensing, instead, is possible only for $\hat{t} \in \{0, \ldots, \hat{t}_{\max} - 1\}$.

In the following, the generic state will be denoted as $Q(\hat{t}) = \{BO_c, BO_s, \hat{t}\}$ and the probability of being in a generic state will be denoted as $\mathbb{P}\{BO_c = c, BO_s = k, \hat{t} = j\} = P\{c, k, j\}$.

## 5.3.2 Formulation of the Mathematical Model

### 5.3.2.1 Steps Followed by the Model

Let us denote as $b^j$ the probability that in the $j$th slot the channel is found to be busy after sensing. This probability will be initially left as parameter, and its computation will be provided at the end of Sect. 5.3.3.4.

The model provides the following metrics:

1. the probability that a node ends the transmission of its packet in a given slot, $j$, denoted as $P\{T^j\}$, with $j \in \{0, \ldots, \hat{t}_{\max} + D - 1\}$;
2. the probability that the sink receives the end of a packet, coming from whatever a node, in a given slot $j$, denoted as $P\{R^j\}$, with $j \in \{0, \ldots, \hat{t}_{\max} + D - 1\}$;

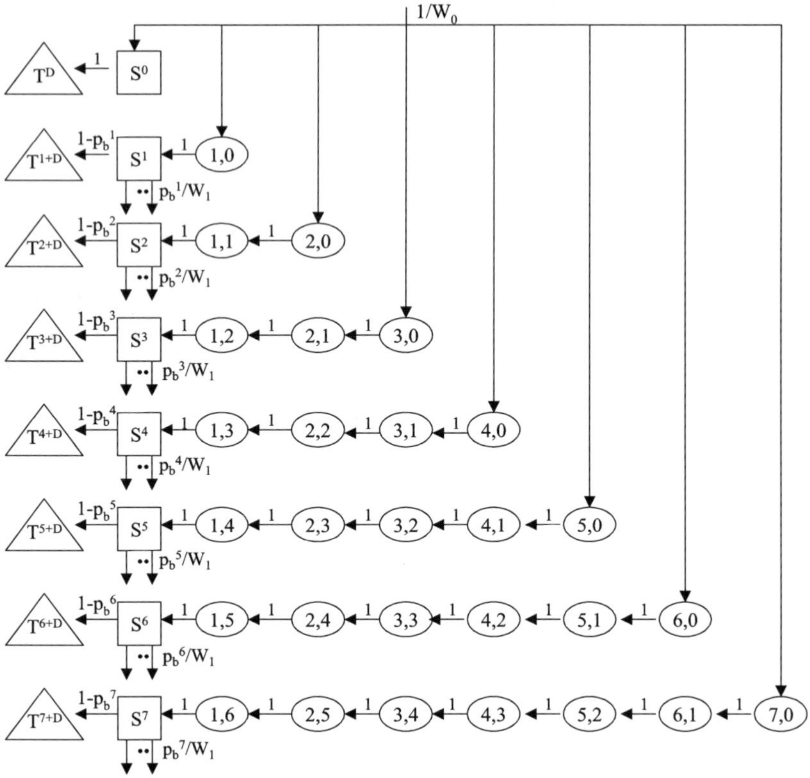


**Fig. 5.3** The state-transition diagram related to the first backoff stage

3. the success probability for a transmission, i.e., the probability that a node succeeds in transmitting its packet in a round whatever the slot, denoted as $p_s$;
4. the mean energy spent by a node in a round, denoted as $E_{\text{mean}}$.

The probability $P\{T^j\}$ depends on the probability of being in sensing state in the slot $j - D$. Because a packet occupies $D$ slots, a node sensing the channel in slot $j - D$ and finding it free, will end its transmission in slot $j$. To determine the sensing probabilities, we model the behavior of a single node, using a state-transition diagram [16], describing the relation between all possible states in which a node can be (Figs. 5.3, 5.4, 5.5, 5.6, and 5.7). From this diagram, we obtain the probability of being in sensing state at the $j$th slot and at the $k$th backoff stage ($BO_s = k$), denoted as $P\{S_k^j\} = P\{0, k, j\}$, whatever $j$ and $k$. This is made in the remainder of this subsection. From these probabilities, we can derive the probability of being in sensing state at the $j$th slot, denoted as $P\{C^j\}$ with $j \in \{0, \ldots, \hat{t}_{\max} - 1\}$, and therefore $P\{T^j\}.P\{T^j\}$, $P\{R^j\}$, $p_s$, $E_{\text{mean}}$, and $b^j$ are derived in Sect. 5.3.3. In this section, the algorithm used to compute all the target performance metrics is also provided.

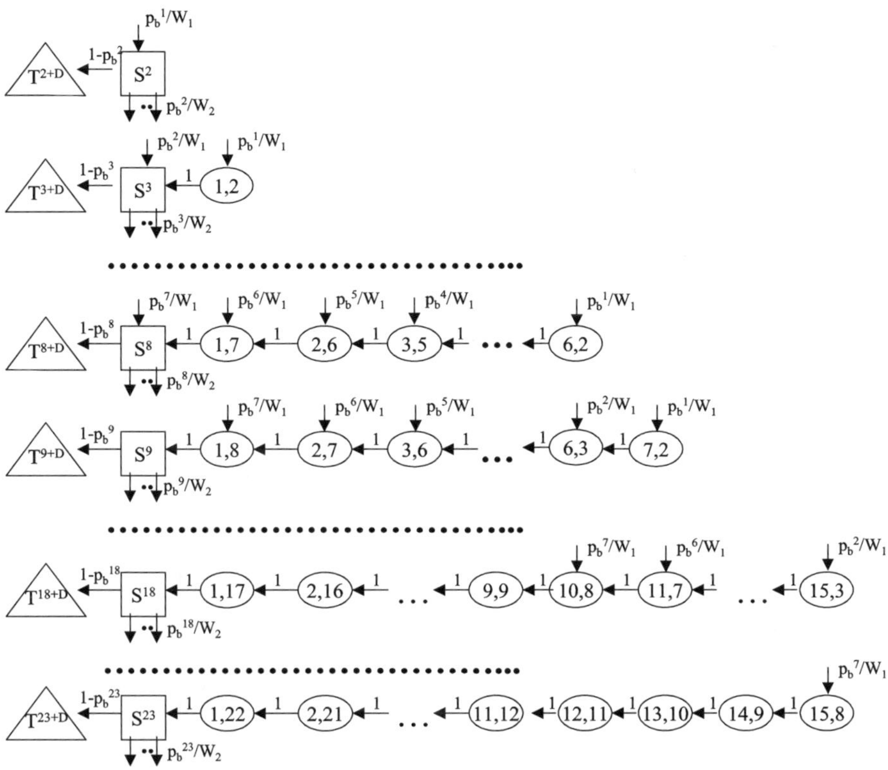

**Fig. 5.4** The state-transition diagram related to the second backoff stage

### 5.3.2.2 Sensing Probabilities

The state-transition diagram of the bidimensional process $Q(\hat{t})$ is presented through different figures: one for each backoff stage. For the sake of clarity, we show here the diagram obtained when the MAC parameters are set to the defaults values: $BE_{\min} = 3$, $BE_{\max} = 5$, $NB_{\max} = 4$. In this case, five backoff stages are present and the related diagrams are shown in Figs. 5.3, 5.4, 5.5, 5.6 and 5.7. In particular, Fig. 5.3 addresses the case with $BO_s = 0$, Fig. 5.4 with $BO_s = 1$, Fig. 5.5 with $BO_s = 2$, Fig. 5.6 with $BO_s = 3$, and Fig. 5.7 with $BO_s = 4$. As will be clarified in the following, they are linked together through transitions that originate from some states of a figure and terminate in the states of the subsequent figure. Since each figure is related to a specific value of $BO_s$, for the sake of simplicity in the drawings, the generic backoff state (ovals in the figures) is simply denoted as $\{c,j\}$, omitting the value of $BO_s$; the sensing states (squares) are denoted as $S^j$ with no pedex $k$. Finally, the transmission states (triangles in the figures) are denoted as $T^j$, with no pedex $k$.

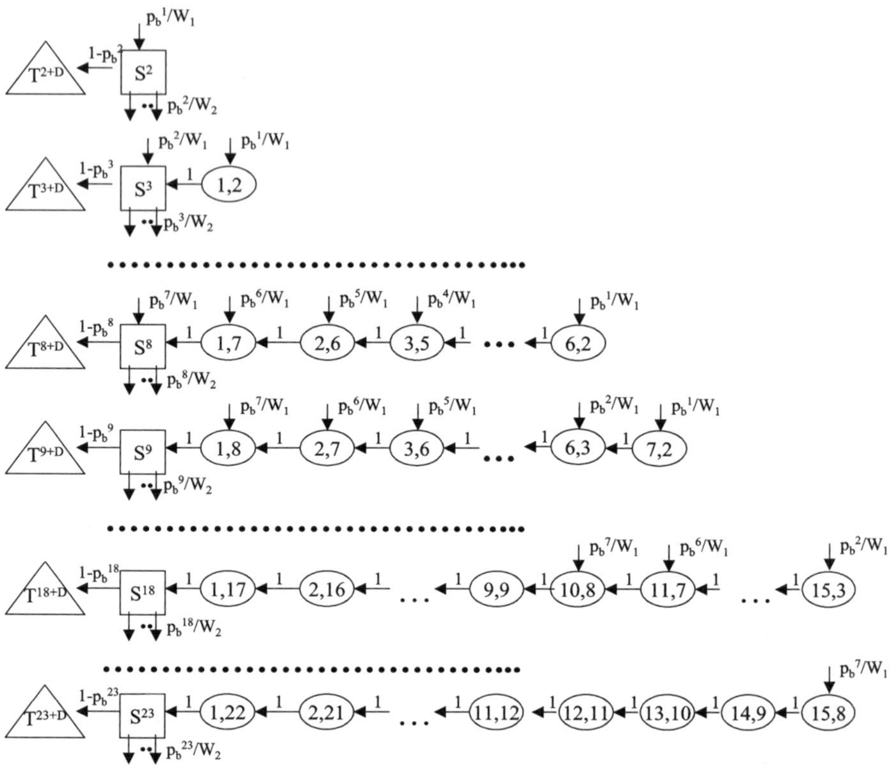

**Fig. 5.5** The state-transition diagram related to the third backoff stage

In the following, the different parts of the state-transition diagram are described and the probabilities of being in the different states of the chain and the transition probabilities between the states are provided.

## First Backoff Stage ($BO_s = 0$)

At the beginning of the backoff algorithm, each node extracts an integer, uniformly distributed between 0 and $W_0 - 1$. At $t = 0$ a node enters, with probability $1/W_0$, one of the states $\{c, 0, 0\}$ with $c \in \{0, \ldots, W_0 - 1\}$. If the extracted value is 0, the node in slot 0 will sense the channel and in slot 1 it will transmit its packet, because no transmission may occur in the first slot ($P\{T^0 = 0\}$) and, therefore, the channel will be certainly found free ($b^0 = 0$). If a value larger than 0 is extracted, the node will decrease its backoff counter at each slot until the counter will reach the zero value, when the node will sense the channel. After the sensing phase, if the channel is found free the node will transmit the packet; otherwise, it will pass to the following backoff stage and another value, uniformly distributed between 0 and $W_1 - 1$, will be extracted. In Fig. 5.3 , the transitions that originated from the

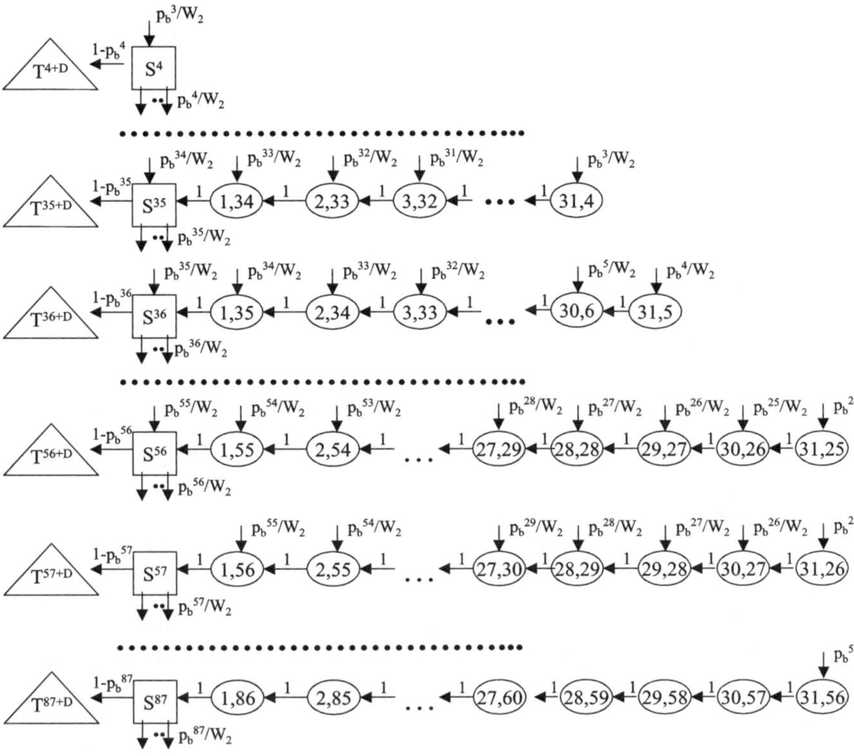


**Fig. 5.6** The state-transition diagram related to the fourth backoff stage

sensing states enter in the states of Fig. 5.4 . For example, if a node is in the state $S_0^1$ and it finds the channel busy, it will enter the state $S_1^2$, or one of the states $\{c, 1, 2\}$, with $c \in \{0, \ldots, W_1 - 1\}$, with the same probability $b^1/W_1$. The arrival state depends on the new backoff counter value extracted. Denoting as $P\{BO_c = c_1, BO_s = k_1, t = j_1 | BO_c = c_0, BO_s = k_0, \hat{t} = j_0\} = P\{c_1, k_1, j_1 | c_0, k_0, j_0\}$ the transition probability from the state $\{c_0, k_0, j_0\}$ to the state $\{c_1, k_1, j_1\}$, the transition probabilities between the backoff states are given by:

$$P\{c, 0, j+1 | c+1, 0, j\} = 1 , \tag{5.2}$$

for $c \in \{0, \ldots, W_0 - 2\}$ and $j \in \{0, \ldots, W_0 - 2\}$.

This equation accounts for the fact that, at the beginning of each time slot, the backoff time counter is decreased by 1 until it reaches the zero value, with probability 1. The probabilities of being in a sensing state are given by:

$$P\{S_0^j\} = \begin{cases} \frac{1}{W_0} & \text{for } j \in \{0, \ldots, W_0 - 1\} \\ 0 & \text{for } j > W_0 - 1. \end{cases} \tag{5.3}$$

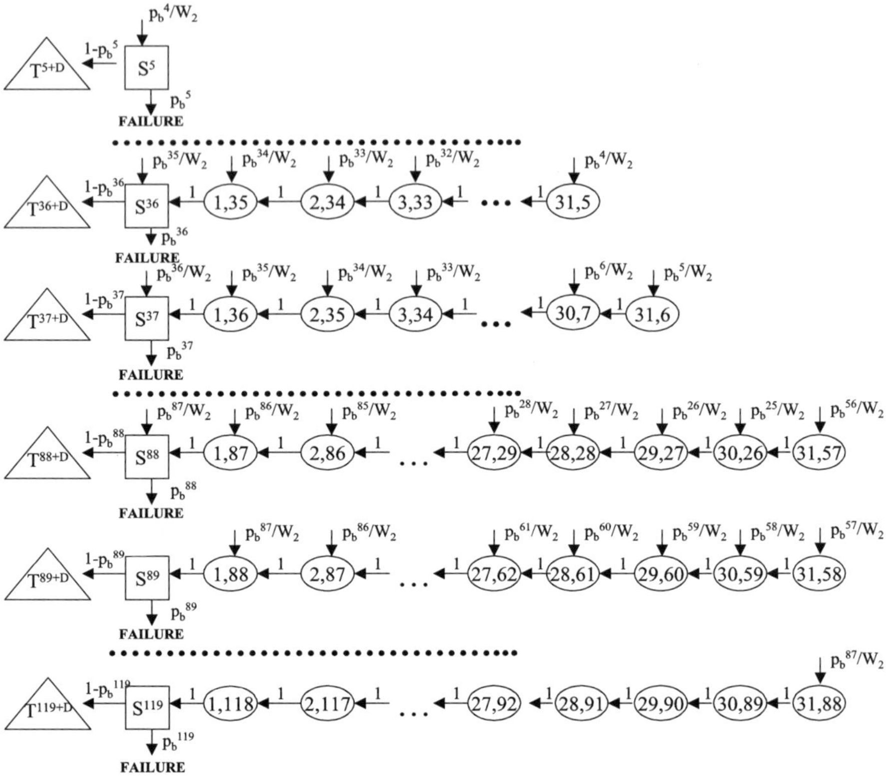

**Fig. 5.7** The state-transition diagram related to the fifth backoff stage

## Other Backoff Stages ($\mathbf{BO_s = 1, \ldots, NB_{max}}$)

We consider here the backoff stages $BO_s = 1, \ldots, NB_{\max}$ and we refer to the parts of the state transition diagram illustrated in Figs. 5.4, 5.5, 5.6, and 5.7, related to the case of $BE_{\min} = 3$, $BE_{\max} = 5$, and $NB_{\max} = 4$. Since a node can arrive in the backoff stage $BO_s = k$ only after it has finished the previous backoff stage, it cannot reach this stage before $j = k + 1$; therefore, $S_k^j$ for $j < k + 1$ do not appear in the diagrams.

As in the previous case, the transition probabilities between backoff states in the $k$th backoff stage are given by:

$$P\{c, k, j + 1 | c + 1, k, j\} = 1 \ , \tag{5.4}$$

for $c \in \{0, \ldots, W_k - 2\}$ and $j \in \{k + 1, \ldots, W_{0,\ldots,k} - 2\}$, where $W_{0,\ldots,k} = W_0 + W_1 + \cdots + W_k$. In the following, we will denote as $W_{x,y,z}$, the sum $W_x + W_y + W_z$.

The transition probabilities between the states of the $(k-1)$th backoff stage and those of the $k$th backoff stage are given by:

$$P\{c,k,j+1|0,k-1,j\} = \frac{b^j}{W_k}\ , \tag{5.5}$$

for $c \in \{0,\ldots,W_k-1\}$ and $j \in \{k,\ldots,W_{0,\ldots,k-1}-1\}$. This equation accounts for the fact that if the channel is found busy at the $j$th slot, the node will go to one of the states $\{c,k,j+1\}$, with $c \in \{0,\ldots,W_k-1\}$, with the same probability $1/W_k$.

In case $W_{0,\ldots,k-1}+1 \leq W_k+k$, the probabilities of being in the sensing state are given by:

$$P\{S_k^j\} = \begin{cases} \sum_{v=k}^{j-1} P\{S_{k-1}^v\} \cdot \frac{b^v}{W_k} & \text{for } j \in \{k+1,\ldots,W_{0,\ldots,k-1}\} \\ P\{S_k^{W_{0,\ldots,k-1}}\} & \text{for } j \in \{W_{0,\ldots,k-1}+1,\ldots,W_k+k\} \\ P\{S_k^{W_{0,\ldots,k-1}}\} - \sum_{v=k}^{j-W_k-1} P\{S_{k-1}^v\} \cdot \frac{b^v}{W_k} & \text{for } j \in \{W_k+k+1,\ldots,W_{0,\ldots,k}-1\} \\ 0 & \text{otherwise}. \end{cases} \tag{5.6}$$

Let us consider the case $BO_s = 1$ shown in Fig. 5.4. The second equation derives from the fact that until $j \leq W_0$, the probability of being in sensing in the second backoff stage depends on the probabilities of being in sensing in the first backoff stage and to find the channel busy. As an example, a node can arrive in $S_1^3$ if it is in $S_0^1$, finds the channel busy, and extracts the value 1 for the second backoff stage; or it is in $S_0^2$, finds the channel busy, and extracts the value 0 for the second backoff stage (see Figs. 5.3 and 5.4). The third equation accounts for the fact that for $j > W_0$, there are no more transitions between the states of $BO_s = 0$ and the ones of $BO_s = 1$, because the last slot in which a node can sense the channel in the first backoff stage is $j = W_0 - 1$. Finally, when $j$ reaches $W_1 + 2$, $P\{{S_1}^{18}\}$ is obtained by subtracting the probability $P\{S_0^1\}\frac{b^1}{W_1}$ from $P\{S_1^{W_0}\}$. Therefore, $P\{{S_1}^{18}\} = \sum_{v=2}^{W_0-1} P\{{S_0}^v\} \cdot \frac{b^v}{W_1}$. In fact, if a node is in $S_0^1$ it moves (in case of channel busy) to states $\{c,1,2\}$ with $c \in \{0,\ldots,15\}$; therefore, the state $\{16,1,2\}$ does not exist (see Fig. 5.4). The last possible sensing state we can have in this part of the chain is $S_1^{23}$, which means that the second backoff stage will be completed by a node at maximum in the 24th slot.

Finally, in the case $W_{0,\ldots,k-1}+1 > W_k+k$, the probabilities of being in sensing are given by:

$$P\{S_k^j\} = \begin{cases} \sum_{v=k}^{j-1} P\{S_{k-1}^v\} \cdot \frac{b^v}{W_k} & \text{for } j \in \{k+1,\ldots,W_k+k\} \\ \sum_{v=k}^{j-1} P\{S_{k-1}^v\} \cdot \frac{b^v}{W_k} - \sum_{v=k}^{j-W_k-1} P\{S_{k-1}^v\} \cdot \frac{b^v}{W_k} & \text{for } j \in \{W_k+k+1,\ldots,W_{0,\ldots,k-1}\} \\ P\{S_k^{W_{0,\ldots,k-1}}\} - \sum_{v=W_{0,\ldots,k-2}}^{j-W_k-1} P\{S_{k-1}^v\} \cdot \frac{b^v}{W_k} & \text{for } j \in \{W_{0,\ldots,k-1}+1,\ldots,W_{0,\ldots,k}-1\} \\ 0 & \text{otherwise}. \end{cases} \tag{5.7}$$

## 5.3.3 Performance Metrics Derived from the Model

### 5.3.3.1 Transmission Probabilities

As stated above, the aim of the model is to evaluate the probability that a generic node ends its packet transmission in slot $j$, denoted as $P\{T^j\}$, with $j \in \{0, \ldots, \hat{t}_{\max} + D - 1\}$.

A node finishes its transmission in slot $j$, if in slot $j - D$ it senses the channel finding it free. Therefore, this probability is given by:

$$P\{T^j\} = P\{C^{j-D}\} \cdot (1 - b^{j-D}) . \tag{5.8}$$

Because a node transmits a packet occupying $D$ slots, we associate $P\{T^j\}$ to the slot in which the transmission terminates; therefore, for $j<D$, $P\{T^j\} = 0$.

The cumulative distribution function $F_T(j)$, corresponding to the probability that a packet coming from whatever a node is transmitted in the channel within slot $j$, is given by:

$$F_T(j) = \sum_{v=0}^{j} P\{T^v\} . \tag{5.9}$$

The probability of being in the sensing state at the instant $j$ is given by:

$$P\{C^j\} = \sum_{k=0}^{NB_{\max}} P\{S_k^j\}. \tag{5.10}$$

### 5.3.3.2 Reception and Success Probability

To evaluate the other target probabilities, we have to model how the number of nodes that compete for accessing to the channel varies with time. We denote as $n_c^j$ the number of nodes which have not transmitted yet at the end of slot $(j - 1)$ and that will compete for slot $j$. In particular, in slot 0, the number of nodes which compete for the channel is equal to $n$ and as none can transmit in slot 0 ($P\{T^0\} = 0$) $n_c^1$ is equal to $n$ too (see Fig. 5.8). Whereas if we set $D = 1$, in slot 1 some nodes may terminate the transmission, each with probability $P\{T^1\}$, and at the end of this slot the number of nodes that still have to transmit their packets, $n_c^2$, will depend on $P\{T^1\}$ and could be lower than $n$. In Fig. 5.8, an illustrative example (considering the case $D = 1$ and five competing nodes) is shown: one of the five competing nodes transmits in slot 1 and, therefore, in the second slot we have four nodes competing for the channel ($n_c^2 = 4$). Therefore, $n_c^j$ is a binomially distributed random variable (r.v.). In the case with $D = 1$, the probability that $k$ over $\hat{n}$ nodes at slot $j$ have not transmitted the packet yet, conditioned on the fact that at the

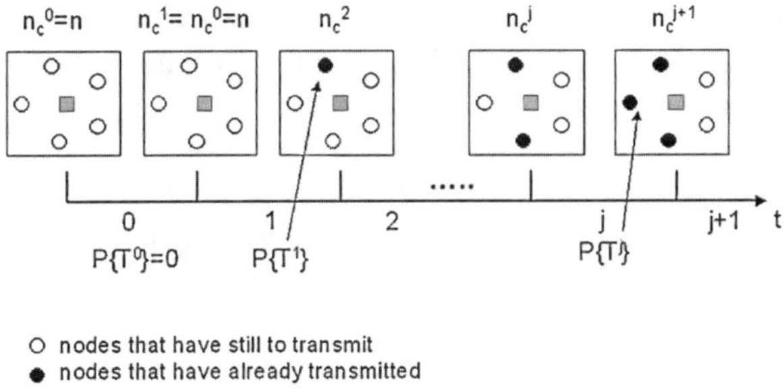


**Fig. 5.8** The behavior of the number of nodes that have still to access the channel in the different time slots (example)

$(j-1)$th slot $\hat{n}$ nodes are competing for the channel ($n_c^{j-1} = \hat{n}$), denoted as $B^j(k, \hat{n})$, is given by:

$$B^j(k,\hat{n}) = P\{n_c^j = k | n_c^{j-1} = \hat{n}\}$$
$$= \binom{\hat{n}}{k}(1 - b^{j-2})\left(P\{C^{j-2}\}\right)^{\hat{n}-k} \cdot \prod_{k=0}^{NB_{\max}}\left(1 - P\{S_k^{j-2}\}\right)^k, \quad (5.11)$$

where $(1 - b^{j-2})(P\{C^{j-2}\})^{\hat{n}-k}$ is the probability that $\hat{n} - k$ nodes transmit in slot $j$ and $\prod_{k=0}^{NB_{\max}}(1 - P\{S_k^{j-2}\})^k$ is the probability that the remaining $k$ nodes do not transmit, as they do not sense the channel in slot $j-2$. In Eq. 5.11, $n_c^{j-1}$ is, in turn, a binomially distributed r.v., having a probability distribution that depends on the probabilities $P\{T^l\}$ with $l \in \{1,\ldots,j-2\}$. Therefore, to find the statistics of $n_c^j$, Eq. 5.11 should be averaged over the statistics of $n_c^{j-1}$, which would depend on the statistics of $n_c^{j-2}$, and so on. By increasing the initial number $n_c^0$ of nodes in the network and the time slot considered (i.e., the value of $j$), the complexity of the evaluation of the statistics of $n_c^{j-1}$ increases exponentially, because of the need to follow all possible combinations of values of $n_c^2$, $n_c^3$, etc. To reduce such complexity we have introduced an approximation: we do not model $n_c^{j-1}$ as a r.v., but we set its value at the value of $k$ that corresponds to a maximum value of the probability $B^{j-1}(k, \hat{n})$. Therefore, $n_c^{j-1}$ is given by:

$$n_c^{j-1} = \underset{k}{\operatorname{argmax}}\, B^{j-1}(k,\hat{n}). \quad (5.12)$$

Moreover, we have also evaluated performance by simply setting $n_c^j = n$, regardless of the value of $j$. In Sect. 5.3.4, simulation results are compared with the mathematical analysis results, considering both models of $n_c^j$. Results show that the two models bring approximatively to the same results and that a good agreement with simulations is obtained in both cases. Therefore, the approximation introduced does not affect significantly the performance.

The modelling of $n_c^j$ in the case $D > 1$ is even more complex than the case $D = 1$, because $n_c^j$ depends on the number of nodes starting their transmission in the last $D$ slots. But as by increasing $D$ each node occupies the channel for a longer time, $n_c^j$ will decrease slowly with time; therefore, it is reasonable to set $n_c^j = n$, regardless of the value of $j$. If, in fact, this approximation is accurate for $D = 1$, it will be even more so for $D > 1$. Therefore, only the case $n_c^j = n$ whatever be $j$ has been considered in Sect. 5.3.4 for $D > 1$.

Now, we can evaluate as follow the probability $p_s$ that a generic packet is transmitted successfully on the channel:

$$p_s = \sum_{j=0}^{\hat{\imath}_{max}+D-1} P\{Z^j\}\,, \tag{5.13}$$

where $P\{Z^j\}$ is the probability that a successful transmission ends in slot $j$, which means that one and only one transmission starts in $j - D + 1$. Because we assume that all nodes can hear each other, if in slot $j - D + 1$ only one node starts its transmission, the sink will receive correctly (i.e., without collisions) the end of the packet in $j$. From the law of total probability we obtain:

$$\begin{aligned} P\{Z^j\} = {} & P\{1 \text{ tx in } (j-D+1)|\text{channel free in } (j-D)\} \cdot P\{\text{channel free in } (j-D)\} \\ & + P\{1 \text{ tx in } (j-D+1)|\text{channel busy in } (j-D)\} \cdot P\{\text{channel busy in } (j-D)\}, \end{aligned} \tag{5.14}$$

where $P\{1 \text{ tx in } (j-D+1)|\text{channel free in } (j-D)\}$ and $P\{1 \text{ tx in } (j-D+1)|$ channel busy in$(j-D)\}$ are the probabilities that one and only one transmission starts in slot $j - D + 1$ conditioned on the fact that the channel in $j - D$ is free or busy, respectively. As only one transmission starts in slot $j - D + 1$ if only one node, over $n_c^{j-D}$, senses the channel in slot $j - D$ and as no transmissions may start in slot $j - D + 1$ if the channel is busy in slot $j - D$, $P\{Z^j\}$ is given by:

$$P\{Z^j\} = (1 - b^{j-D})P\{C^{j-D}\} \cdot \prod_{k=0}^{NB_{max}} \left(1 - P\{S_k^{j-D}\}\right)^{n_c^{j-D}-1}, \tag{5.15}$$

where $P\{C^{j-D}\}$ is the probability that one node senses the channel in $j - D$ and $\prod_{k=0}^{NB_{max}} (1 - P\{S_k^{j-D}\})^{n_c^{j-D}-1}$ is the probability that the remaining $n_c^{j-D} - 1$ nodes do not sense the channel in slot $j - D$.

Finally, the probability $P\{R^j\}$ that in slot $j$ the sink receives the end of a packet, coming from any node, is given by:

$$P\{R^j\} = n_c^j \cdot P\{Z^j\}\,. \tag{5.16}$$

#### 5.3.3.3 The Energy Consumption

Here, the mean energy consumed by a node during a round is derived. A node spends energy when it receives or transmits a packet and also when it is in backoff state. After the transmission of the packet, the node switches off and does not consume energy. The node will stay in the off state till the reception of the following query.

Let $P_s = 82.5$ mW be the power spent in receiving and sensing states; $P_{bo} = 50$ mW be the power spent in backoff state and $P_t = 75.8$ mW the power spent during transmission (see Freescale IEEE 802.15.4 devices [17]). The mean energy spent by a node in a round, is given by:

$$E_{mean} = \sum_{j=0}^{\hat{t}_{max}+D-1} E_t^j + E_s^j + E_{bo}^j, \tag{5.17}$$

where $E_t^j$, $E_s^j$, and $E_{bo}^j$ are the different energy contributions spent in transmission, sensing and backoff, respectively, for a node ending its transmission in slot $j$.

Since no retransmission is performed, each node will transmit only one packet per round. Therefore,

$$E_t^j = P_t \cdot \frac{D \cdot N_{bit}}{R_b} \cdot P\{T^j\}, \tag{5.18}$$

where $N_{bit} = 10$ bytes is the number of bits transmitted in one slot (having duration $d_{bo}$). The energy spent in the sensing state depends on how many slots are used by the node for sensing the channel. A node transmitting in slot $j$ could have sensed the channel for one slot, in the case that it found the channel free at the end of the first backoff stage, for two slots in case it has found the channel free at the end of the second backoff stage, etc. This energy is given by

$$E_s^j = P_s \cdot \frac{N_{bit}}{R_b} \cdot (1 - b^{j-D}) \sum_{k=0}^{NB_{max}} (k+1) \cdot P\{S_k^{j-D}\}, \tag{5.19}$$

where $b^j$ is the probability to find the channel busy in slot $j$, and $(1 - b^{j-D}) \cdot P\{S_k^{j-D}\}$ is the probability that a node at the end of the $k$th backoff stage, finds the channel free and ends transmitting in slot $j$. Finally, the energy spent in the backoff state depends on how many slots are occupied by the node for the backoff procedure. This number depends, in turn, on the number of backoff stages performed. Therefore, we have

$$E_{bo}^j = P_{bo} \cdot \frac{N_{bit}}{R_b} \cdot (1 - b^{j-D}) \sum_{k=0}^{NB_{max}} (j - k - D) \cdot P\{S_k^{j-D}\}, \tag{5.20}$$

where $j - k - D$ is the number of slots during which a node that has finished the $k$th backoff stage has performed backoff. This value is the same no matter what values of backoff counter are extracted at each backoff stage.

#### 5.3.3.4 Derivation of the Probability that the Channel is Found Busy

By denoting as $f^j$ the probability that the channel in $j$ is free, the probability to find the channel busy in $j$ is given by:

$$b^j = 1 - f^j \,. \tag{5.21}$$

From the law of total probability we can express $f^j$ as:

$$\begin{aligned} f^j = {} & P\{\text{no tx in } j|\text{channel free in } (j-1)\} \cdot P\{\text{channel free in } (j-1)\} + \\ & + P\{\text{no tx in } j|\text{channel busy in } (j-1)\} \cdot P\{\text{channel busy in } (j-1)\}. \end{aligned} \tag{5.22}$$

where $P\{\text{no tx in } j|\text{channel free in } (j-1)\}$ and $P\{\text{no tx in } j|\text{channel busy in } (j-1)\}$, are the probabilities that no transmissions occur in slot $j$ conditioned to the fact that the channel in $j-1$ is free or busy, respectively.

When $D = 1, P\{\text{no tx in } j|\text{channel free in } (j-1)\} = \prod_{k=0}^{NB_{\max}} (1 - P\{S_k^{j-1}\})^{n_c^{j-1}-1}$, as if the channel in $j-1$ is free, no transmissions occur in $j$ if no nodes sense the channel in $j-1$. When, instead, the channel in $j-1$ is busy, no transmissions may certainly occur in $j$. Therefore, in this case, $f^j$ is given by:

$$f^j = \left(1 - b^{j-1}\right) \prod_{k=0}^{NB_{\max}} \left(1 - P\{S_k^{j-1}\}\right)^{n_c^{j-1}-1} + b^{j-1} \,. \tag{5.23}$$

When, instead, $D > 1$ the second term of Eq. 5.22 coincides with the probability that in slot $j-1$ a transmission ends, i.e., the probability that at least one transmission starts in slot $j - D$, given by: $(1 - b^{j-D-1}) \cdot [1 - \prod_{k=0}^{NB_{\max}} (1 - P\{S_k^{j-D-1}\})^{n_c^{j-D-1}-1}]$.

Therefore, in this case, $f^j$ is given by:

$$\begin{aligned} f^j = {} & \left(1 - b^{j-1}\right) \prod_{k=0}^{NB_{\max}} \left(1 - P\{S_k^{j-1}\}\right)^{n_c^{j-1}-1} \\ & + \left(1 - b^{j-D-1}\right) \cdot \left[1 - \prod_{k=0}^{NB_{\max}} \left(1 - P\{S_k^{j-D-1}\}\right)^{n_c^{j-D-1}-1}\right]. \end{aligned} \tag{5.24}$$

Obviously, when $j \leq D$ the second term of Eq. 5.24 becomes null.

#### 5.3.3.5 The Algorithm

The algorithm that allows the evaluation of all the aformentioned performance metrics follows. The simplest case with $n_c^j = n, \forall j$ is considered, but the algorithm can be used for characterizing how $n_c^j$ evolves with time (i.e., depends on $j$), by simply substituting the formula to derive $n_c^j$ in the sequence of steps below (see first instruction of For cycle).

Initialisation of the parameters for $j = 0$:

- set $n_c^0 = n$;
- set $b^0 = 0$;
- set $P\{S_0^0\} = 1/W_0$, $P\{S_k^0\} = 0$ for $k \in \{1, \ldots, NB_{\max}\}$;
- set $P\{C^0\} = P\{S_0^0\}$;
- set $P\{T^0\} = 0$, $P\{Z^0\} = 0$, $P\{R^0\} = 0$.

For ( $j = 1; j <= \hat{t}_{\max} + D - 1; j++$)

- {
- set $n_c^j = n$;
- compute $b^j$ according to Eq. 5.21, by using Eq. 5.23 in the case $D = 1$ and Eq. 5.24 in the case $D > 1$;
- compute $P\{S_0^j\}$ according to Eq. 5.3
- compute $P\{S_1^j\}$ according to Eq. 5.6
- compute $P\{S_2^j\}$ according to Eq. 5.6
- compute $P\{S_3^j\}$ according to Eq. 5.7
- compute $P\{S_4^j\}$ according to Eq. 5.7
- compute $P\{C^j\}$ according to Eq. 5.10
- compute $P\{T^j\}$ according to Eq. 5.8
- compute $P\{Z^j\}$ according to Eq. 5.15
- compute $P\{R^j\}$ according to Eq. 5.16
- }

compute $p_s$ through Eq. 5.13 and $E_{\text{mean}}$ through Eq. 5.17.

### 5.3.4 Numerical Results

For the purpose of numerical comparison, a dedicated simulation tool written in C language has been developed. The simulator generates a network composed of $n$ nodes and a sink, sending queries and waiting for the data from nodes. Upon reception of the query, nodes start the CSMA/CA algorithm trying to transmit their packets. The CSMA/CA protocol described above is implemented. Ideal channel conditions are assumed; therefore, all nodes can "hear" each other and can receive correctly the query at each round. No capture effect is considered: in case two or more packets collide, they are all lost. Finally, no ACK and retransmission mechanisms are performed. In the simulator, nodes could be in four different states: backoff, sensing, transmission, and idle. When backoff is performed, only the backoff counter is updated (decreased by one in each slot). During sensing the state of the channel is evaluated: if there are nodes in the transmission state, the channel is considered busy. Finally, during a transmission the presence of other transmissions must be checked. If two or more transmissions are overlapped (totally or partially) the packet is considered lost. Finally if the node, at the reception of the subsequent query has not still accessed the channel, the packet is considered lost. We consider $10^4$ transmissions in our simulator, meaning that $10^4$ queries are simulated.

In the following, we set $z_{\mathrm{B}} = 60$ bytes, $BE_{\mathrm{min}} = 3$, $BE_{\mathrm{max}} = 5$, $NB_{\mathrm{max}} = 4$, $T_{\mathrm{q}} = 100$ ms, if not otherwise specified. Figures 5.9, 5.10, and 5.11 show the probability $P\{T^j\}$ as a function of time $\hat{t} = j$, representing the time slot, for $n = 3$, 5, and 7, respectively, having fixed $D = 1$. Even if these probabilities could be larger than zero for $j \in \{0, \ldots, \hat{t}_{\mathrm{max}} + D - 1\}$—as, in all the three case, for $j > 26$ it holds that $P\{T^j\}$ tends to zero—the curves are shown for $j \leq 26$.

Both mathematical analysis and simulation results are reported, considering the two models: $n_{\mathrm{c}}^j$ variable according to Eq. 5.12 and $n_{\mathrm{c}}^j = n$. As one can see, the two mathematical models bring approximatively the same results, and both do not present relevant differences with respect to simulations; therefore, the model is validated. Owing to its simplicity, all the other results shown here have been obtained by considering $n_{\mathrm{c}}^j = n$, whatever is $j$.

It can be seen in the figures that, in all cases, no traffic is present in the first slot, because no transmission may occur: a node that extracts the 0 value at the first backoff stage will sense the channel in slot 0 and will transmit in slot 1. This happens with probability $1/W_0 = 1/8$, whatever be $n$, and this is also the maximum value that $P\{T^j\}$ can assume. If a node extracts the value 0 at the first backoff stage, in fact, it will certainly transmit in slot 1, whereas if a larger value is extracted there is a certain probability that the channel is found busy. Therefore, $P\{T^j\}$ assumes lower values for $j > 1$. When a node tries to access the channel for the first time, it will delay the transmission for a random number of slots in the range $\{0, \ldots, 7\}$. As the network is composed of few nodes, the probability to find the channel busy is low; therefore, $P\{T^j\}$ for $j \in \{1, \ldots, 8\}$, which correspond to the cases in which the node extracts the value 0, or 1, … , or 7 respectively, are the largest. Lower probabilities are associated to the slots from 9 to 23, in which only nodes that have found the channel busy and are performing the second backoff stage (plus some nodes already performing the third or fourth or fifth backoff stage, which are a minority) transmit. From slot 24, the probabilities show a further decrease, because in these slots there are only transmissions of nodes that have ended the second backoff too and are running the third or fourth or fifth backoff stage; once again the probability that the channel is found busy for two or three times is very low, and few nodes will transmit after slot 24. If we compare

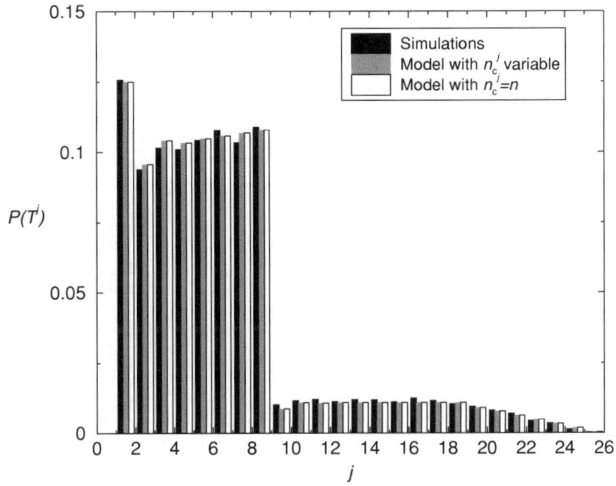


**Fig. 5.9** The probabilities $P\{T^j\}$, obtained through simulation and through the mathematical model, considering $n_c^j = n_c^0$ for every $j$ and $n_c^j$ variable, in the case $n = 3$ and $D = 1$

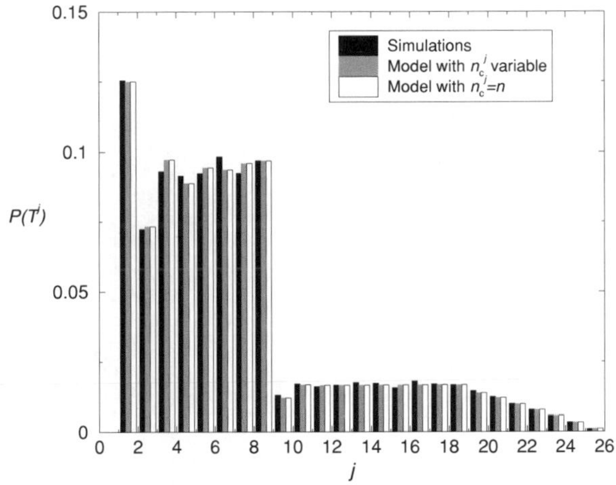


**Fig. 5.10** The probabilities $P\{T^j\}$, obtained through simulation and through the mathematical model, considering $n_c^j = n_c^0$ for every $j$ and $n_c^j$ variable, in the case $n = 5$ and $D = 1$

Figs. 5.9, 5.10, and 5.11 we can note that by increasing $n$ the probabilities of having transmissions in slots from 2 to 8 decrease and, consequently, the transmission probabilities in slots from 9 to 23 increase. The reason is that by increasing the number of nodes, the probability to find the channel busy at the end of the first backoff increases. Finally, it can be noted that in all cases we have two relative minima in slot 2: as the probability to transmit a packet in slot 1 is large, the probability to find the channel busy in this slot is also large; therefore, the probability to transmit in the following slot is quite small (see Eq. 5.8).

In Figs. 5.12 and 5.13, the probabilities $P\{Z^j\}$ and $P\{R^j\}$ are shown, as functions of $j$, representing the time slot, for $n = 3$, 5, and 7. Once again, the model is validated by simulations: the values obtained through the analysis and simulations are very similar. The differences are due to the approximation we have made in modeling $n_c^j$ (set equal to $n$, whatever be $j$). The trends are very similar to those

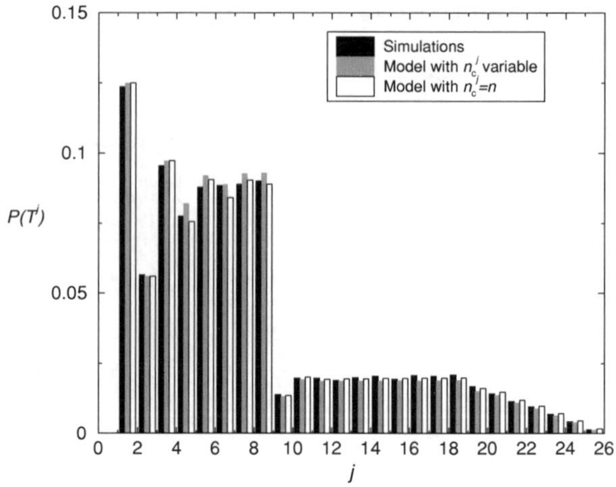


**Fig. 5.11** The probabilities $P\{T^j\}$ obtained through simulation and through the mathematical model considering $n_c^j = n_c^0$ for every $j$ and $n_c^j$ variable, in the case $n = 7$ and $D = 1$.

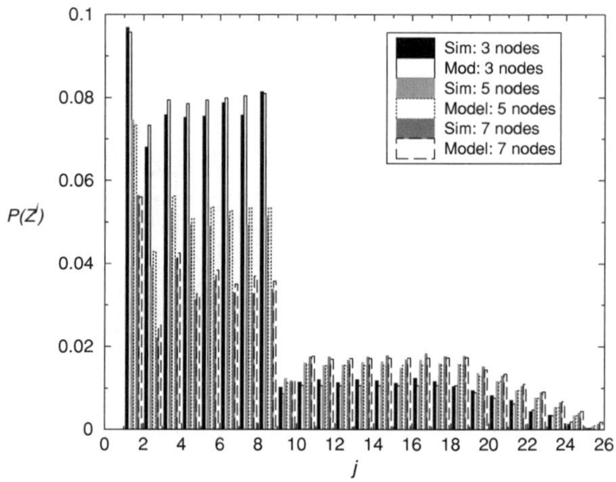


**Fig. 5.12** The probabilities $P\{Z^j\}$ obtained through simulation and through the mathematical model for $n = 3$, 5, and 7, having fixed $D = 1$

obtained for $P\{T^j\}$: the largest values of probabilities are for slots 1 to 8; lower probabilities are present for slots from 9 to 23, and then the probabilities tend to zero. Moreover, we have a maximum in slot 1 and a relative minimum in slot 2.

To validate the model for the cases $D > 1$, in Fig. 5.14 $F_T(j)$ is shown, as a function of $j$, for different values of $n$ and $D$. Both mathematical analysis (lines) and simulation results (symbols) are reported to validate the model: a good agreement between the two results can be found in all cases. The non linear behavior of the curves for small values of the ordinate are due to the sudden changes in values of $P\{T^j\}$ already commented before. In all cases, no traffic is present for $j < D$, and when $j = D$ then $P\{T^j\}$ assumes its maximum value, equal

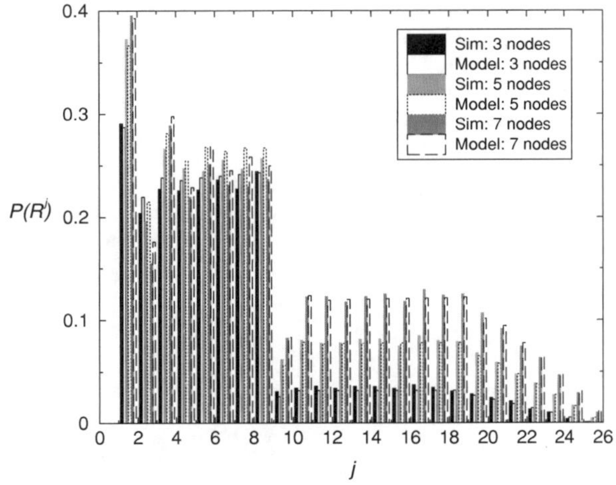


**Fig. 5.13** The probabilities $P\{R^j\}$ obtained through simulation and through the mathematical model for $n = 3$, 5 and 7, having fixed $D = 1$

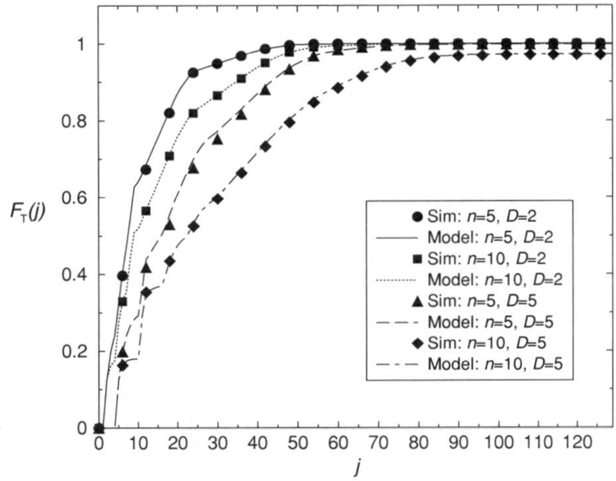


**Fig. 5.14** The cumulative function $F_T(j)$, as a function of $j$, obtained through simulations (*symbols*) and through the mathematical model (*lines*), for different values of $n$ and $D$

to 1/8, whatever be $n$ (as in the case $D = 1$). By increasing $n$ and $D$, the delay in accessing the channel increases. Moreover, in the case of $n = 10$ and $D = 10$, we can note that $F_T(j)$ does not reach 1, as there is a certain probability that a node cannot succeed in accessing the channel.

In Fig. 5.15, $p_s$ is shown, as function of $n$, for different values of $D$. Results obtained through simulation (symbols) and the mathematical model (lines) are reported. Once again, simulations validate the model. As one can see, $p_s$ decreases monotonically by increasing $n$, because the number of nodes competing for the channel increases. There exists an optimum value of $D$, denoted as $D_{\text{opt}}$, maximizing $p_s$, and this value depends on $n$. For the sake of readability of the drawings, here only the curves obtained for $D = 1$, 3, and 5, are shown. However, the model has been validated for $1 \leq D \leq 10$ and $1 \leq n \leq 50$. From these results, we have found also that for $1 < n < 12$, $D_{\text{opt}} = 7$; for $12 < n < 18$,

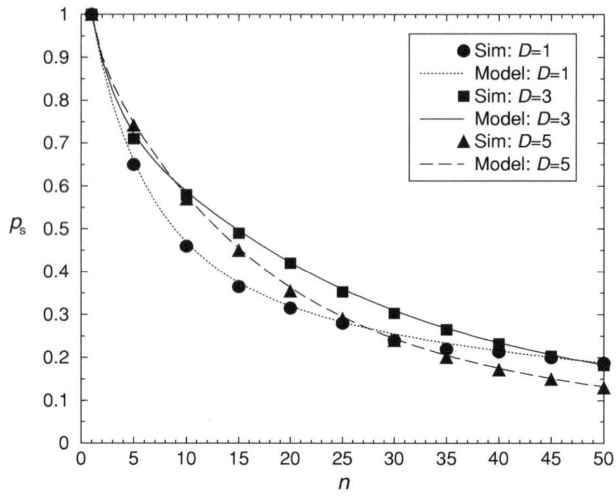


**Fig. 5.15** The success probability, $p_s$, obtained through simulations (*symbols*) and through the mathematical model (*lines*), as a function of $n$, for different values of $D$

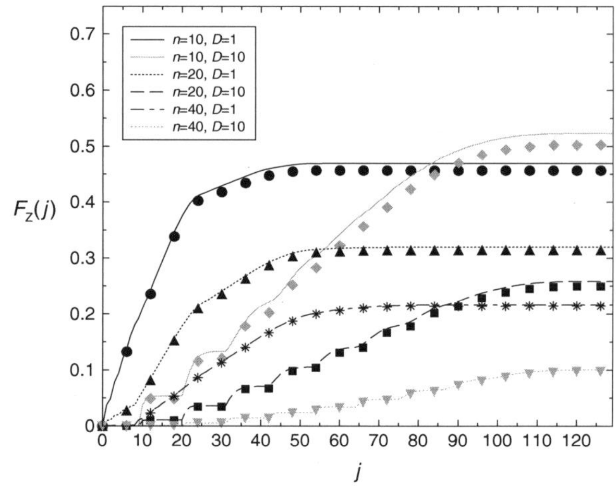


**Fig. 5.16** The cumulative function, $F_Z(j)$, as a function of $j$, obtained through simulations (*symbols*) and through the mathematical model (*lines*), for different values of $n$ and $D$

$D_{opt} = 5$; for $n > 68$, $D_{opt} = 2$. Therefore, it clearly appears that $D_{opt}$ decreases when augmenting $n$.

To better understand how the distribution of the traffic varies when low, medium, and high offered load are present, in Fig. 5.16, the cumulative function of $P\{Z^j\}$, denoted as $F_Z(j)$, is shown, as a function of $j$, for different values of $n$ and $D$ ($F_Z(j)$ is obtained by substituting $P\{T^v\}$ with $P\{Z^v\}$ in Eq. 5.9). Both mathematical analysis (lines) and simulation (symbols) results are reported to validate the model: a good agreement is found in almost all cases. As expected, once we fix $D$, by decreasing $n$ (therefore the offered load), $F_Z(j)$ decreases; once we set $n$, instead, the value of $D$ maximizing $F_Z(j)$ depends on $n$ as stated earlier. As an example, for $n = 10$, to obtain the largest value of $F_Z(j)$ we have to fix $D = 10$, whereas for $n = 40$, the largest value is reached for $D = 1$. However, if we set

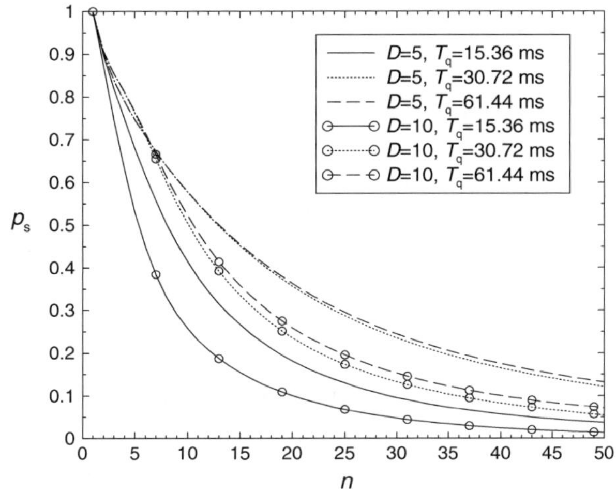


**Fig. 5.17** The success probability, $p_s$, as a function of $n$, for different values of $D$ and $T_q$

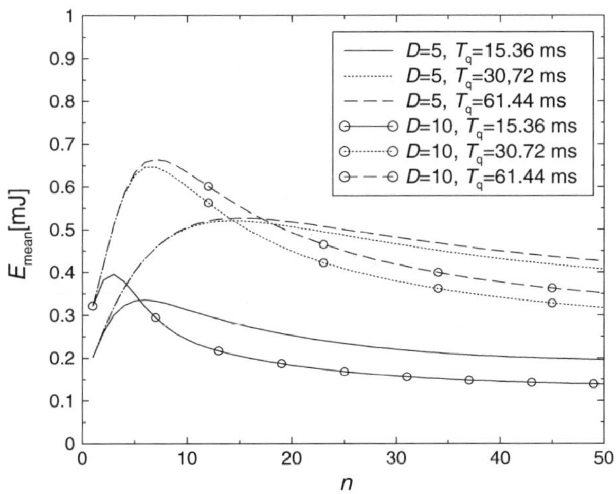


**Fig. 5.18** The mean energy consumed per round, $E_{mean}$, as a function of $n$, for different values of $D$ and $T_q$

$D = 10$, the maximum value of $F_Z(j)$ is reached with a higher delay. As can be seen, in fact, the curves with $D = 1$ have a higher slope and reach the maximum value with shorter delays.

The behavior of $p_s$ and $E_{mean}$ is now investigated, by varying the time interval $T_q$ between two successive queries, having fixed $D = 5$ and 10.

In Fig. 5.17, $p_s$ is shown as a function of $n$. As expected, by increasing $T_q$, $p_s$ gets larger, because nodes have more time to access the channel. However note that the increase of $p_s$ is obtained at the cost of longer delays, resulting also in an increasing of $E_{mean}$. In Fig. 5.18, $E_{mean}$ is shown, as a function of $n$, for the same set of parameters $D$ and $T_q$ considered in Fig. 5.17. As it can be noted, here the increase of $T_q$ results in an increasing of $E_{mean}$, since nodes will stay on for a longer time, and also have higher probability to transmit their packets. For small values of $n$, by increasing $D$, $E_{mean}$ gets larger, because of the greater amount of energy spent for transmitting larger packets. Conversely, for high $n$, the larger $D$, the lower will be the probability that a node succeeds in accessing the channel, decreasing the energy spent by the node.

By comparing Figs. 5.17 and 5.18, we can deduce that a tradeoff between energy consumption and success probability should be found.

## 5.4 The Beacon-Enabled Model

In this section, the beacon-enabled model, considering both star and tree-based topologies, is derived. The section is structured as follows: Sect. 5.4.1 introduces the metrics derived from the model, whereas in Sect. 5.4.2 and Sect. 5.4.3 the mathematical model of the CSMA/CA algorithm and the performance metrics related to the CAP portion of the superframe are derived. All the above cited subsections are related to the star topology, and the related results are discussed in Sect. 5.4.4. Sect. 5.4.5, instead, is devoted to the tree-based topology.

### *5.4.1 Performance Metrics Derived from the Model*

Recall that in case of star topologies, we set $BI = SD = T_q$ and that we also denote as $s$ the $s$th slot (i.e., backoff period) in the superframe. Note that, here, we introduce a new variable for denoting the generic slot $s$, which is different from the value $j$ previously defined; the relationship between $s$ and $j$ will be explained in the following.

The model provides the following metrics:

- the probability that a node ends the transmission of its packet in a given slot $s$, denoted as $P\{T^s\}$, with $s \in \{0, \ldots, T_q/d_{bo} - 1\}$;
- the probability that the sink receives the packet tail, coming from a node, in a given slot $s$, denoted as $P\{Z^s\}$, with $s \in \{0, \ldots, T_q/d_{bo} - 1\}$;
- the success probability for a transmission, that is the probability that a node succeeds in transmitting its packet in the superframe whatever the slot, denoted as $p_s$;
- the average delay, $D_{mean}$, with which a packet is received by the sink.

We denote as $N_{GTS}$ the number of Guaranteed Time Slots (GTSs) allocated (see Fig. 5.2, upper part). According to the standard each GTS must have a duration multiple of $60 \cdot 2^{SO} \cdot T_s$; we denote this duration as $d_{GTS}$, equal to $D_{GTS} \cdot 60 \cdot 2^{SO} \cdot T_s$, with $D_{\mathrm{GTS}}$ integer (see Fig. 5.2, above part). Since an inter-frame space between two successive packets received by the sink must be guaranteed, $D_{\mathrm{GTS}}$ is chosen such that the GTS contains the packet and the inter-frame space. The inter-frame space duration depends on the size of the MAC protocol data unit (MPDU): for MPDU lower than 18 bytes an inter-frame space of 12 $T_s$ must be present, whereas 40 $T_s$ are needed for MPDU larger than 18 bytes [10]. We set the GTS duration equal to the minimum possible duration which allows to contain the packet and the inter-frame space. Therefore, by denoting as $d_{\mathrm{ifs}}$ the duration of the inter-frame space, we have $D_{\mathrm{GTS}} = \lceil (D \cdot d_{\mathrm{bo}} + d_{\mathrm{ifs}})/(60 \cdot 2^{SO} T_s) \rceil$, and the number of backoff periods occupied by each GTS is equal to $D_{\mathrm{GTS}} \cdot 3 \cdot 2^{SO}$.

We assume that when $N_{\mathrm{GTS}}$ GTSs are allocated, the sink selects randomly the $N_{\mathrm{GTS}}$ nodes to which the GTSs are allocated. Therefore, no resource allocation strategies are accounted for. For the scenario considered, this assumption is reasonable, since all nodes transmit packets of the same size and no priority policy between nodes is needed. We also recall here that each node has only a packet to be transmitted per superframe: therefore, it will use the CAP or the CFP (but not both). In these conditions, the probability that a node has a specific GTS allocated is $1/n$, whereas the probability that a node has any GTS allocated (that is the probability that a node can use the CFP) is $N_{\mathrm{GTS}}/n$. However, note that the model could be applied to any GTS allocation strategy, by simply changing the probability $1/n$.

To simplify the formulas in the following, we will indicate with the integer $j$ the slots in the CAP portion, and with $P\{T^j\}_{\mathrm{CAP}}$ and $P\{Z^j\}_{\mathrm{CAP}}$, the probabilities that a node succeeds in accessing the channel and in transmitting its packet in slot $j$ of the CAP portion, being $j \in \{0, \ldots, T_{\mathrm{CAP}}/d_{\mathrm{bo}} - 1\}$, where $T_{\mathrm{CAP}}$ is the duration of the CAP portion given by: $T_{\mathrm{CAP}} = T_q - d_B - N_{\mathrm{GTS}} \cdot d_{\mathrm{GTS}}$. Therefore, we simply set $j = s - d_B/d_{\mathrm{bo}}$ (see Fig. 5.2).

Therefore, the probabilities $P\{T^s\}$ and $P\{Z^s\}$ in the CAP portion are given by:

$$P\{T^s\} = P\{T^j\}_{\mathrm{CAP}} \cdot \frac{n - N_{\mathrm{GTS}}}{n}, \tag{5.25}$$

for $s \in \{d_B/d_{\mathrm{bo}}, \ldots, T_{\mathrm{CAP}}/d_{\mathrm{bo}} + d_B/d_{\mathrm{bo}} - 1\}$ and $j \in \{0, \ldots, T_{\mathrm{CAP}}/d_{\mathrm{bo}} - 1\}$; and null otherwise.

$$P\{Z^s\} = P\{Z^j\}_{\mathrm{CAP}} \cdot \frac{n - N_{\mathrm{GTS}}}{n}, \tag{2.26}$$

for $s \in \{d_B/d_{\mathrm{bo}}, \ldots, T_{\mathrm{CAP}}/d_{\mathrm{bo}} + d_B/d_{\mathrm{bo}} - 1\}$ and $j \in \{0, \ldots, T_{\mathrm{CAP}}/d_{\mathrm{bo}} - 1\}$; and null otherwise.

In the CFP, $P\{T^s\} = P\{Z^s\} = 1/n$ for $s = T_{\mathrm{CAP}}/d_{\mathrm{bo}} + d_B/d_{\mathrm{bo}} + k \cdot D \cdot 3 \cdot 2^{SO} + D - 1$, with $k \in \{0, \ldots, N_{\mathrm{GTS}} - 1\}$; and null otherwise. Recall that transmissions are referred to the last slot in which the transmission occurs and that no collisions happen in GTSs.

We can also evaluate the cumulative functions, $F_T(s)$ and $F_Z(s)$, defined as the probabilities that a node transmits its packet within slot $s$, and that a node transmits correctly its packet within $s$, respectively. They can be expressed as $F_T(s) = \sum_{v=0}^{s} P\{T^v\}$ and $F_Z(s) = \sum_{v=0}^{s} P\{Z^v\}$.

The probability $p_s$ of successful packet transmission by a node in a network composed of $n$ nodes organised in a star topology is:

$$p_s(n) = p_{s_{CAP}}(n - N_{GTS}) \cdot \frac{n - N_{GTS}}{n} + \frac{N_{GTS}}{n}, \tag{5.27}$$

where $p_{s_{CAP}}(n - N_{GTS})$ is the success probability for a packet transmitted in the CAP portion, through the CSMA/CA algorithm, when $n - N_{GTS}$ nodes compete for the channel. The success probability for a packet transmitted in the CFP, instead, is equal to one. Finally the average delay, $D_{mean}$, is given by:

$$D_{mean} = \sum_{s=0}^{T_q/d_{bo}-1} (s+1)\frac{P\{Z^s\}}{p_s}, \tag{5.28}$$

where $s + 1$ is the delay, in backoff periods, of a packet correctly received in slot $s$, and $\frac{P\{Z^s\}}{p_s}$ is the probability that the packet tail is received in slot $s$, given that the packet has been correctly received.

The probabilities $P\{T^j\}_{CAP}$, $P\{Z^j\}_{CAP}$ and $p_{s_{CAP}}$, related to the CAP portion, are derived in the following subsections where the mathematical model of the CSMA/CA algorithm is introduced.

## 5.4.2 Formulation of the Mathematical Model of the CSMA/CA Algorithm

### 5.4.2.1 Node States

As in the non beacon-enabled case, a node accessing the channel during the CAP portion of the superframe can be in one of four states: backoff, sensing, transmission, or idle. Given the scenario under consideration, we need to model only the backoff and sensing states.

The node state is modeled as a three-dimensional process $Q(\hat{t}) = \{BO_c(\hat{t}), BO_s(\hat{t}), CW(\hat{t})\}$, where $\hat{t}$ is an integer, representing the time, expressed in number of slots, having set the origin of this time axis ($\hat{t} = 0$) at the instant in which nodes receive the beacon. Therefore, $\hat{t} = j$ denotes the $j$th slot (from $j \cdot d_{bo}$ to $(j+1) \cdot d_{bo}$), after the reception of the beacon, that is the interval of time between $d_B + j \cdot d_{bo}$ and $d_B + (j+1) \cdot d_{bo}$.

$BO_c(\hat{t})$ and $BO_s(\hat{t})$ represent, once again, the backoff time counter and the backoff stage at time $\hat{t}$, respectively, and $CW(\hat{t})$ is the value of $CW$ at time $\hat{t}$.

As in the non beacon-enabled case, we can identify $NB_{\max} + 1$ different backoff stages obtained by considering the different possible combinations of the pair $(NB, BE)$, shown in Table 5.1.

Since there exists a maximum value for $NB$, there will be also a maximum delay affecting the transmission of a packet. This maximum is reached in case a node extracts at every backoff stage the higher backoff time counter and at the end of each backoff stage it always finds the channel busy. Therefore, the last slot in which a transmission can start is $\hat{t}_{\max} = \sum_{k=0}^{NB_{\max}} W_k + k + 1$, and the last slot in which a transmission can finish is $(\hat{t}_{\max} + D - 1)$.

In the following, the generic state will be denoted as $Q(\hat{t}) = \{BO_{\mathrm{c}}, BO_{\mathrm{s}}, CW, \hat{t}\}$ and the probability of being in a generic state will be denoted as $P\{BO_{\mathrm{c}} = c, BO_{\mathrm{s}} = k, CW = w, \hat{t} = j\} = P\{c, k, w, j\}$. In particular, the probability of being in a backoff state will be denoted as $P\{c, k, 2, j\}$, since in these states $CW$ is equal to 2. The probabilities of being in the first sensing phase (i.e., when $CW = 2$) and in the second sensing phase (i.e., when $CW = 1$), at the $j$th slot and in the $k$th backoff stage, will be denoted as $P\{S2_k^j\} = P\{0, k, 2, j\}$ and $P\{S1_k^j\} = P\{0, k, 1, j\}$, respectively. Note that when a node is in sensing state, then $BO_{\mathrm{c}}$ is equal to zero.

#### 5.4.2.2 Steps Followed by the Model

Let us denote by $b_w^j$ the probability that in the $j$th slot when $CW = w$ the channel is found to be busy after sensing. Since $CW$ is equal to 2 when a node performs the first sensing phase and to 1 when it performs the second sensing phase, we will denote as $b_2^j$ the probability to find the channel busy in the first phase and as $b_1^j$ the probability to find the channel busy in the second phase. Finally, we will denote as $f^j$ the joint probability to find the channel free in slot $j$ and in slot $j - 1$ (i.e., the probability that a node starting sensing in slot $j - 1$ finds the channel free for two subsequent slots). These probabilities will be initially left as parameters, and their computation will be performed in Sect. 5.4.3.3. The model provides $P\{T^j\}_{\mathrm{CAP}}$ and $P\{Z^j\}_{\mathrm{CAP}}$, with $j \in \{0, \ldots, T_{\mathrm{CAP}}/d_{\mathrm{bo}} - 1\}$, and $p_{s_{\mathrm{CAP}}}$.

The probability $P\{T^j\}_{\mathrm{CAP}}$ depends on the probability of being in sensing state in the slot $j - D - 1$ (since a packet occupies $D$ slots) and to find the channel free for two subsequent slots. To determine the sensing probabilities, the behavior of a single node is modeled, using a state-transition diagram [16], describing the relation between all possible states in which a node can be (see the following subsection). From this diagram, we obtain the probabilities $P\{S1_k^j\}$ and $P\{S2_k^j\}$, whatever be $j$ and $k$. This is made in the remainder of this subsection. From these probabilities, we can derive the probabilities $P\{T^j\}_{\mathrm{CAP}}$, $P\{Z^j\}_{\mathrm{CAP}}$, $p_{s_{\mathrm{CAP}}}$, which are derived in Sect. 5.4.3.1. At the end of this subsection $b_1^j$, $b_2^j$, and $f^j$ are also given. The algorithm used to compute all the target

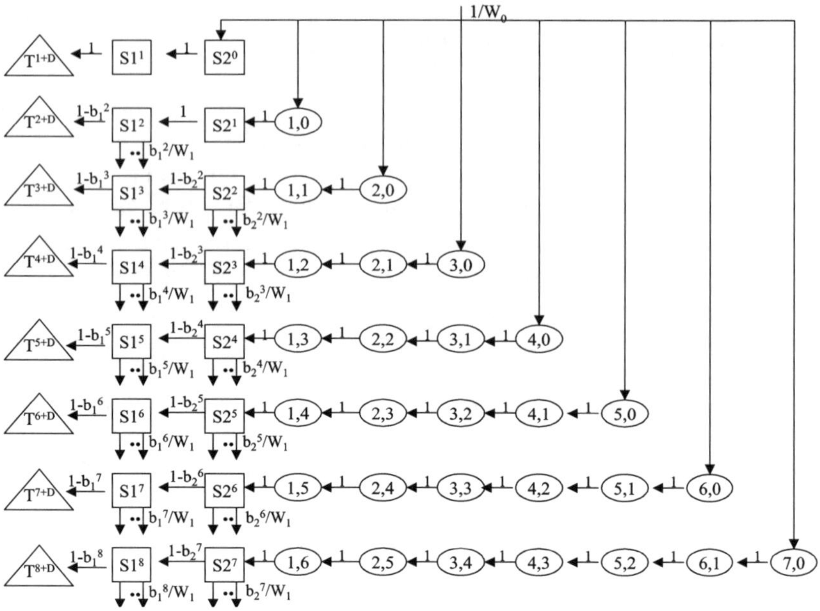


**Fig. 5.19** The state-transition diagram of the first backoff stage ($BO_s = 0$)

performance metrics is not reported owing to its similarity with the one illustrated in Sect. 5.3.3.5.

### 5.4.2.3 Sensing Probabilities

The state-transition diagram of the three-dimensional process $Q(\hat{t})$ is presented through different figures: one for each backoff stage. The part of the diagram related to the first backoff stage ($BO_s = 0$), obtained when the MAC parameters are set to the default values ($BE_{min} = 3$, $BE_{max} = 5$, $NB_{max} = 4$), is reported in Fig. 5.19. The part of the diagram related to the generic backoff stage ($BO_s = k$) is, instead, reported in Figs. 5.20 and 5.21.

As in the non beacon-enabled case, for the sake of simplicity in the drawings, the generic backoff state is simply denoted as $\{c, j\}$, omitting the value of $BO_s$, and also the value of $CW$ (equal to 2 for all the backoff states), and the sensing states are denoted as $S1^j$ and $S2^j$ with no subscript $k$. Finally the transmission states are denoted as $T^j$, with no pedex $k$.

In the following, the state-transition diagram is described. The probabilities of being in the different states of the chain and the transition probabilities between the states are provided.

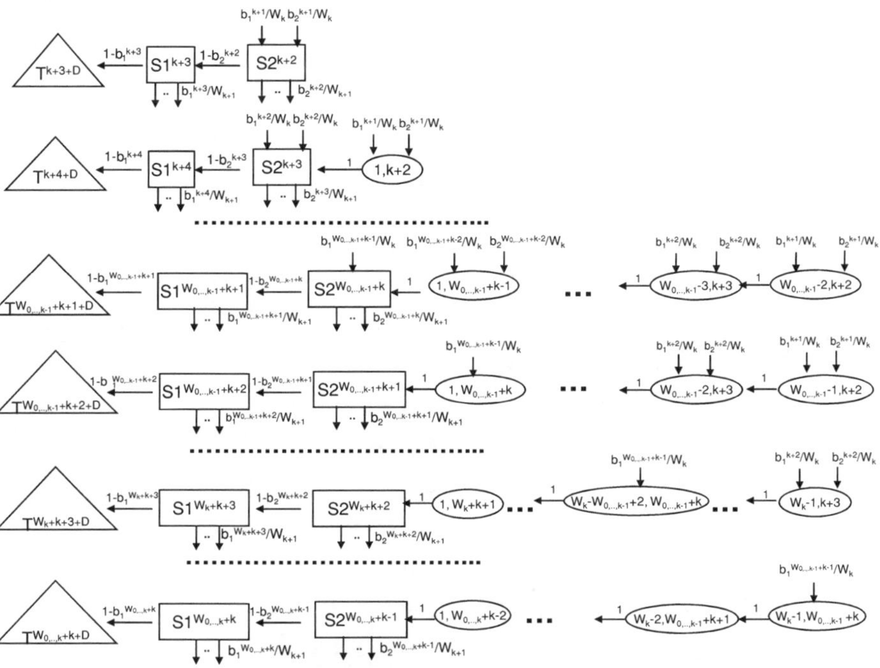


**Fig. 5.20** The state-transition diagram of the $k$th backoff stage when $W_{0,\ldots,k-1} \leq W_k$

## First Backoff Stage($\mathrm{BO}_\mathrm{s} = 0$)

At the beginning of the backoff algorithm, each node extracts an integer value uniformly distributed between 0 and $W_0 - 1$. At $\hat{t} = 0$ a node enters, with probability $1/W_0$, one of the states $\{c, 0, 2, 0\}$ with $c \in \{0, \ldots, W_0 - 1\}$. If the extracted value is 0, in slot 0 and 1 the node will sense the channel and in slot 2 it will transmit its packet, because no transmission may occur in the first two slots and, therefore, the channel will be certainly found free ($f^1 = 1$). In case a value larger than 0 is extracted, the node will decrease its backoff counter at each slot until the counter will reach the zero value, when the node will start sensing. In case the channel is found free for two subsequent slots the node will transmit the packet; otherwise, it will pass to the following backoff stage and another value, uniformly distributed between 0 and $W_1 - 1$, will be extracted. In Fig. 5.19, the transitions that originated from the sensing states enter in the states of Fig. 5.20. For example, if a node is in the state $S2_0^1$ and finds the channel busy, it will enter the state $S2_1^2$ or one of the states $\{c, 1, 2, 2\}$, with $c \in \{1, \ldots, W_1 - 1\}$, with the same probability $b_2^1/W_1$. The state of arrival depends on the new backoff counter value extracted.

Denoting as $P\{BO_\mathrm{c} = c_1, BO_\mathrm{s} = k_1, CW = w_1, \hat{t} = j_1 | BO_\mathrm{c} = c_0, BO_\mathrm{s} = k_0, CW = w_0, \hat{t} = j_0\} = P\{c_1, k_1, w_1, j_1 | c_0, k_0, w_0, j_0\}$ the transition probability from

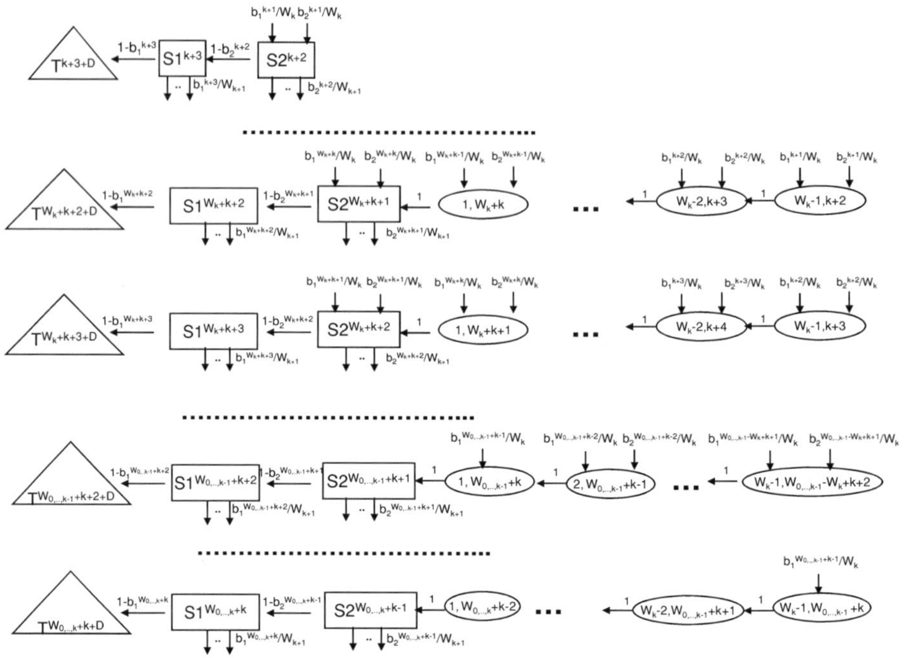

**Fig. 5.21** The state-transition diagram of the $k$th backoff stage when $W_{0,\dots,k-1} > W_k$

the state $\{c_0, k_0, w_0, j_0\}$ to the state $\{c_1, k_1, w_1, j_1\}$, the transition probabilities between the backoff states are given by:

$$P\{c, 0, 2, j+1 | c+1, 0, 2, j\} = 1 \ , \tag{5.29}$$

for $c \in \{0, \dots, W_0 - 2\}$ and $j \in \{0, \dots, W_0 - 2\}$. This equation accounts for the fact that, at the beginning of each time slot, the backoff time counter is decreased by 1, until it reaches the zero value, with probability 1.

The probabilities of being in a sensing state when $CW = 2$ are given by:

$$P\{S2_0^j\} = \begin{cases} \frac{1}{W_0} & \text{for } j \in \{0, \dots, W_0 - 1\} \\ 0 & \text{otherwise} \ . \end{cases} \tag{5.30}$$

The probabilities of being in a sensing state when $CW = 1$ are given by:

$$P\{S1_0^j\} = \begin{cases} P\{S2_0^{j-1}\} \cdot (1 - b_2^{j-1}) & \text{for } j \in \{1, \dots, W_0\} \\ 0 & \text{otherwise} \ . \end{cases} \tag{5.31}$$

A node, in fact, will sense the channel for the second time if and only if it finds the channel free during the first sensing phase.

### Other Backoff Stages ($\mathbf{BO_s = 1, \ldots, NB_{max}}$)

We consider here the backoff stages $BO_s = 1, \ldots, NB_{max}$ and we refer to the parts of the state transition diagram illustrated in Fig. 5.20 for the cases $W_{0,\ldots,k-1} \leq W_k$ and in Fig. 5.21 for the cases $W_{0,\ldots,k-1} > W_k$. Note that in the case of defaults MAC parameters ($BE_{min} = 3$, $BE_{max} = 5$ and $NB_{max} = 4$), Fig. 5.20 shows the cases $BO_s = 1$ and 2 and Fig. 5.21 the cases $BO_s = 3$ and 4.

As in the case with $BO_s = 0$, the transition probabilities between backoff states in the $k$th backoff stage are given by:

$$P\{c,k,2,j+1|c+1,k,2,j\} = 1 \,, \tag{5.32}$$

for $c \in \{0,\ldots,W_k - 2\}$ and $j \in \{k+2,\ldots,W_{0,1,\ldots,k} + k - 2\}$, where $W_{0,1,\ldots,k} = W_0 + W_1 + \cdots + W_k$. In the following we will denote as $W_{x,y,z}$ the sum $W_x + W_y + W_z$.

The transition probabilities between the sensing states at $CW = 2$ of the backoff stage $k$ and those of the backoff stage $k+1$ are given by:

$$P\{c,k,2,j+1|0,k-1,2,j\} = \frac{b_2^j}{W_k} \,, \tag{5.33}$$

for $c \in \{0,\ldots,W_k - 1\}$ and $j \in \{k+1,\ldots,W_{k-1} + k - 2\}$. This equation accounts for the fact that in case a node is in the $(k-1)$st backoff stage and the channel at slot $j$ is found busy, the node will reach one of the states $\{c,k,2,j+1\}$, with $c \in \{0,\ldots,W_k - 1\}$, with the same probability $1/W_k$.

The transition probabilities between the sensing states of two subsequent backoff stages when $CW = 1$ are given by:

$$\{c,k,2,j+1|0,k-1,1,j\} = \frac{b_1^j}{W_k} \,, \tag{5.34}$$

for $c \in \{0,\ldots,W_k - 1\}$ and $j \in \{k+1,\ldots,W_{k-1} + k - 1\}$.

If $W_{0,\ldots,k-1} \leq W_k$ (see Fig. 5.20), the probabilities of being in sensing when $CW = 2$ are given by:

$$P\{S2_k^j\} = \begin{cases} \sum_{v=k+1}^{j-1} (P\{S1_{k-1}^v\} \cdot \frac{b_1^v}{W_k} + P\{S2_{k-1}^v\} \cdot \frac{b_2^v}{W_k}) & \text{for } j \in \{k+2,\ldots,W_{0,\ldots,k-1} + k\} \\ P\{S2_k^{W_{0,\ldots,k-1}+k}\} & \text{for } j \in \{W_{0,\ldots,k-1} + k + 1,\ldots,W_k + k + 1\} \\ \sum_{v=j-W_k}^{W_{0,\ldots,k-1}+k-1} (\{S1_{k-1}^v\} \cdot \frac{b_1^v}{W_k} + \{S2_{k-1}^v\} \cdot \frac{b_2^v}{W_k}) & \\ & \text{for } j \in \{W_k + k + 2,\ldots,W_{0,\ldots,k} + k - 1\} \\ 0 & \text{otherwise.} \end{cases} \tag{5.35}$$

Let us consider the case $BO_s = 1$, when $BE_{min} = 3$ and $BE_{max} = 5$. The first equation derives from the fact that until $j \leq W_0$, the probability of being in sensing in the second backoff stage depends on the probabilities of being in sensing in the first backoff stage and to find the channel busy the first or the second time. As an example, a node can arrive in $S2_1^3$ if it is in $S1_0^2$ or in $S2_0^2$, finds the channel busy, and extracts the value 0 for the second backoff stage (see Figs. 5.19 and 5.20). The second equation accounts for the fact that for $j > W_0 + 1$, there are no more transitions between the states of $BO_s = 0$ and the ones of $BO_s = 1$, because the last slot in which a node can sense the channel in the first backoff stage is $j = W_0 = 8$. Finally, when $j$ reaches $W_1 + 3 = 19$, the sum starts with $v = 3$ and not 2, since if a node is in $S1_0^2$ (or in $S2_0^2$) it moves (in case of channel busy) to states $\{c, 1, 2, 3\}$ with $c \in \{0, \ldots, 15\}$; therefore the state $\{16, 1, 2, 3\}$ does not exist (see the figure).

Whereas, if $W_{0,\ldots,k-1} > W_k$ (see Fig. 5.21), the probabilities of being in sensing when $CW = 2$ are given by:

$$P\{S2_k^j\} = \begin{cases} \sum_{v=k+1}^{j-1} P(\{S1_{k-1}^v\} \cdot \frac{b_1^v}{W_k} + P\{S2_{k-1}^v\} \cdot \frac{b_2^v}{W_k}) & \text{for } j \in \{k+2, \ldots, W_k + k + 1\} \\ \sum_{v=j-W_k}^{j-1} P(\{S1_{k-1}^v\} \cdot \frac{b_1^v}{W_k} + P\{S2_{k-1}^v\} \cdot \frac{b_2^v}{W_k}) & \text{for } j \in \{W_k + k + 2, \ldots, W_{0,\ldots,k-1} + k\} \\ \sum_{v=j-W_k}^{W_{0,\ldots,k-1}+k-1} P(\{S1_{k-1}^v\} \cdot \frac{b_1^v}{W_k} + P\{S2_{k-1}^v\} \cdot \frac{b_2^v}{W_k}) & \\ & \text{for } j \in \{W_{0,\ldots,k-1} + k + 1, \ldots, W_{0,\ldots,k} + k - 1\} \\ 0 & \text{otherwise.} \end{cases} \tag{5.36}$$

Finally, the probabilities of being in sensing when $CW = 1$, $BO_s = k$, with $k > 0$, are given by:

$$P\{S1_k^j\} = \begin{cases} P\{S2_k^{j-1}\} \cdot (1 - b_2^{j-1}) & \text{for } j \in \{k+3, \ldots, W_{0,\ldots,k} + k\} \\ 0 & \text{otherwise.} \end{cases} \tag{5.37}$$

## 5.4.3 *Performance Metrics Related to the CAP Portion*

### 5.4.3.1 Transmission Probabilities

As stated before, the aim of our model is to evaluate the probability that a generic node ends its packet transmission in slot $j$, $P\{T^j\}_{\text{CAP}}$, with $j \in \{0, \ldots, T_{\text{CAP}}/d_{\text{bo}} - 1\}$.

A node finishes its transmission in slot $j$, if in slot $j - D - 1$ it starts sensing the channel finding it free for two subsequent slots. The probability that a node starts sensing in slot $j$ is the sum of the probabilities of starting sensing in the $j$th slot and at the $k$th backoff stage, considering all the possible backoff stages. Therefore, we obtain:

$$P\{T^j\}_{\text{CAP}} = \begin{cases} f^{j-D} \cdot \sum_{k=0}^{NB_{\max}} P\{S2_k^{j-D-1}\} & \text{for } j \in \{D+1, \ldots, \hat{t}_{\max} + D - 1\} \\ 0 & \text{otherwise.} \end{cases} \tag{5.38}$$

Because a node transmits a packet occupying $D$ slots, we associate $P\{T^j\}_{\mathrm{CAP}}$ to the slot in which the transmission terminates.

$P\{T^j\}_{\mathrm{CAP}}$ obtained from Eq. 5.38 is used in Eq. 5.25 to derive the statistics in the whole superframe.

#### 5.4.3.2 Reception and Success Probability

To evaluate the other target probabilities, we have to model how the number of nodes that compete for the access to the channel varies with time. We denote as $n_\mathrm{c}^j$ the number of nodes which have not transmitted yet at the end of slot $j-1$ and that will compete for slot $j$. As time goes by, some nodes in the network may access the channel, so that the number of nodes competing for the channel decreases for increasing values of $j$. In particular, as shown in [18], $n_\mathrm{c}^j$ is a r.v., binomially distributed. However, a precise evaluation of the statistics of this variable is complex from the computational viewpoint, since it depends on the statistics of $n_\mathrm{c}^{j-1}$, whose determination would depend on the statistics of $n_\mathrm{c}^{j-2}$ and so on. To reduce such complexity, in [18] different approximations have been introduced and their impact have been compared. Results show that these approximations bring approximatively to the same results, therefore here we use the simplest approximation, according to which we set $n_\mathrm{c}^j = n_\mathrm{c}$, whatever $j$ is. In the case of star topologies, $n_\mathrm{c}$ is the number of nodes using the CAP, therefore $n - N_{\mathrm{GTS}}$. In Sect. 5.4.4, simulations are compared with the mathematical approach. Results show that a very good agreement with simulations is obtained through the model, despite the approximation introduced.

The probability $p_{s_{\mathrm{CAP}}}$ that a generic packet is transmitted successfully on the channel is given by:

$$p_{s_{\mathrm{CAP}}} = \begin{cases} \sum_{j=0}^{\hat{t}_{\max}+D-1} P\{Z^j\}_{\mathrm{CAP}} & \text{if } \hat{t}_{\max} + D - 1 \le T_{\mathrm{CAP}}/d_{\mathrm{bo}} - 1 \\ \sum_{j=0}^{T_{\mathrm{CAP}}/d_{\mathrm{bo}}-1} P\{Z^j\}_{\mathrm{CAP}} & \text{otherwise,} \end{cases} \tag{5.39}$$

where $P\{Z^j\}_{\mathrm{CAP}}$ is the probability that a successful transmission ends in slot $j$, which means that one and only one transmission starts in slot $j-D+1$.

As only one transmission starts in slot $j-D+1$ if only one node, over $n_\mathrm{c}$, senses the channel in slot $j-D$ and if the channel is free in slots $j-D$ and $j-D-1$, $P\{Z^j\}_{\mathrm{cAP}}$ is given by:

$$P\{Z^j\}_{\mathrm{CAP}} = f^{j-D} \sum_{k=0}^{NB_{\max}} P\{S2_k^{j-D-1}\} \cdot \prod_{k=0}^{NB_{\max}} (1 - P\{S2_k^{j-D-1}\})^{n_\mathrm{c}-1}, \tag{5.40}$$

where the second factor gives the probability that one node senses the channel in slot $j-D$, whatever the backoff stage, and the third factor gives the probability that the remaining $n_\mathrm{c}-1$ nodes do not sense the channel in slot $j-D$.

#### 5.4.3.3 Probability of Finding the Channel Busy

The channel is found busy in slot $j$ if a transmission starts in slot $j$, or in slot $j-1$, up to slot $j-D+1$, since each node transmits a packet occupying $D$ slots. Therefore, by denoting as $P\{T_1^j\}$ the probability that at least one transmission starts in slot $j$, the probability of finding the channel busy during the first sensing phase ($CW=2$) is given by:

$$b_2^j = \sum_{v=j-D+1}^{j} P\{T_1^v\}. \tag{5.41}$$

On the other hand, $b_1^j$ is the probability to find the channel busy conditioned to the fact that the channel in slot $j-1$ was free, since a node performs the second sensing phase only if it has found the channel free in the first slot. Therefore, it is the probability that slot $j-2$ is free and that there is at least one node starting sensing in this slot:

$$b_1^j = (1-b_2^{j-2}) \cdot \left[1 - \prod_{k=0}^{NB_{\max}} (1 - P\{S2_k^{j-2}\})^{n_c-1}\right], \tag{5.42}$$

where the second factor (between the brackets), is the probability that at least one node starts sensing in slot $j-2$.

The channel will be jointly free in slots $j$ and $j-1$ if no transmissions start in slot $j$, $j-1$, up to $j-D$, therefore, the probability $f^j$ is given by:

$$f^j = 1 - \sum_{v=j-D}^{j} P\{T_1^v\}. \tag{5.43}$$

Finally, the probability that at least one transmission starts in slot $j$ is given by:

$$P\{T_1^j\} = f^{j-1} \cdot \left[1 - \prod_{k=0}^{NB_{\max}} (1 - P\{S2_k^{j-2}\})^{n_c-1}\right]. \tag{5.44}$$

### 5.4.4 *Numerical Results: The Star Topology*

We consider the same simulator, written in C, used in Sect. 5.3.4 to validate the non beacon-enabled model. The CSMA/CA protocol has been changed to account for the two subsequent sensing phases and also the superframe structure has been implemented. If GTSs are used, the sink randomly selects the $N_{\mathrm{GTS}}$ nodes to which it allocates GTSs, whereas the remaining nodes will use the CAP portion.

According to the application scenario, nodes have only one packet per superframe to transmit and, in case the CAP is used, they start the CSMA/CA algorithm at the same time. We consider $10^4$ transmissions in our simulator, meaning that $10^4$ queries and superframes are simulated.

In the following, we set $d_{\text{ifs}} = 12T_s$ for the case $D = 2$, $d_{\text{ifs}} = 40T_s$ for the cases $D > 2$ and $z_B = 60$ bytes assuming that a payload is present in the beacon packet.

In Figs. 5.22 and 5.23, the cumulative function $F_T(s)$ is shown, as a function of time $s$, for different values of $n$ and with $D = 2$, when no GTSs and seven GTSs are allocated, respectively. Both mathematical (lines) and simulation (symbols) results are reported to validate the model: an excellent agreement between the two cases can be found in all cases. Results are obtained by setting $SO = 1$, therefore $T_q = 1920 \cdot T_s = 30.72$ ms. There is no traffic toward the sink in the first part of the superframe owing to the transmission of the beacon and to the sensing phases. As expected, by increasing $n$, the delay with which a node accesses the channel increases. The curves do not reach the value 1, since some nodes do not succeed in accessing the channel by the end of the superframe. In Fig. 5.23, we can observe the statistics of the traffic in the CFP, characterised by steps in each GTS. Being $d_{\text{ifs}} = 12T_s$, we have $D_{\text{GTS}} = 1$. The reason why the curves in the CAP portion of the superframe are downscaled, with respect to those of Fig. 5.22, is that here the traffic in the CAP is due to the presence of $n - N_{\text{GTS}}$ nodes instead of $n$. Moreover, nodes use the CAP only if they do not have a GTS allocated (i.e., with probability $(n - N_{\text{GTS}})/n$). For these reasons the traffic in the CAP portion decreases.

Figure 5.24 reports $F_Z$, as a function of $s$, for different values of $n$, having set $D = 2$. The behavior of the curves is similar to that of Figs. 5.22 and 5.23; the main difference is that, owing to collisions, some transmitted packets are not correctly received by the sink and, therefore, the curves are down shifted. As one can see, a good agreement between simulation and analytical results is obtained also for this metric.

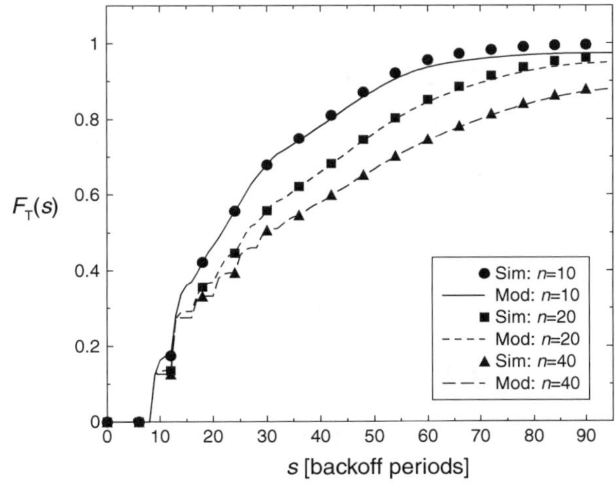


**Fig. 5.22** The cumulative function, $F_T(s)$, when no GTSs are allocated, having fixed $D = 2$

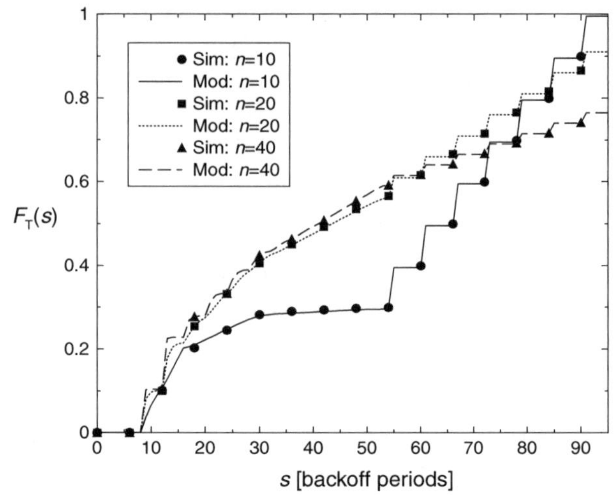


**Fig. 5.23** The cumulative function, $F_T(s)$, when seven GTSs are allocated, having fixed $D = 2$

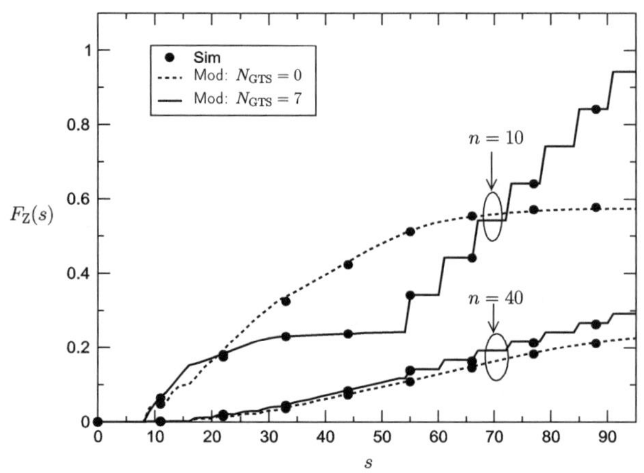


**Fig. 5.24** The cumulative function $F_Z(s)$, as a function of $s$, with and without allocation of GTSs

In Figs. 5.25 and 5.26, $p_s$ is shown, as a function of $n$, for different values of $SO$, setting $D = 5$ and $D = 10$, respectively. The cases of no GTSs and $N_{GTS}$ equal to the maximum number of GTSs allocable are considered. As explained above, this maximum number depends on the values of $D$ and $SO$. As we can see, $p_s$ decreases monotonically (for $n > 1$ when $N_{GTS} = 0$ and for $n > N_{GTS}$ when $N_{GTS} > 0$), by increasing $n$, since the number of nodes competing for the channel increases. As expected, the use of GTSs improves the performance, since less nodes compete for the channel. By increasing $SO$, $p_s$ gets larger, since the CAP duration increases and nodes have more time to access the channel.

We now introduce the concepts of throughput, denoted as $S$, and offered traffic, denoted as $G$. We define the throughput as the number of bytes per unit of time successfully transmitted to the sink, and the offered traffic $G$ as the maximum number of bytes the network was deployed to deliver per unit of time, i.e., the amount of traffic that nodes offer to the sink. More precisely, $G$ can be expressed as

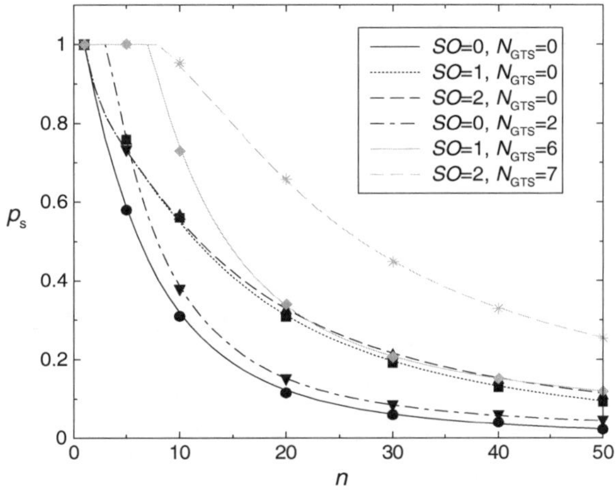


**Fig. 5.25** The success probability, $p_s$, as a function of $n$, having fixed $D = 5$

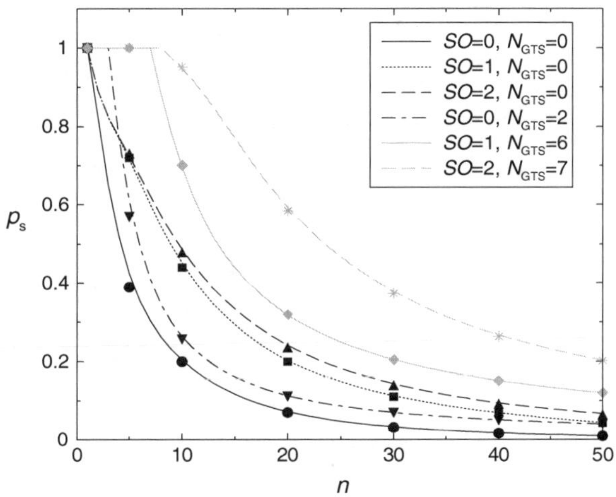


**Fig. 5.26** The success probability, $p_s$, as a function of $n$, having fixed $D = 10$

$$G = \frac{n \cdot z}{T_q} [\text{bytes/s}]. \tag{5.45}$$

The throughput $S$ can be expressed as

$$\begin{aligned} S = p_s \cdot G &= \frac{n \cdot z}{T_q} \left( p_{s_{CAP}} \cdot \frac{n - N_{GTS}}{n} + \frac{N_{GTS}}{n} \right) \\ &= \frac{z}{T_q} [p_{s_{CAP}} \cdot (n - N_{GTS}) + N_{GTS}] [\text{bytes/s}]. \end{aligned} \tag{5.46}$$

Recall that, in this case, $z = D \cdot 10$ bytes and $T_q = BI = SD$.

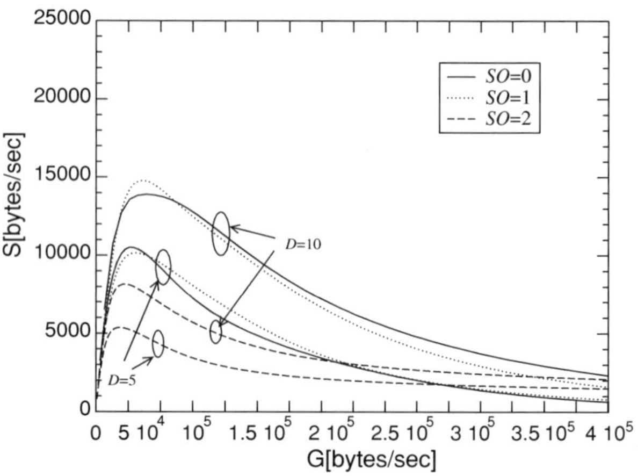


**Fig. 5.27** The throughput $S$, as a function of $G$, when no GTSs are allocated

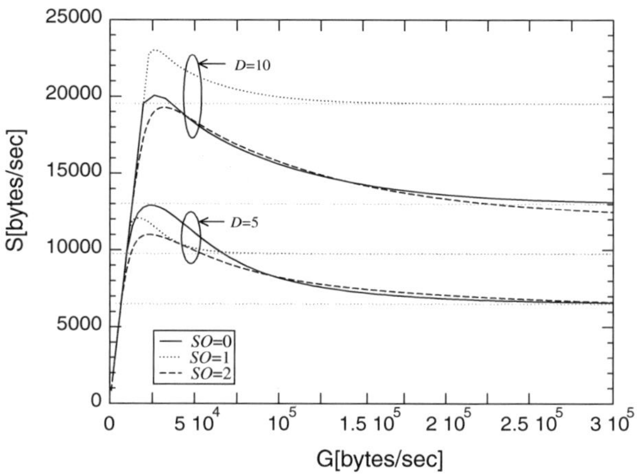


**Fig. 5.28** The throughput $S$, as a function of $G$, when the maximum number of GTSs is allocated

In Figs. 5.27 and 5.28, $S$ is shown, as a function of $G$, when varying $SO$ and $D$, when no GTSs and when the maximum number of GTSs is allocated, respectively. When few nodes are distributed in the network, by increasing $G$, $S$ gets larger. With many nodes an increase of $G$ results in a decrease of $S$, since many nodes are competing for the channel. This means that in star topologies it is not convenient to increase too much $n$ (i.e., the cost of the network), since many packets will be lost and also that when $n$ gets larger star topologies are not suitable (this outcome was, in fact, expected). Moreover, we can note that there exists a value of $SO$ maximising $S$, which depends on $G$ and $D$. As an example, for $D = 5$ when $G$ is small, an increase of $SO$, even though $p_s$ increases, results in a decrement of $S$, since $S$ also depends on $1/T_q$. When, instead, the offered traffic gets larger, collisions increase and larger values of $SO$ are required. On the other hand, when $D = 10$, the optimum value of $SO$ is 1, for low $G$. This is due to the fact that, having large packets, when $SO = 0$ too many packets are lost, owing to the short duration of the superframe.

When no GTSs are allocated (Fig. 5.27) $S$ decreases monotonically since $\lim_{G\to\infty} p_{\mathrm{sCAP}} = 0$. When, instead, GTSs are allocated (Fig. 5.28), there exists an horizontal asymptote, derived as follows:

$$\lim_{G\to\infty} S = \lim_{G\to\infty}\left(\frac{z}{T_{\mathrm{q}}}\cdot p_{\mathrm{sCAP}}\cdot(n - N_{\mathrm{GTS}}) + \frac{z\cdot N_{\mathrm{GTS}}}{T_{\mathrm{q}}}\right) = \frac{z\cdot N_{\mathrm{GTS}}}{T_{\mathrm{q}}}. \tag{5.47}$$

As an example, when $SO = 1$ and $D = 10$, the maximum number of GTSs allocable is $N_{\mathrm{GTS}} = 6$ and the horizontal asymptote is $S = 19531.25$ bytes/s.

### *5.4.5 The Tree-Based Topology*

As stated above, when the number of nodes in the PAN increases, star topologies are not suitable, and peer-to-peer or tree-based topologies should be used [2]. The tree-based topology defined by the Zigbee Alliance [2] for IEEE 802.15.4 networks is considered here.

We consider a $(T + 1)$-level tree-based topology, where the tree is rooted at the sink (namely, at level zero), and level $i$ nodes receive data from level $i + 1$ nodes and forward them to level $i - 1$ nodes, towards the sink (see Fig. 1.16).

As stated in Chap. 1, the tree formation procedure is started by the sink, which broadcasts beacons to nodes. A candidate node receiving the beacon may request to join the network at the sink. If the sink allows the node to join, it will begin transmitting periodic beacons such that other candidate nodes may join the network. In particular, each router in the tree, after the reception of the beacon coming from the parent, will select the instant when transmitting its beacon. Each child node tracks the beacon of its parent and transmits its own beacon at a predefined offset with respect to the beginning of its parent beacon. Obviously, the beacon packets are sent only by the routers in the tree.

We assume that all active parts of the superframes generated by the routers and by the sink have the same duration (i.e., we set a unique value of $SO$). In these conditions, once we set the value of $BO$, the number of routers (including the sink) that will have a portion of superframe available for receiving data from their children will be equal to $2^{BO-SO}$ (see Fig. 5.29). If the number of routers in the network is larger than $2^{BO-SO}$, some routers will not have a portion of superframe available, their children cannot access the channel, and their packets will be lost. We denote as $p_{\mathrm{frame}_i}$ the probability that a level $i$ router has a portion of superframe allocated in which it could receive data from its level $i + 1$ children.

We denote as $p_i$ the probability that a node is at level $i$ of the hierarchy and as $p_{\mathrm{s}}(n_i)$ the success probability for a level $i$ node competing for the channel with the other $n_i - 1$ nodes, connected to the same parent at level $i - 1$. A packet coming from a level $i$ node will be correctly received by the sink, in case it is successfully

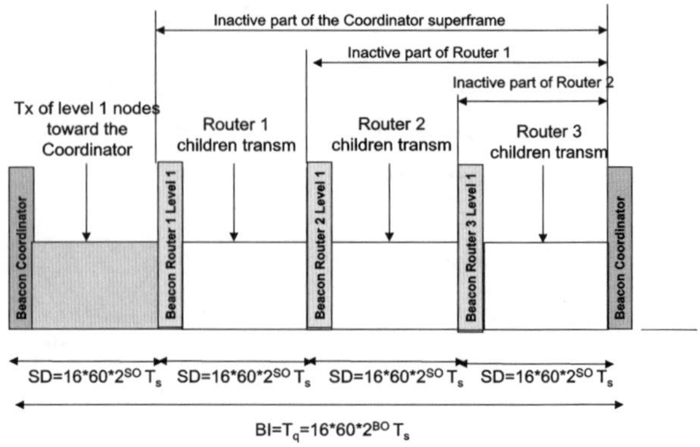


**Fig. 5.29** The superframe structure used in the tree-based topology

transmitted by the level $i$ node from which it is generated, and by all the routers from level $i-1$ till level one, forwarding it toward the sink. The success probability for a node accessing the channel in the tree is therefore:

$$p_{s_{tree}} = \sum_{i=1}^{T-1} p_i \cdot p_{frame_{i-1}} \cdot \prod_{k=1}^{i} \overline{p_{s_k}}. \tag{5.48}$$

where $\overline{p_{s_k}}$ is the average success probability for a node at level $k$, given by:

$$\overline{p_{s_k}} = \sum_{n_k=0}^{N_k} p_s(n_k) \cdot \mathbb{P}\{n_k\} \tag{5.49}$$

where $\mathbb{P}\{n_k\}$ is the probability to have $n_k$ nodes at level $k$ connected to the same parent at level $k-1$, being $N_k$ the total number of nodes at level $k$. According to the channel access strategy defined above, only the children of a given parent compete for the channel and, therefore, the tree could be seen as a series of stars, each having a parent and its children, operating independently (i.e., without collisions). Therefore, $p_s(n_k)$ is given by Eq. 5.13, by simply setting $n = n_k$. On the other hand, $\mathbb{P}\{n_k\}$ depends on the strategy used to realise the tree.

Note that Eq. 5.48 could be used to evaluate the success probability for a node accessing the channel when a $T$-level tree-based topology is established, whatever the strategy used to implement the tree.

Now the success probability $p_{s_{tree}}$ is evaluated in the particular case of a three-level tree ($T = 2$). We denote as $N_i = p_i \cdot n$, the number of level $i$ nodes. We assume that each level one router performs data aggregation: the received packets are aggregated to that generated by the router itself, resulting in a packet of the same size of the single aggregated packets. We also assume that level two nodes select randomly the level one parent and that the active part of the sink superframe is used by level one nodes to transmit toward the sink. The remaining $2^{BO-SO} - 1$ superframe portions are randomly allocated to level one routers for receiving data from their children. Under such assumption $p_{\text{frame}_0} = 1$, whereas there exists a certain probability, denoted as $p_{\text{frame}_1}$, that a level one router has not a portion of the superframe available. This probability can be expressed as

$$p_{\text{frame}_1} = \frac{2^{BO-SO} - 1}{N_R}, \tag{5.50}$$

where $N_R$ is the mean number of level one routers, that is the number of level one nodes with at least one child, given by:

$$N_R = \sum_{i=0}^{N_1} \binom{N_1}{i} (p_{\text{child}})^i \cdot (1 - p_{\text{child}})^{N_1 - i}. \tag{5.51}$$

where $p_{\text{child}} = 1 - (1 - \frac{1}{N_1})^{N_2}$ is the probability that a level one node has at least a child, and $1/N_1$ is the probability that a level two node is connected to a given level one node.

Since level two nodes randomly select level one nodes to which transmit to, the number of level two nodes connected to the same level one node will be binomially distributed:

$$\mathbb{P}\{n_2 = i\} = \binom{N_2}{i} \left(\frac{1}{N_1}\right)^i \cdot \left(1 - \frac{1}{N_1}\right)^{N_2 - i}. \tag{5.52}$$

Therefore, the average success probability for a node being at level two will be:

$$\overline{p_{s_2}} = \sum_{i=0}^{N_2} p_s(i) \cdot \binom{N_2}{i} \left(\frac{1}{N_1}\right)^i \cdot \left(1 - \frac{1}{N_1}\right)^{N_2 - i}, \tag{5.53}$$

where $p_s(i)$ is the success probability given by Eq. 5.13 when $i$ nodes at level two are competing for transmitting to the same level one node. Finally, according with Eq. 5.48, we achieve

$$p_{s_{\text{tree}}} = p_1 \cdot \overline{p_{s_1}} + p_2 \cdot p_{\text{frame}_1} \cdot \overline{p_{s_1}} \cdot \overline{p_{s_2}}, \tag{5.54}$$

where $p_{\text{frame}_1}$ is given by Eq. 5.50, and $\overline{p_{s_1}} = p_s(N_1)$. The average delay, denoted as $D_{\text{mean}_{\text{tree}}}$ in the case of tree, depends on the average delays of the packets coming from level one and level two nodes, denoted as $D_{\text{mean}_1}$ and $D_{\text{mean}_2}$, respectively.

As stated above, level one nodes use the first portion of the superframe defined by the sink to transmit packets (see Fig. 5.29). Since these nodes could be seen as nodes of a star topology transmitting their packets directly to the sink, according to Eq. 5.28, we obtain:

$$D_{\text{mean}_1} = \sum_{s=0}^{SD/d_{\text{bo}}-1} (s+1) \frac{P\{Z^s\}(N_1)}{p_s(N_1)}, \tag{5.55}$$

where $\frac{P\{Z^s\}(N_1)}{p_s(N_1)}$ is the probability that the packet tail is correctly received by the sink in slot $s$, when $N_1$ nodes compete for the channel, given that the packet is correctly received. $SD$ is the duration of the active part of the superframe defined by the sink.

A packet coming from a level two node, instead, must be transmitted toward the level one parent and then from the latter to the sink. Therefore, two superframes are needed, one for each hop. Level one routers, in fact, always transmit to the sink the packets received by their children in the previous superframe. Therefore, the total average delay suffered by a level two node packet will be equal to $T_q = BI$ plus the average delay of its parent (i.e., the average delay of a level one node packet): $D_{\text{mean}_2} = D_{\text{mean}_1} + T_q$. Note that $D_{\text{mean}_2}$ does not depend on the instant in which the parent receives the packet within its superframe.

Finally the average delay suffered by a packet coming from whatever a node in a tree is given by

$$D_{\text{mean}_{\text{tree}}} = p_1 \cdot D_{\text{mean}_1} + p_2 \cdot D_{\text{mean}_2} \tag{5.56}$$

### 5.4.6 Numerical Results: Tree-Based Topology

Numerical results obtained in the three-level tree are now discussed and compared with results obtained in the star topology case. Since $p_{s_{\text{tree}}}$, $D_{\text{mean}_{\text{tree}}}$, and $P\{Z^s\}$ depend on $p_s$ obtained in the star topology case, that have been validated in Sect. 5.4.4, simulation results are not reported here. In Fig. 5.30, $p_{s_{\text{tree}}}$ is shown as a function of $N_1$, for different values of $n$, $D$ and $SO$, having set $BO = 5$. There exists an optimum value of $N_1$ maximising $p_{s_{\text{tree}}}$, and this value obviously increases by increasing $n$ and is approximatively equal to $\frac{-1+\sqrt{1+4n}}{2}$. Therefore, it is independent of $D$ and $SO$. This means that, once $n$ is fixed, there exists an optimum split between level 1 and level 2 nodes, maximising the success probability.

In Fig. 5.31, results related to the two topologies, showing the success probability, as a function of $n$, for different values of $SO$ and $BO$ by setting $D = 5$, are compared. For a fair comparison, the success probability is computed by fixing the same value of $T_B$, i.e., by giving to nodes the same time to access the channel. To this end, we set $SO = BO$ for the star topology, and we compare the case "star" with $SO = BO = 1$ with the case "tree" with $BO = 1$ and $SO = 0$. Whereas the case "star" with $SO = BO > 1$ (note that the cases $SO = BO = 2, 3$, etc. bring to the same $p_s$) are compared with the cases "tree" with $BO > 1$, whatever $SO$ is. In the tree case $N_1$ is set to the optimum value maximising $p_{s_{\text{tree}}}$ obtained from Fig. 5.30. As we can see, when $BO = 1$, the star is preferable, since in the tree only one router has a part of the superframe allocated, therefore, many packets of level 2 nodes will be lost. For $BO > 1$, instead, the tree outperforms the star. The difference between the star and the tree, obviously, increases by increasing $BO$ and $SO$, resulting in an increase of $p_{\text{frame}}$ and $p_s$, respectively.

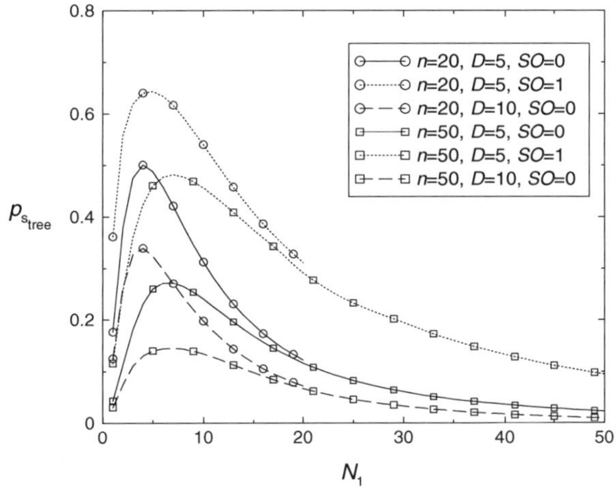


**Fig. 5.30** The success probability, $p_{s_{tree}}$, as a function of $N_1$, for a tree-based topology

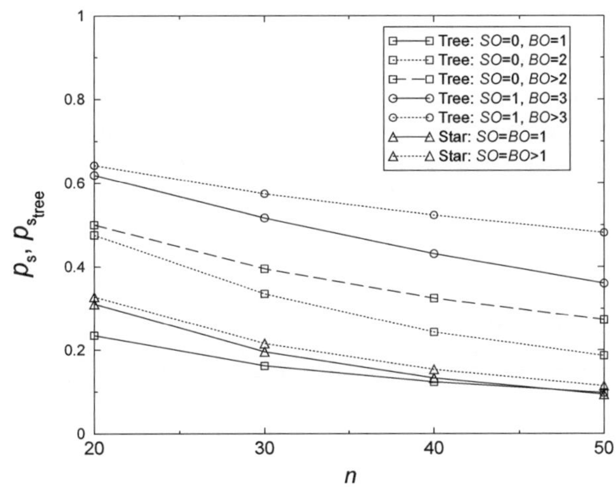


**Fig. 5.31** The success probability, as a function of $n$, when a star and tree topologies are used

In Fig. 5.32, the average delays obtained in case of star and tree-based topologies as a function of $n$ are shown. The curves are obtained by setting $D = 5$ and $N_1 = 3$ in the case of trees. The delays increase by increasing $n$, since the probability to find the channel busy and to delay the transmission increases. An horizontal asymptote is also present due to the maximum delay a packet may incur, equal to the superframe duration $SD$ in the "star" case and to $SD + BI$ in the "tree" case. As expected, the delays are larger for trees, since packets coming from level two nodes need two superframes to reach the sink. Also note that by increasing $BO$ delays get significantly larger. The curves "tree" with $SO = 0$, $BO = 3$ and "tree" with $SO = 1$ and $BO = 3$ are overlapped since $BI$ assumes the same value and the delays of level one nodes are approximatively the same (in fact also the curves "star" with $SO = BO = 0$ and $SO = BO = 1$ are approximatively the same).

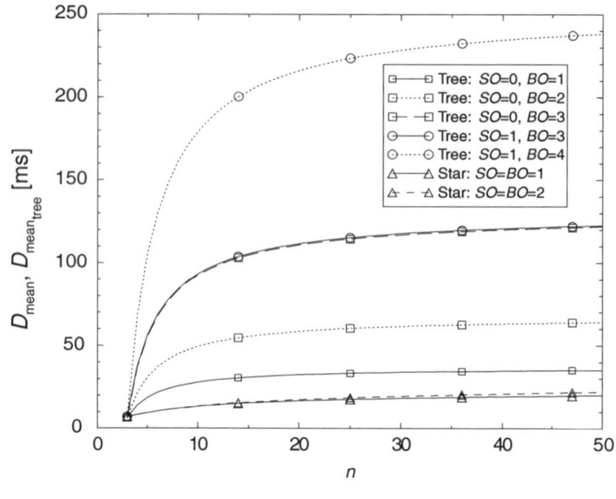


**Fig. 5.32** The average delay, as a function of $n$, when star and tree topologies are used

By comparing Figs. 5.31 and 5.32, we can, finally, deduce that the choice of the topology depends on the application requirements. If the application requires a high success probability and can tolerate large delays, trees are preferable; if, instead, more stringent constraints in terms of delays are imposed, star topologies are better. However, as widely shown in Chap. 4, trees allow to realise larger networks distributed in wide areas, whereas the number of nodes that may reach directly the sink is limited by connectivity problems.

## 5.5 Comparison Between the Beacon- and Non Beacon-Enabled Modes

The comparison is performed in terms of success probability, $p_s$, and throughput, $S$ (defined in Sect. 5.4.4), considering the star topology (since trees may not be formed in the non beacon-enabled case [2]). Have in mind that $T_q$ is given by Eq. 1.1 for the beacon one, and can be set whatever a value in the non beacon-enabled case [18].

Figure 5.33 compares the values of $p_s$ obtained in the two modes, when no GTS are allocated. Results are obtained through the mathematical models. Here $T_q = 61.44$ ms, which corresponds to $SO = 2$ in the beacon-enabled case and is larger than the maximum possible delay in the non beacon-enabled mode. This means that the largest values of $p_s$ that could be obtained in both the modalities are considered. A logarithmic scale is used to better visualise the differences between the curves. It can be seen that there are no relevant differences between the two modalities, when no GTSs are allocated. When instead GTSs are used, relevant differences are present.

If we compare Figs. 5.25 and 5.26, related to the beacon-enabled mode (with $D = 5$ and 10, respectively) with results of the non beacon-enabled mode shown in

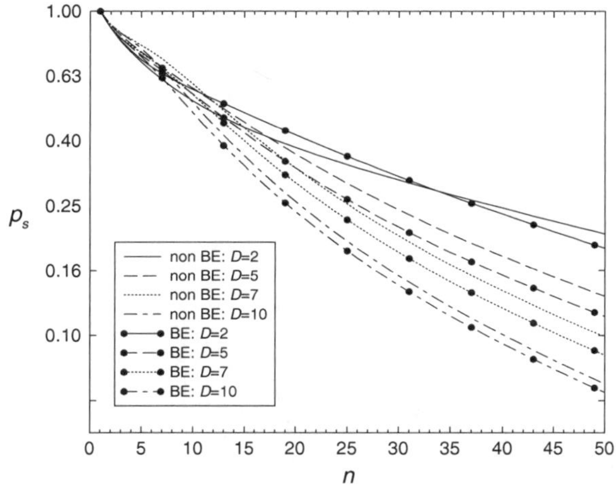


**Fig. 5.33** The success probability, $p_s$, for the beacon- and non beacon-enabled modes, as a function of $n$

Fig. 5.17, we can note that in both cases, $D = 5$ and 10, once we fix the round or superframe duration, results are approximatively the same if no GTSs are allocated, whereas, there is a notable increasing of $p_s$ in the beacon-enabled case when GTSs are allocated. Note that the cases $T_q = 15.36$ ms, $T_q = 30.72$ ms and $T_q = 61.44$ ms correspond to $SO = 0$, 1 and 2, respectively.

In Figs. 5.34 and 5.35, $S$ as a function of $G$, when varying $T_q$, for $D = 2$ and $D = 10$, respectively, are shown. Both beacon- and non beacon-enabled modes are considered. In both figures, once $G$ is fixed there exists a value of $SO$ (i.e., $T_q$) maximising $S$. For $D = 2$ (Fig. 5.34) when $G$ is low, an increase of $T_q$ results in a decrement of $S$ since, even though $p_s$ gets greater (since nodes have more time to transmit their packets), the query interval is longer and, therefore, the number of bytes per second received by the sink decreases. When, instead, the offered traffic gets larger, collisions increase and larger values of $T_q$ are required, to increase $S$. On the other hand, when $D = 10$ in the beacon-enabled case, the optimum value of $SO$ is 1, for low $G$. This is due to the fact that, having large packets, when $T_q = 15.36$ ms too many packets are lost, owing to the short duration of the superframe. If the curves obtained in the two modalities are compared, we see that (especially in the case $D = 2$) the non beacon-enabled mode outperforms the beacon-enabled mode. The differences in terms of $p_s$, in fact, are few when $T_q = 61.44$ ms, but increase when $T_q$ gets lower, since, on average, the delays in the beacon-enabled case are larger [18]. This fact results in notably larger values of $S$ in the non beacon-enabled case, for low $T_q$.

Finally, for comparison of Figs. 5.34 and 5.35, we note that, once $G$ is fixed, $S$ gets notably larger when $D$ increases, since more bytes/s are correctly transmitted toward the sink.

In Figs. 5.34 and 5.35 simulation results are not reported, since $S$ depends on $p_s$, validated in many figures of this chapter.

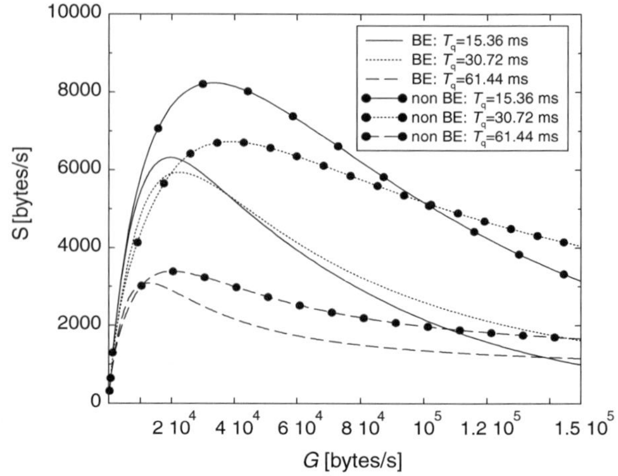


**Fig. 5.34** The throughput, $S$, as a function of $G$, for $D = 2$, for the beacon- and non beacon-enabled modes

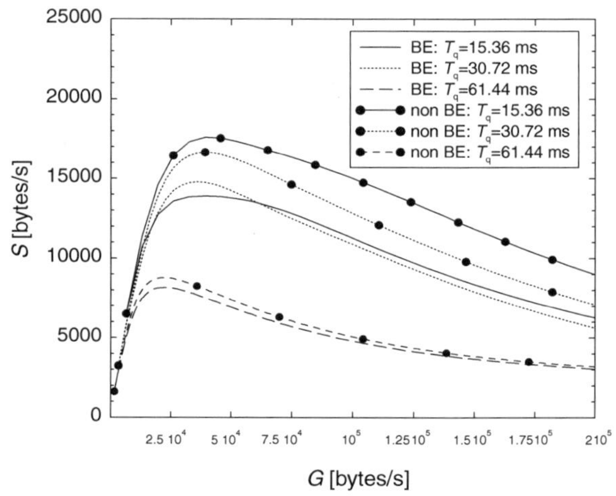


**Fig. 5.35** The throughput, $S$, as a function of $G$, for $D = 10$, for the beacon- and non beacon-enabled modes

## 5.6 Concluding Remarks

A novel analytical model for the IEEE 802.15.4 MAC protocol, considering both non beacon- and beacon-enabled PAN, where nodes are organised in a star topology (or possibly, a tree, in the beacon-enabled case), is provided. The model does not suffer from the limitations shown by related works in the literature; however, more importantly, it also introduces a very new challenge in the modelling of CSMA-based MAC protocols for WSN. This challenge regards the application scenario considered here: the sink periodically triggers nodes and waits for replies. This implies that each node has one and only one packet to be transmitted at each query received, and also that the number of nodes competing for the channel decreases by passing time. Therefore, this scenario imposes the use of a new approach in modeling the MAC protocol, different from that developed by Bianchi [9] and followed by almost all the successive literature from 2000 till now. As stated above, in fact, in [9] the network is studied in saturated conditions, or, anyway, in conditions in which the statistics of the traffic generated by node is defined a priori. The other relevant issue of this model is that it allows the evaluation of the statistical distribution of the traffic generated by nodes toward the sink, never investigated before analytically.

## 5.7 Further Readings

In the literature, there are several works devoted to the study of IEEE 802.15.4 networks. Performance evaluation of the IEEE 802.15.4 MAC protocol has been carried out by means of simulation for small and low-load networks in [11] and for dense networks in [12]. In [13], the performance of the beacon-enabled mode is evaluated through OPNET for different network settings to understand the impact of the protocol attributes on network performance. [19] presents simulation results for the MAC protocol and addresses the coexistence between IEEE 802.11 and IEEE 802.15.4, through experimental measurements. Also some studies have tried to describe analytically the behavior of the IEEE 802.15.4 MAC protocol. Few works devoted their attention to non beacon-enabled mode [4]; most of the analytical models are related to beacon-enabled networks [5, 6, 8, 14, 15]. In [4], the authors try to model the unslotted CSMA/CA protocol for non beacon-enabled networks, but they do not capture correctly the protocol, because they include in the Markov chain two subsequent sensing phases, and not one, as fixed in the standard (see Sect. 5.1). The model described in [5] fails to match simulation results, as described in [8], as the authors use the same Markov formulation and assumptions made by Bianchi in [9], where the 802.11 MAC protocol is considered. This protocol, in fact, is significantly different from the one defined by the IEEE 802.15.4 standard [18]. A better model is proposed in [6] where, however, the probabilities of being in sensing in the two subsequent slots are not correctly captured by the Markov chain (see also [8]). In [7], the events of finding the channel free in the first and in the second slots are considered independent and the probabilities that those two events happen are assumed to be equal. Another model is presented in [20] where, however, different simplifying assumptions are made: the uniform distribution of the backoff counter within the backoff windows are assumed to be geometrically distributed, so that the backoff algorithm becomes memoryless. In [21], the analysis made in [20] has been extended from the one-hop case to the multi-hop case, by using the same assumption made in [20]. Finally, regarding the use of GTSs, there are some papers in the literature studying performance achieved when GTSs are used [13] or GTSs allocation strategies [22], but these studies have been carried out through simulations.

## References


1. A. Koubaa, M. Alves, E. Tovar, Modeling and worst-case dimensioning of cluster-tree wireless sensor networks, in *Proceedings of IEEE International Real-Time Systems Symposium (RTSS)*, Rio de Janeiro, Brazil, December 2006, pp. 412–421 (2006)
2. Zigbee Alliance, http://www.zigbee.org
3. C. Buratti, R. Verdone, A hybrid hierarchical architecture: from a wireless sensor network to the fixed infrastructure, in *Proceedings of IEEE European Wireless (EW)*, Prague, Czech Republic, June 2008, pp. 1–7 (2008)
4. T.O. Kim, H. Kim, J. Lee, J.S. Park, B.D. Choi, Performance analysis of the IEEE 802.15.4 with non beacon-enabled CSMA/CA in non-saturated condition, in *Proceedings of International Conference on Embedded and Ubiquitous Computing (EUC)*, Seoul, Korea, August 2006, pp. 884–893 (2006)
5. J. Misic, S. Shafi, V.B. Misic, Maintaining reliability through activity management in an 802.15.4 sensor cluster. IEEE Trans. Veh. Technol. **55**(3), 779–788 (2006)
6. T. Park, T. Kim, J. Choi, S. Choi, W. Kwon, Throughput and energy consumption analysis of IEEE 802.15.4 slotted CSMA/CA. IEEE Electron. Lett. **41**(18), 1017–1019 (2005)
7. Z. Chen, C. Lin, H. Wen, H. Yin, An analytical model for evaluating IEEE 802.15.4 CSMA/CA protocol in low rate wireless application, in *IEEE International Conference on Advanced Information Networking and Applications Workshop (AINAW)*, vol. 2. Niagara Falls, Ontario, Canada, May 2007, pp. 899–904 (2007)
8. S. Pollin, M. Ergen, S. Ergen, B. Bougard, L.D. Perre, I. Moerman, A. Bahai, P. Varaiya, F. Catthoor, Performance analysis of slotted carrier sense IEEE 802.15.4 medium access layer. IEEE Trans. Wirel. Commun. **7**(9), 3359–3371 (2008)
9. G. Bianchi, Performance analysis of the IEEE 802.11 distributed coordination function. IEEE J. Select. Areas Commun. **18**(3), 535–547 (2000)
10. IEEE 802.15.4 Std, Wireless Medium Access Control (MAC) and Physical Layer (PHY) Specifications for Low-Rate Wireless Personal Area Networks (LR-WPANs) (IEEE Computer Society Press, 2003), pp. 1–679, ISBN:0-7381-3677-5
11. G. Lu, B. Krishnamachari, C.S. Raghavendra, Performance evaluation of the IEEE 802.15.4 MAC for low-rate low-power wireless networks, in *Workshop on Energy-Efficient Wireless Communications and Networks (EWCN)*, Phoenix, AZ, USA, April 2004, pp. 701–706 (2004)
12. B. Bougard, F. Catthoor, D.C. Daly, A. Chandrakasan, W. Dehaene, Energy efficiency of the IEEE 802.15.4 standard in dense wireless microsensor networks: Modeling and improvement perspectives, in *Proceedings of Design, Automation and Test in Europe (DATE)*, vol. 1, Messe Munich, Germany, March 2005, pp. 196–201 (2005)
13. A. Koubaa, M. Alves, E. Tovar, A comprehensive simulation study of slotted CSMA/CA for IEEE 802.15.4 wireless sensor networks, in *IEEE International Workshop on Factory Communication Systems (WFCS)*, Turin, Italy, June 2006, pp. 183–192 (2006)
14. J. Misic, V.B. Misic, S. Shafi,. Performance of IEEE 802.15.4 beacon-enabled PAN with uplink transmissions in non-saturation mode—access delay for finite buffers, in *Proceedings of the International Conference on Broadband Networks (BroadNets)*, San Jose, CA, USA, October 2004, pp. 416–425 (2004)
15. J. Misic, S. Shafi, V.B. Misic, The impact of MAC parameters on the performance of 802.15.4 PAN. Elsevier Ad hoc Netw. J. **3**(5), 509–528 (2005)
16. L. Kleinrock, *Queueing Systems* (Wiley, New York, 1975)
17. Freescale, *Freescale Semiconductor's MC13192 Developer's Kit* http://www.freescale.com/webapp/sps/site/prod_summary.jsp?code=MC13192
18. C. Buratti, R. Verdone, Performance analysis of IEEE 802.15.4 non-beacon enabled mode. IEEE Trans. Veh. Technol. **58**(7), 3480–3493 (2009)
19. M. Petrova, J. Riihijarvi, P. Mahonen, S. Labella, Performance study of IEEE 802.15.4 using measurements and simulations, in *Proceedings of IEEE Wireless Communications and Networking Conference (WCNC)*, Las Vegas, NV, USA, April 2006, pp. 487–492 (2006)

20. I. Ramachandran, A.K. Das, S. Roy, Analysis of the contention access period of IEEE 802.15.4 MAC. ACM Trans. Sens. Netw. **3**(1), 29 (2007)
21. M. Martalò, S. Busanelli, G. Ferrari, Markov chain-based performance analysis of multihop IEEE 802.15.4 wireless networks. Elsevier Perform. Eval. **66**(12), 722–741 (2009)
22. P. Jurcik, A. Koubaa, M. Alves, E. Tovar, Z. Hanzalek, A simulation model for the IEEE 802.15.4 protocol: delay/throughput evaluation of the GTS mechanism, in *IEEE International Symposium on Modeling, Analysis, and Simulation of Computer and Telecommunication Systems (MACOTS)*, Instanbul, Turkey, October 2007, pp. 109–116 (2007)

# Chapter 6
# Area Throughput for Multi-Sink Wireless Sensor Networks

In this chapter, the models for the evaluation of connectivity properties in multi-sink Wireless Sensor Networks (WSNs) and of the IEEE 802.15.4 Medium Access Control (MAC) protocol, described in Chaps. 4 and 5 respectively, have been integrated in a unique framework. The concept of *area throughput*, that is the amount of data per second successfully transmitted to the sinks from a given area, is introduced. This performance metric is strictly related to both connectivity and MAC issues: it depends, in fact, on the probability that a given sensor node is not isolated and that it succeeds in transmitting its packet (i.e., without collisions).

In our scenario we assume that sensors and sinks are distributed according to a Poisson Point Process (PPP) in bounded or unbounded regions and we develop a mathematical tool to derive the area throughput. Note that even if the model is thought for Carrier-Sense Multiple Access (CSMA)-based MAC protocols, with particular attention toward the IEEE 802.15.4 protocol, it could be easily applied to any MAC protocol. The link model considered is that described in Chap. 4, taking into consideration random channel fluctuations (see Eq. 3.17).

This chapter is structured as follows. In Sect. 6.1, the motivations behind this chapter are presented, whereas Sect. 6.2 illustrates the reference scenario. In Sect. 6.3, the analytical framework for the derivation of the area throughput is presented. Corresponding numerical results are shown in Sect. 6.4. Finally, concluding remarks are given in Sect. 6.5.

## 6.1 Aims of the Model

In this chapter we consider a multi-sink WSN, collecting data from the environment through the sampling of some physical entities and sending them to an external user.

C. Buratti et al., *Sensor Networks with IEEE 802.15.4 Systems*, Signals and Communication Technology, DOI: 10.1007/978-3-642-17490-2_6,

Being the acquisition of samples from the target area the main issue for our application scenario, a new metric for studying the behavior of the WSN, namely the *area throughput*, denoting the amount of data per unit of time successfully transmitted to the centralised unit originating from the target area, is defined.

As expected, the area throughput is larger if the density of sensor nodes is larger, but, on the other hand, if a contention-based MAC protocol is used, the density of nodes significantly affects the ability of the protocol to avoid packet collisions (i.e., simultaneous transmissions from separate sensors toward the same sink). If, in fact, the number of sensor nodes per cluster is very large, collisions and backoff procedures can make data transmission impossible under time-constrained conditions, and the samples taken from sensors do not reach the sinks and, consequently, the centralised unit. Therefore, the optimisation of the area throughput requires proper dimensioning of the density of sensors, in a framework model where both MAC and connectivity issues are considered.

Even if the model described above could be applied to any MAC protocol, we particularly refer to CSMA-based protocols, and in particular to the IEEE 802.15.4 air interface, being the reference interface of this book. In this case, sinks will act as Personal Area Network (PAN) coordinators, periodically transmitting queries to sensors and waiting for replies. According to the standard, the different PAN coordinators, and therefore the PANs, use different frequency channels (see the scan functionality performed by the PAN coordinator for establishing a PAN, described in Chap. 1). Therefore, no collisions may occur between nodes belonging to different PANs; however, nodes belonging to the same cluster, will compete to try to transmit their packets to the sink. An infinite area where sensors and sinks are uniformly distributed at random, is considered. Then, a specific portion of space, of finite size and given shape (without loss of generality, we consider a square), is considered as target area (see Fig. 6.1), where sensors and sinks are distributed according to the two distributions accounted for.

The frequency of the queries transmitted by the sinks is denoted as $f_q = 1/T_q$. Each sensor takes, upon reception of a query, one sample of a given phenomenon

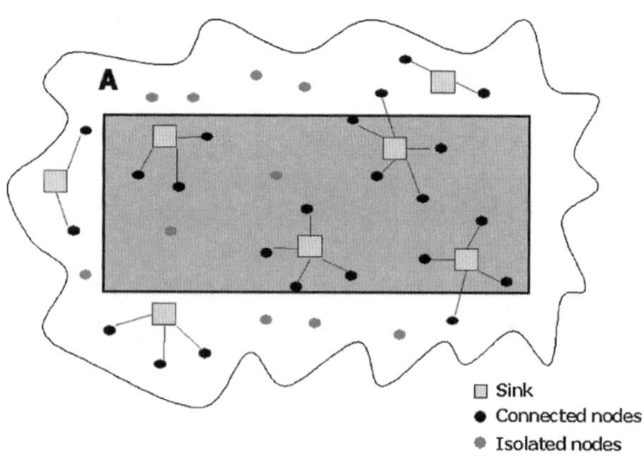


**Fig. 6.1** The reference scenario considered

and forwards it through a direct link to the sink. Once transmission is performed, it switches to an idle state until the next query is received. We denote the interval between two successive queries as *round*.

The amount of data available from the sensors deployed in the area per unit of time, is denoted as *offered load*. The basic objective of this chapter is thus to determine how the area throughput depends on the offered load for different scenarios and system parameters.

In general terms, it might be said that the aim is to define a picture showing how throughput varies with load, as done for many years in the literature for different types of MAC protocols. However, here connectivity and the plurality of sinks are accounted for.

As stated above, the increasing of sensors density in the area, aiming at increasing the quantity of samples taken from the area (i.e., improving the estimation of the process), also causes many data losses, due to MAC failures. One solution can be found in the decimation of the sensor nodes to respond. Other improvements might be introduced by letting the sensor nodes apply a form of aggregation procedure, responding only sporadically to queries, with a single data packet composed of all samples taken since the previous transmission: fewer access attempts are performed, but with longer packets. Such decimation process, or the aggregation strategy, must be driven by an optimisation procedure that, by taking into account the density of sensor nodes and sinks, the frequency of queries, and the randomness of node locations, the radio channel behaviour, and CSMA mechanisms, determines the optimum number of nodes that should respond to any query, and whether aggregating samples provides advantages.

This chapter first addresses such optimisation problem, by showing the behaviour of the area throughput for different aggregation strategies and considering the IEEE 802.15.4 MAC protocol in the non beacon-enabled mode. For the sake of completeness, also an example of results obtained by applying to the framework a very simple CSMA-based MAC protocol is shown. Finally, performance obtained with the beacon-enabled and the non beacon-enabled modes of IEEE 802.15.4, are compared.

## 6.2 Assumptions and Reference Scenario

The reference scenario considered consists of an area of finite size and given shape, where sensors and sinks are both distributed according to an homogeneous PPP. The sensors and sinks densities, are denotes as $\rho_s$ and $\rho_0$, respectively, whereas $A$ is the area of the target domain. Denoting by $k$ the number of sensor nodes in $A$, $k$ is Poisson distributed with mean $\bar{k} = \rho_s \cdot A$ and probability mass function (p.m.f.)

$$g_k = \frac{\bar{k}^k e^{-\bar{k}}}{k!}. \tag{6.1}$$

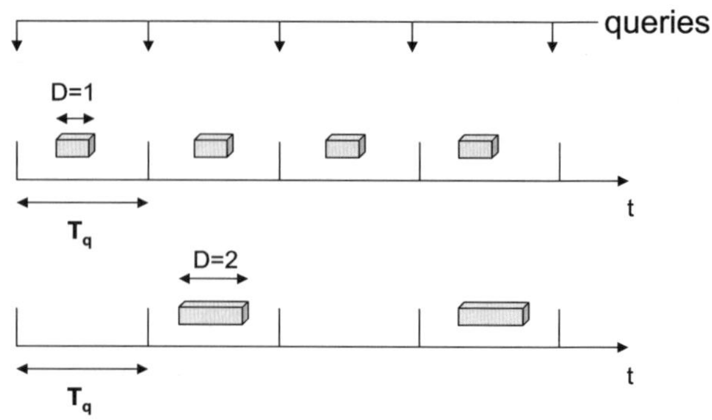


**Fig. 6.2** The aggregation strategy

The average number of sinks in $A$ is denoted as $I = \rho_0 \cdot A$.

Sinks periodically send queries to sensors and wait for replies. In case a sensor node receives a query from more than one sink, it selects the one providing the largest received power and responds to it. It is assumed that sensors may perform some data aggregation before transmitting their packets. For instance, they perform sampling from the environment upon each query, but transmit data only when a given number of samples have been collected. By doing so, transmissions do not occur at each query.

The time needed to transmit a unit of data, that is one sample, is denoted as $T$, whereas $T_{\mathrm{D}}$ is the time needed to transmit a packet. The frequency of the queries transmitted by the sinks is denoted as $f_{\mathrm{q}} = 1/T_{\mathrm{q}}$. $T_{\mathrm{q}}$ is the time interval between two consecutive queries. It is assumed that sensors transmit packets composed of $D$ samples every $D$ queries. At each query sensors take one sample and when $D$ samples are taken, data is aggregated and transmitted. We assume that the aggregation process generates a packet whose transmission requires a time $T_{\mathrm{D}} = D \cdot T$, when $D$ units of data are aggregated. In Fig. 6.2, the aggregation strategies in the cases $D = 1, 2$ are shown as examples.

## 6.3 Evaluation of the Area Throughput

The area throughput is mathematically derived through an intermediate step: first the probability of successful data transmission by an arbitrary sensor node, when $k$ nodes are present in the monitored area, is considered. Then, the overall area throughput is evaluated based on this result.

### *6.3.1 Joint MAC/Connectivity Probability of Success*

Let us consider an arbitrary sensor node that is located in the observed area $A$ at a certain time instant. The aim is computing the probability that it can connect to one of the sinks deployed in $A$ and successfully transmit its data sample to the infrastructure. Such an event is clearly related to connectivity issues (i.e., the sensor must employ an adequate transmitting power in order to reach the sink and not be isolated) and to MAC problems (i.e., the number of sensors which attempt at connecting to the same sink strongly affects the probability of successful transmission). For this reason, we define $P_{s|k}(x, y)$ as the probability of successful transmission conditioned on the overall number, $k$, of sensors present in the monitored area, which also depends on the position $(x, y)$ of the sensor relative to a reference system with origin centered in $A$. This dependence is due to the well-known border effects in connectivity [1].

In particular,

$$\begin{aligned} P_{s|k}(x, y) &= \mathbb{E}_n[p_{\mathrm{MAC}}(n) \cdot p_{\mathrm{CON}}(x, y)] \\ &= \mathbb{E}_n[p_{\mathrm{MAC}}(n)] \cdot p_{\mathrm{CON}}(x, y) \end{aligned} \tag{6.2}$$

where $\mathbb{E}[\cdot]$ is the expectation operator and the impact of connectivity and MAC on the transmission of samples are separated. A packet will be successfully received by a sink if the sensor node is connected to at least one sink and if no MAC failures occur. The two terms that appear in (6.2) are now analysed.

$p_{\mathrm{CON}}(x, y)$ represents the probability that the sensor is not isolated (i.e., it receives a sufficiently strong signal from at least one sink). This probability decreases as the sensor approaches the borders (border effects). $p_{\mathrm{CON}}$ for multi-sink single-hop WSNs, in bounded and unbounded regions, has been computed in Chap. 4. In particular, for unbounded regions, $p_{\mathrm{CON}}(x, y) \simeq p_{\mathrm{CON}}$, that is equal to $q_\infty$, given by Eq. 4.21. Whereas, when bounded regions are considered, $p_{\mathrm{CON}}(x, y)$ is equal to $q(x, y)$ given by Eq. 4.26.

Specifically, since the position of the sensor is in general unknown, $P_{s|k}(x, y)$ of (6.2) can be deconditioned as follows:

$$\begin{aligned} P_{s|k} &= \mathbb{E}_{x,y}[P_{s|k}(x, y)] \\ &= \mathbb{E}_{x,y}[p_{\mathrm{CON}}(x, y)] \cdot \mathbb{E}_n[p_{\mathrm{MAC}}(n)]. \end{aligned} \tag{6.3}$$

$\mathbb{E}_{x,y}[p_{\mathrm{CON}}(x, y)]$ is equal to $\overline{q}$ given by Eq. 4.27 in Chap. 4, when bounded regions are accounted for. When, instead border effects are negligible, $\mathbb{E}_{x,y}[p_{\mathrm{CON}}(x, y)] = \mathbb{E}_{x,y}[p_{\mathrm{CON}}] = p_{\mathrm{CON}}$, given by Eq. 4.21.

Given the channel model described in Chap. 4 (see Eq. 4.1), the average connectivity area of the sensor, that is the average area in which the sinks audible to the given sensor are contained, can be defined as

$$A_{\sigma_s} = \pi e^{\frac{2(L_{\mathrm{th}} - k_0)}{k1}} e^{\frac{2\sigma_s^2}{k_1^2}}. \tag{6.4}$$

In [2], it is also shown that border effects are negligible when $A_{\sigma_s} < 0.1A$. In the following, only this case will be accounted for. In this case:

$$p_{\mathrm{CON}}(x, y) \simeq p_{\mathrm{CON}} = 1 - e^{-\mu_0}, \tag{6.5}$$

where $\mu_0 = \rho_0 A_{\sigma_s} = IA_{\sigma_s}/A$ is the mean number of audible sinks on an infinite plane from any position [3].

$p_{\text{MAC}}(n)$, $n \geq 1$, is the probability of successful transmission when $n-1$ interfering sensors are present. It accounts for MAC issues. When the IEEE 802.15.4 MAC protocols are considered, the models to derive $p_{\text{MAC}}(n)$ are given in Chap. 5. In particular, $p_{\text{MAC}}(n)$ is the success probability, $p_{\text{s}}$, derived in Chap. 5, when $n$ nodes are competing for the channel. In particular, $p_{\text{s}}$ is given by Eq. 5.13, for the non beacon-enabled case and by Eq. 5.27, for the beacon-enabled case.

In general, when CSMA-based MAC protocols are considered, $p_{\text{MAC}}(n)$ is a monotonic decreasing function of the number, $n$, of sensors which attempt to connect to the same serving sink. This number is in general a random variable in the range $\{0, \ldots, k\}$. In fact, note that in (6.2) there is no explicit dependence on $k$, except for the fact that $n \leq k$ must hold. Moreover in our case we assume $1 \leq n \leq k$, as there is at least one sensor competing for access with probability $p_{\text{CON}}$ (6.5).

In [4], Orriss et al. showed that the number of sensors uniformly distributed on an infinite plane that hear one particular sink as the one with the strongest signal power (i.e., the number of sensors competing for access to such sink), is Poisson distributed with mean

$$\bar{n} = \mu_{\text{s}} \frac{1 - e^{-\mu_0}}{\mu_0}, \tag{6.6}$$

with $\mu_{\text{s}} = \rho_{\text{s}} A_{\sigma_s}$ being the mean number of sensors that are audible by a given sink. Such a result is relevant toward our goal even though it was derived on the infinite plane. In fact, when border effects are negligible (i.e., $A_{\sigma_s} < 0.1A$) and $k$ is large, $n$ can still be considered Poisson distributed. The only two things that change are:

- $n$ is upper bounded by $k$ (i.e., the p.m.f. is truncated);
- the density $\rho_{\text{s}}$ is to be computed as the ratio $k/A$ [$\text{m}^{-2}$], thus yielding $\mu_{\text{s}} = k\frac{A_{\sigma_s}}{A}$.

Therefore, we assume $n \sim \text{Poisson}(\bar{n})$, with

$$\bar{n} = \bar{n}(k) = k\frac{A_{\sigma_s}}{A}\frac{1 - e^{-\mu_{\text{sink}}}}{\mu_{\text{sink}}} = k\frac{1 - e^{-IA_{\sigma_s}/A}}{I}. \tag{6.7}$$

Finally, by making the average in (6.3) explicit and neglecting border effects (see (6.5)), we get

$$P_{\text{s}|k} = (1 - e^{-IA_{\sigma_s}/A}) \cdot \frac{1}{M}\sum_{n=1}^{k} p_{\text{MAC}}(n)\frac{\bar{n}^n e^{-\bar{n}}}{n!}, \tag{6.8}$$

where

$$M = \sum_{n=1}^{k} \frac{\bar{n}^n e^{-\bar{n}}}{n!} \tag{6.9}$$

is a normalizing factor.

### 6.3.2 Area Throughput

The area throughput has been defined as the amount of data, successfully transmitted toward the sinks, per unit of time. The data here is identify with the sample, being a sample the unit of data transmitted (packet when $D = 1$); therefore, the metric will be expressed in [samples/s].

According to the aggregation strategy described in the previous subsection, the amount of samples generated by the network as response to a given query is equal to the number of sensors, $k$, that are present and active when the query is received. As a consequence, the average number of data samples-per-query generated by the network is the mean number of sensors, $\bar{k}$, in the observed area.

Now, denote by $G$ the average number of data samples generated per unit of time, given by

$$G = \bar{k} \cdot f_{\mathrm{q}} = \rho_{\mathrm{s}} \cdot A \cdot \frac{1}{T_{\mathrm{q}}}[\text{samples/s}]. \tag{6.10}$$

From (6.10) we have $\bar{k} = GT_{\mathrm{q}}$.

The average amount of data received by the infrastructure per unit of time (area throughput), $S$, is given by:

$$S = \sum_{k=0}^{+\infty} S(k) \cdot g_k[\text{samples/s}], \tag{6.11}$$

where

$$S(k) = \frac{k}{T_{\mathrm{q}}} P_{\mathrm{s}|k}, \tag{6.12}$$

$g_k$ as in (6.1), and $P_{\mathrm{s}|k}$ as in (6.8).

Finally, by means of (6.8–6.10), Eq. 6.11 may be rewritten as

$$S = \frac{1 - e^{-IA_{\sigma_s}/A}}{T_{\mathrm{q}}} \cdot \sum_{k=1}^{+\infty} \frac{\sum_{n=1}^{k} p_{\mathrm{MAC}}(n) \frac{\bar{n}^n e^{-\bar{n}}}{n!}}{\sum_{n=1}^{k} \frac{\bar{n}^n e^{-\bar{n}}}{n!}} \cdot \frac{(GT_{\mathrm{q}})^k e^{-GT_{\mathrm{q}}}}{(k-1)!}. \tag{6.13}$$

## 6.4 Numerical Results

A square area, having area $A = 10^6\,\mathrm{m}^2$, where an average number of ten sinks are distributed according to a PPP ($I = 10$), is considered. We also set $k_0 = 40$ dB, $k_1 = 13.03$, $\sigma_s = 4$ dB (the values are taken from experimental measurements made on the field with Freescale devices [5]), and $L_{\mathrm{th}} = 107$ dB.

In this section, the behavior of the area throughput, $S$, as a function of the offered load, $G$, is shown.

First, the optimal aggregation strategy is investigated, showing results for a single-sink scenario with no connectivity problems, with the purpose of motivating the use of the aggregation strategy, then the multi-sink scenario is considered. Then, a comparison of the area throughput obtained with the two modalities, beacon- and non beacon-enabled, is provided.

Note that the results shown in the following, obtained by applying the aggregation strategy described above, are also valid for a more general scenario, where nodes transmit packets of duration $T_D$ every query, and no aggregation strategy is performed. In this case $S$, expressed in [samples/s] is still given by Eq. 6.13, but now a sample coincides with a packet (i.e., it has duration $D \cdot T$ and not $T$). If, instead, we are interested in $S$ in [bytes/s], to take into account the quantity of information contained in each packet, we have simply to multiply $S$ given by Eq. 6.13 by $D \cdot 10$.

## 6.4.1 The Optimum Aggregation Strategy

### 6.4.1.1 The Single-Sink Scenario, without Connectivity Problems

Here results obtained through the non beacon-enabled mode of the IEEE 802.15.4, related to a single-sink scenario with $n$ sensors and no connectivity problems, are shown. These results are interesting because they motivate the choice of the above described aggregation strategy. It is shown indeed, that given $n$, there exists an optimum value of $D$, $D_{opt}$, maximising the throughput, $S$. Therefore, if sensors are aware of the size $n$ of the cluster they belong to, they could select $D = D_{opt}$, obtained through our results, and transmit the aggregated packet every $D_{opt}$ queries.

The interval of time $T$ needed to transmit a unit of data will be equal to the backoff period, $d_{bo} = 320\,\mu s$, defined in Chap. 5. It is assumed that the sinks allow sensors to try to access the channel for all the time they need. Therefore, by setting the query size equal to 10 bytes (i.e., the query is transmitted in $T$), we fix $T_q = (\hat{t}_{max} + D + 1)T = (121 + D)T$, being $(\hat{t}_{max} + D)T$ the maximum delay with which a node can transmit a packet having size $D \cdot 10$ bytes (see Eq. 5.1).

Since here we have ensured connectivity, a single sink and a deterministically fixed number, $k = n$, of sensors competing for access, we have $p_{CON} = 1$ and $P_{s|k} = p_{MAC}$. Hence, the area throughput is simply:

$$S = \frac{n}{(121 + D)T} \cdot p_{MAC}(n). \tag{6.14}$$

In Fig. 6.3, $S$ is shown, as a function of $n$, for different values of $D$. As we can see, $S$ presents a maximum. In fact, for small $n$, $p_{MAC}$ approaches zero slower then $1/n$ and thus by increasing $n$, $S$ also increases. On the contrary, for large $n$, $p_{MAC}$

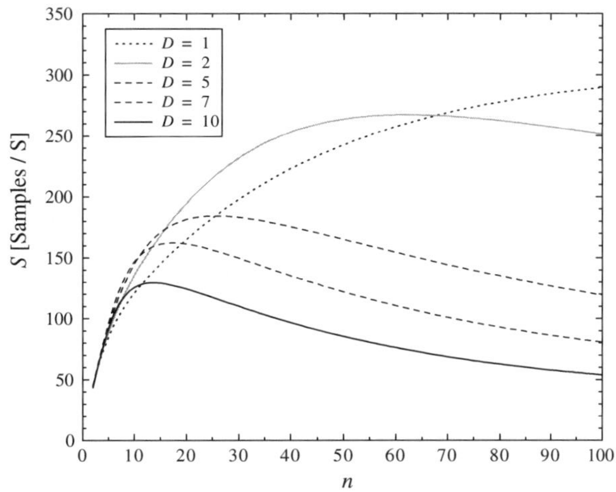


**Fig. 6.3** $S$, as a function of $n$, for different values of $D$, in a single sink connected case

approaches zero faster then $1/n$ and thus by increasing $n$, the product $n \cdot p_{\mathrm{MAC}}(n)$ decreases, and so does $S$. The physical interpretation is that too many packet losses occur when traffic is too heavy. The maximum values of $S$ depend on $D$ and are obtained for different values of $n$. As we can see, for $1 < n < 12$, $D_{\mathrm{opt}} = 7$; for $12 < n < 18$, $D_{\mathrm{opt}} = 5$; for $18 < n < 68$, $D_{\mathrm{opt}} = 2$ and for $n > 68$ $D_{\mathrm{opt}} = 1$. Therefore, it clearly appears that $D_{\mathrm{opt}}$ decreases when increasing $n$.

The aggregation strategy proposed here is achievable only in case sensors know $n$. This parameter could be estimated by sensors, for example, by computing the number of times the channel is found busy in a given interval of time. The probability to find the channel busy, in fact, is strictly related to $n$.

#### 6.4.1.2 The Multi-Sink Scenario

Once again the IEEE 802.15.4 in non beacon-enabled mode, is considered; therefore, $T = 320\,\mu\mathrm{s}$ and $T_{\mathrm{q}} = (121 + D)T$. Since a typical IEEE 802.15.4 air interface is considered, a limit on the number of sensors that could be connected to a given sink should be imposed [6, 7]. To this end, we denote as $n_{\max}$ the maximum number of sensors that could be served by a sink and define a new probability (to replace $p_{\mathrm{MAC}}(n)$ in (6.13)) $p'_{\mathrm{MAC}}(n)$ given by:

$$p'_{\mathrm{MAC}}(n) = \begin{cases} p_{\mathrm{MAC}}(n), & n_{\max} \\ p_{\mathrm{MAC}}(n_{\max}) \cdot n_{\max}/n, & n > n_{\max} \end{cases} \qquad (6.15)$$

where $p_{\mathrm{MAC}}(n)$ is obtained through the model described in Chap. 5, and $1 - n_{\max}/n$ is the probability that a sensor is not served by the sink it is connected to, owing to the capacity constraint. Performance curves are obtained by setting $n_{\max} = 20$. Moreover, the case of negligible border effects is considered.

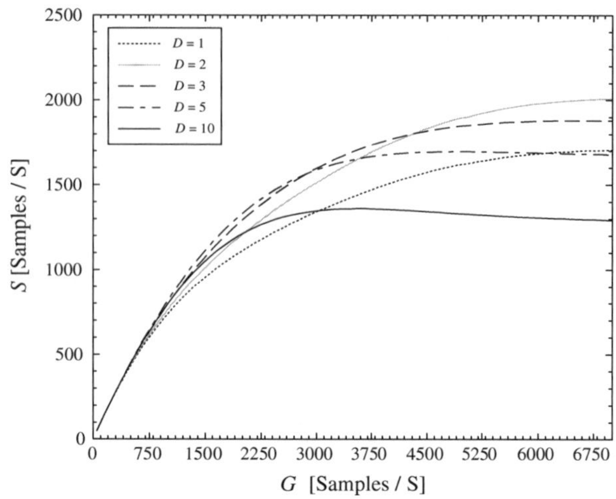


**Fig. 6.4** $S$, as a function of $G$, for different values of $D$, having fixed $p_{\mathrm{CON}}(x,y) = 1$

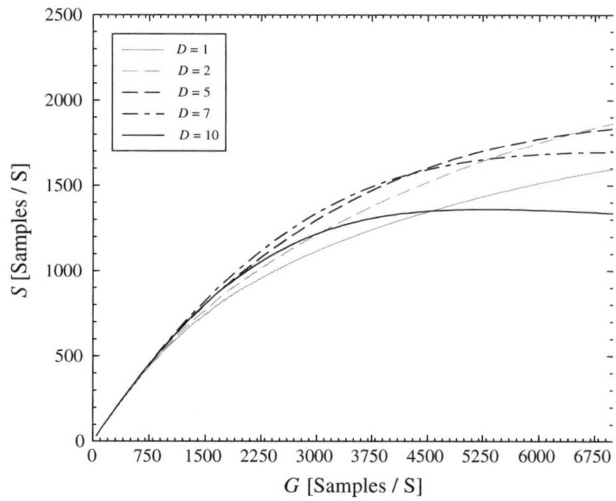


**Fig. 6.5** $S$, as a function of $G$, for different values of $D$, having fixed $p_{\mathrm{CON}}(x,y) = 0.67$

In Figs. 6.4 and 6.5, $S$, as a function of $G$, for different values of $D$ when $p_{\mathrm{CON}} = 1$ and 0.67 respectively, is shown. In both figures, there exists a value $D_{\mathrm{opt}}$ which decreases by increasing $G$. Moreover, from Fig. 6.4 we can see that for $0 < G < 3000$ samples/s (when $I = 10$, $G = 3000$ samples/s corresponds to $n = 12$) $D_{\mathrm{opt}} = 7$; for $3000 < G < 4500$ samples/s ($G = 4500$ samples/s corresponds to $n = 18$) $D_{\mathrm{opt}} = 5$; and for $G > 4500$ samples/s $D_{\mathrm{opt}} = 2$. Therefore, the behavior of $D_{\mathrm{opt}}$, as a function of $G$, is exactly the same shown in Fig. 6.3.

By comparing Figs. 6.4 and 6.5, we can observe the effects of connectivity on $S$. Once $D$ is fixed, the values of $S$ reached for large offered load are approximatively the same reached when $p_{\mathrm{CON}} = 1$. The decrease of $p_{\mathrm{CON}}$, in fact, results in a lower mean number of sensors per sink, therefore the decreasing of $p_{\mathrm{CON}}$ is compensated by an increasing of $p_{\mathrm{MAC}}(n)$. However, the behavior of the curves for low values of $G$ is different (the curves have different slopes). If we fix $D = 5$ and we want to obtain $S = 1500$ samples/s, when $p_{\mathrm{CON}} = 0.67$, we need to deploy on average 158 sensors, whereas, when $p_{\mathrm{CON}} = 1$, 106 sensors on average are sufficient. Therefore, the loss of connectivity brings to a larger cost in terms of number of sensors that must be deployed to obtain the desired $S$.

To increase the values of $S$, instead, we need to increase $I$. In fact, given a value of $G$, by increasing $I$ the connectivity improves and also the losses due to MAC decrease, since $n$ decreases.

Finally, an example of results obtained by considering a simpler MAC protocol model where the probability of success, $p''_{\mathrm{MAC}}(n)$ (to be included in (6.13)), is a linear function of $n$, is shown. In [8], it is shown, in fact, that in some cases the success probability for a non-persistent CSMA protocol, decreases linearly with the number of nodes. Therefore, we model $p''_{\mathrm{MAC}}(n)$ as:

$$p''_{\mathrm{MAC}}(n) = m \cdot n + 1, \tag{6.16}$$

and we denote by $n^*$ the value such that $p''_{\mathrm{MAC}}(n^*) = 0$.

In Fig. 6.6, three cases are accounted for: $m = -0.01$, corresponding to $n^* = 100$; $m = -0.02$, corresponding to $n^* = 50$; and $m = -0.04$, corresponding to $n^* = 25$. By decreasing $n^*$, the maximum of $S$ is reached for lower values of $G$. Therefore, for a given value of $G$, by increasing the slope of $p''_{\mathrm{MAC}}(n)$, $S$ increases. The maximum value of $S$ obtained with $n^* = 50$ is approximately twice as large as the one obtained with $n^* = 25$, but it is reached for an offered load that is twice

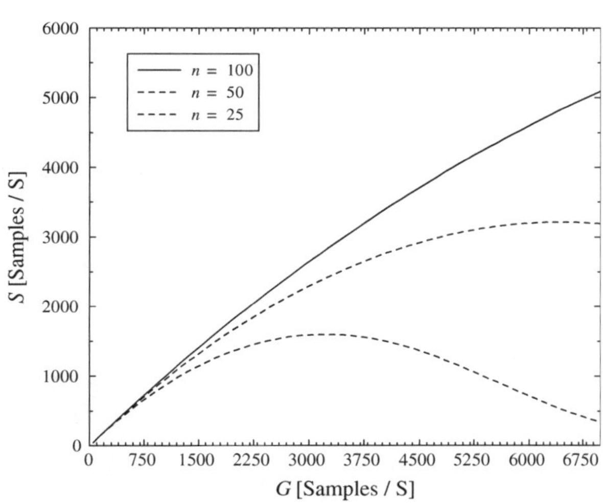


**Fig. 6.6** $S$, as a function of $G$, in the case of a CSMA based protocol, having a $p_{\mathrm{MAC}}(n)$ decreasing linearly with $n$, for different values of $n^*$

over. Therefore, this increase in the maximum value is reached at the cost of deploying more sensors.

### *6.4.2 Comparing Beacon- and Non Beacon-Enabled Modes*

In this subsection, area throughput obtained with the two modalities beacon- and non beacon-enabled, considering different values of $D$, $SO$, $N_{\mathrm{GTS}}$, $T_{\mathrm{q}}$, and different connectivity levels, is shown.

The query packet size is set equal to 60 bytes, therefore, it is transmitted in $6 \cdot T$ seconds, and $T_{\mathrm{q}} = (126 + D)T$ for the non beacon-enabled mode, once again to allow sensors to access the channel for all the time needed.

Here, a limit on the number of sensors that could be connected to the same sink is not imposed, therefore, Eq. 6.15 is not used.

In Figs. 6.7 and 6.8, $S$, as a function of $G$, when varying $SO$, $N_{\mathrm{GTS}}$, and $T_{\mathrm{q}}$, for $D = 2$ and $D = 10$, is shown, respectively. The input parameters that we entered give a connection probability $p_{\mathrm{CON}} = 0.89$. Both beacon- and non beacon-enabled modes are considered. In both figures it can be noted that, once $SO$ is fixed (beacon-enabled case), an increase of $N_{\mathrm{GTS}}$ results in an increment of $S$, since $p_{\mathrm{MAC}}$ increases. Moreover, once $N_{\mathrm{GTS}}$ is fixed, there exists a value of $SO$ maximising $S$. When $D = 2$, an increase of $SO$ results in a decrement of $S$ since, even though $p_{\mathrm{MAC}}$ gets greater, the query interval is longer and, therefore, the number of samples per second received by the sink decreases. On the other hand, when $D = 10$ and all possible GTSs are allocated, the optimum value of $SO$ is 1. This is due to the fact that, having large packets, when $SO = 0$ too many packets are lost,

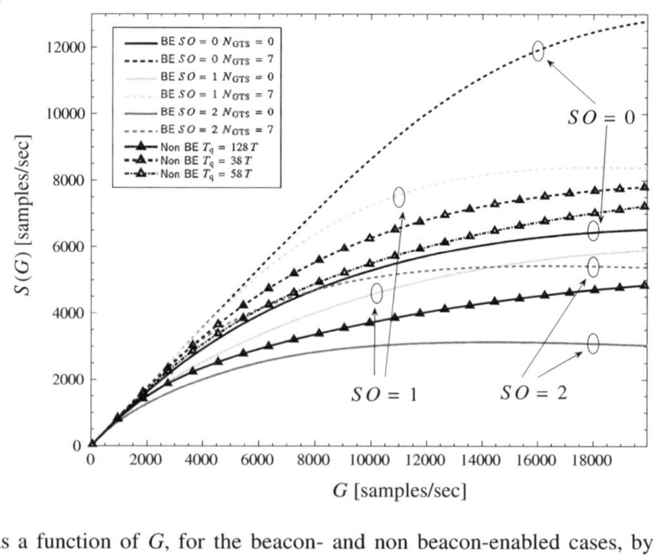


**Fig. 6.7** $S$ as a function of $G$, for the beacon- and non beacon-enabled cases, by varying $SO$, $N_{\mathrm{GTS}}$, and $T_{\mathrm{q}}$, having fixed $D = 2$

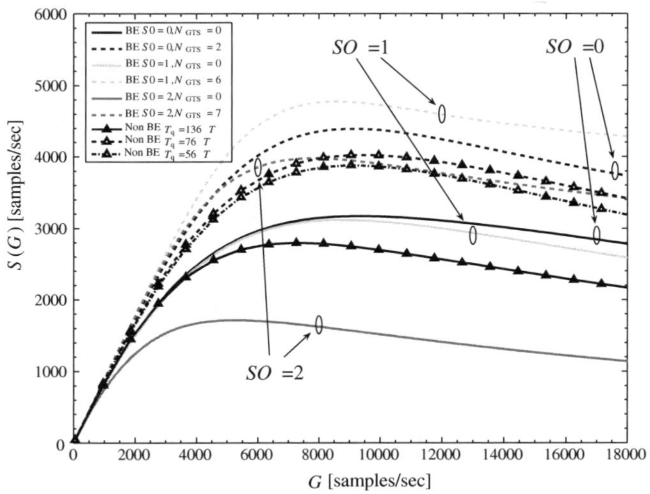


**Fig. 6.8** $S$ as a function of $G$, for the beacon- and non beacon-enabled cases, by varying $SO$, $N_{\text{GTS}}$, and $T_{\text{q}}$, having fixed $D = 10$

owing to the short duration of the superframe. However, when $N_{\text{GTS}} = 0$ the best case is, once again, $SO = 0$, since in this case MAC losses are approximately the same obtained in the case $SO = 1$ (see Fig. 5.26), which, however, brings to a higher query interval. In conclusion, we can deduce that the use of GTSs is always advantageous, and that there exists an optimum value of $SO$ maximising $S$, which depends on $D$ and $N_{\text{GTS}}$.

Concerning the non beacon-enabled case, in both figures it can be noted that, by decreasing $T_{\text{q}}$, $S$ gets larger even though $p_{\text{MAC}}$ decreases, since, once again, the MAC losses are balanced by larger values of $f_{\text{q}}$.

By comparison of Figs. 6.7 and 6.8, we note that, once the offered load, $G$, is fixed, $S$ gets notably smaller when $D$ increases. $S$, in fact, is expressed in terms of samples/s received by the sink, and not in bytes/s. Therefore, once $T_{\text{q}}$ is fixed, by increasing $D$, $p_{\text{MAC}}$ gets smaller. On the other hand, by increasing $D$, the maximum value of $S$ is reached for lower values of $G$. This means that, when $D$ is small, the maximum value of $S$ is reached at the cost of deploying more sensors.

Finally, we show the effects of connectivity on the area throughput. When $p_{\text{CON}}$ is less than 1, only a fraction of the deployed nodes has a sink in its vicinity. In particular, an average number, $\bar{k} = p_{\text{CON}} G T_{\text{q}}/I$, of sensors compete for access at each sink. In Fig. 6.9, we consider the non beacon-enabled case with $D = 2$, $T_{\text{q}} = 128\,T$ and $D = 10$, $T_{\text{q}} = 136\,T$. When $D = 10$, $T_{\text{q}} = 136\,T$, for high offered loads the area throughput tends to decay, since packet collisions dominate. Hence, by moving from $p_{\text{CON}} = 1$ to $p_{\text{CON}} = 0.89$, we observe a slight improvement due to the fact that a smaller average number of sensors tries to connect to the same sink. Conversely, when $D = 2$, $T_{\text{q}} = 128\,T$, $S$ is still increasing with $G$, then by moving from $p_{\text{CON}} = 1$ to $p_{\text{CON}} = 0.89$, we just reduce the useful traffic. Furthermore, when $p_{\text{CON}} = 0.15$, the offered load is very light, so that we are working

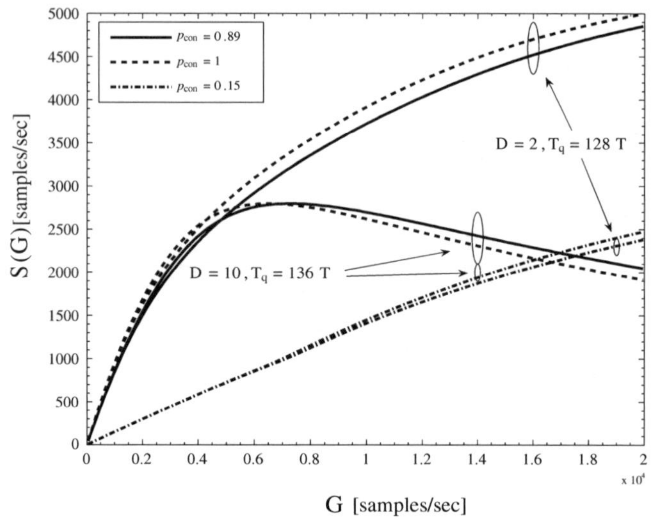


**Fig. 6.9** $S$, as a function of $G$, in the non beacon-enabled case, for different values of $D$ and $p_{CON}$, having fixed $T_q$ to the maximum delay

in the region where $p_{MAC}(D = 2, T_q = 128\,T) < p_{MAC}(D = 10, T_q = 136\,T)$ (see Fig. 5.17), resulting in a slightly better performance of the case with $D = 2$. Thus we conclude that the effect of lowering $p_{CON}$ results in a stretch of the curves reported in the previous plots.

## 6.5 Concluding Remarks

A multi-sink WSN where sensor nodes transmit their packets to a sink selected among many, by using a CSMA-based MAC protocol, is studied. A new performance metric, accounting for connectivity and MAC issues jointly, namely the area throughput, has been defined. This new concept allows the study of this kind of networks under a new perspective, even if, in general terms, the aim is to define a picture showing how throughput varies with load, as done for many years in the literature. However, here, connectivity issues and the presence of multiple sinks are accounted for. This implies, mainly, that performance depends not only on the number of nodes in the network, and on the packet size, but also on sinks density and on transmit power (i.e., $L_{th}$). In fact, in case the application fixes the minimum value of $S$, from the figures we could obtain not only the number of nodes that must be distributed in the network (i.e., the offered load, $G$), but also (once $G$ is fixed), the number of sinks that must be distributed, or the transmit power (from which depends $L_{th}$, and, therefore, $p_{CON}$). Other minor outcomes could be derived from this chapter: (i) the model developed allows the evaluation of an optimum aggre-gation strategy, maximising $S$; (ii) a comparison in terms of area throughout between the beacon- and non beacon-enabled modes of the IEEE 802.15.4, is provided.

## References

1. C. Bettstetter, On the minimum node degree and connectivity of a wireless multihop network in *Proceedings of ACM Symposium on Mobile Ad Hoc Networks and Computing*. (Mobihoc), Lausanne, Switzerland, June 2002, pp. 80–91
2. F. Fabbri, R. Verdone, A statistical model for the connectivity of nodes in a multi-sink wireless sensor network over a bounded region, in *Proceedings of IEEE European Wireless* (EW), Prague, Czech Republic, June 2008, pp. 1–6
3. J. Orriss, S.K. Barton, Probability distributions for the number of radio transceivers which can communicate with one another. IEEE Trans. Commun. **51**(4), 676–681 (2003)
4. J. Orriss, A. Zanella, R. Verdone, S. Barton, Probability distributions for the number of radio transceivers in a hot spot with an application to the evaluation of blocking probabilities, in *Proceedings of IEEE International Symposium on Personal, Indoor, and Mobile Radio Communications*, vol. 2, Lisboa, Portugal, September 2002, pp. 654–657
5. Freescale, *Freescale Semiconductor's MC13192 Developer's Kit* http://www.freescale.com/webapp/sps/site/prod_summary.jsp?code=MC13192
6. Zigbee Alliance, Website: http://www.zigbee.org
7. A. Koubaa, M. Alves, E. Tovar, Modeling and worst-case dimensioning of cluster-tree wireless sensor networks, in *Proceedings of IEEE International Real-Time Systems Symposium (RTSS)*, Rio de Janeiro, Brazil, December 2006, pp. 412–421
8. Y.C. Tay, K. Jamieson, H. Balakrishnan, Collision-minimizing CSMA and its applications to wireless sensor networks. IEEE J. Sel. Areas Commun. **22**(6), 1048–1057 (2004)

# Part IV
# Cross-Layer Design

# Chapter 7
# Decentralized Detection in IEEE 802.15.4 Wireless Sensor Networks

In this chapter, we present a mathematical model to study the interplay between a decentralized detection task in clustered Wireless Sensor Networks (WSNs) and the IEEE 802.15.4 Medium Access Control (MAC) protocol. In particular, the models introduced in Chaps. 2 and 5 are integrated, to develop a unique framework jointly accounting for MAC and distributed detection issues.

The scenario described in Chap. 2 is considered: sensors and Fusion Centers (FCs) are distributed with the aim of detecting an event of interest. Sensors are organized in clusters, with FCs acting as cluster heads, and are supposed to observe the same common binary phenomenon. As in Chap. 5, a query-based application is accounted for: FCs send periodic queries and wait for replies coming from sensors. After reception of data, FCs perform data fusion with a majority-like fusion rule and send their decisions to an Access Point (AP), where a final data fusion is carried out and an estimate of the phenomenon is obtained. Decentralized detection and MAC issues are jointly investigated through analytical modelling. The proposed framework allows the derivation of the probability of decision error at the AP, when accounting for packets' losses due to possible collisions. The impact of different clustering configurations and of noisy communications is also investigated.

This chapter is structured as follows. In Sect. 7.1, a few preliminaries are provided: the mathematical framework for decentralized detection presented in Part II is briefly recalled in Sect. 7.1.1, whereas Sect. 7.1.2 describes how the access to the channel is managed in the considered scenario. In Sect. 7.2, a new mathematical model, accounting for decentralized detection and MAC issues jointly, is derived. Sections 7.3 and 7.4 report numerical results and concluding remarks, respectively.

C. Buratti et al., *Sensor Networks with IEEE 802.15.4 Systems*,
Signals and Communication Technology, DOI: 10.1007/978-3-642-17490-2_7,

## 7.1 Preliminaries

### 7.1.1 Decentralized Detection

In this subsection, we recall a few preliminaries on decentralized detection of a common binary phenomenon in the presence of an ideal (collision-less) MAC protocol [1]. A complete analysis of this problem is provided in Chap. 2. As in Sect. 2.1.1, we consider a network scenario where $n$ sensors observe a common binary phenomenon with probability $p_0$. The sensors are clustered into $n_c < n$ groups, and each sensor can communicate only with its local FC. The groups may be either *uniform* or *non-uniform*. For the sake of simplicity, only one level of FCs is considered in this chapter. We consider the reference scenario shown in Fig. 7.1: FCs act as Personal Area Network (PAN) coordinators gathering data from sensors belonging to their clusters and transmitting their decisions to the final destination, that is the AP. We assume that a different network (e.g., an infrastructure-based network where radio resources are scheduled) is used for the communication between the FCs and the AP and there are no MAC losses (i.e., there is a contention-free access). In this case, FCs will act as gateways between two different networks: the IEEE 802.15.4 (sensor) network and the infrastructure-based network. A logical representation of the overall considered architecture is shown in Fig. 7.1a, whereas Fig. 7.1b shows a more detailed view of a scenario with a single cluster and the AP. Note that the representation of Fig. 7.1 is equivalent to that in Fig. 2.1. In a scenario with noisy communication links, they are simply modeled as Binary Symmetric Channels (BSCs), with cross-over probability of the BSC model denoted as $p$ [2]. This simplified model is accurate only in the presence of additive communication noise, without other impairments, such as inter-symbol interference and path loss, are accounted for. For instance, when additive Additive White Gaussian Noise (AWGN) and binary phase shift keying are considered, it holds that $p = Q(\sqrt{\gamma_b})$, where $\gamma_b$ is the channel Signal-to-Noise Ratio (SNR) and

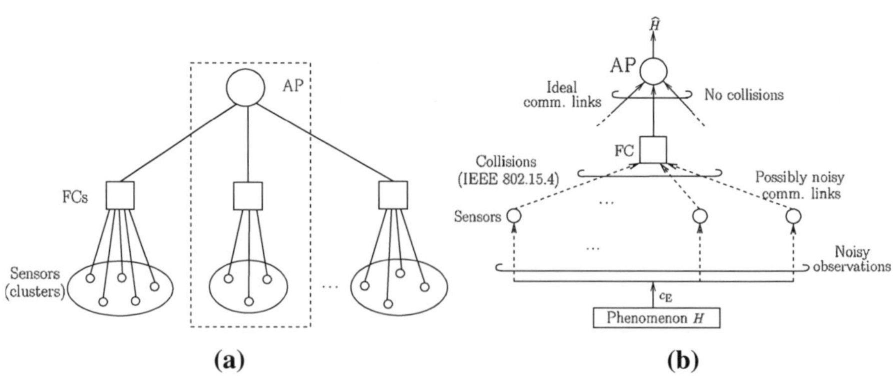


**Fig. 7.1** Logical representation of a clustered sensor network (case **a**) and detailed representation of a specific cluster highlighted above (case **b**)

$Q(x) \triangleq \int_x^{+\infty} \frac{1}{\sqrt{2\pi}} \exp(-y^2/2) \mathrm{d}y$. When an ideal (collision-less) MAC protocol is considered, the number of data packets received at an FC is equal to the number of sensors in the corresponding cluster, since no losses at MAC level occur. Recalling $\mathscr{D}$ as the $n_\mathrm{c}$-dimensional vector containing the number of decisions received at the $n_\mathrm{c}$ FCs, the same analytical approach for the probability of decision error given in Sect. 2.1.2.2 can thus be applied. Therefore, the probability of decision error in a generic scenario with noisy communication links, given a clustering configuration (i.e., $\mathscr{D}$), becomes that in (2.14):

$$
\begin{aligned}
P_\mathrm{e}(\mathrm{SNR}, p|\mathscr{D}) = p_0 \sum_{i=k}^{n_\mathrm{c}} \sum_{j=1}^{n_\mathrm{c}} \prod_{\ell=1}^{n_\mathrm{c}} \Big\{ c_{i,j}(\ell) p_{\ell,}^{1|0}\left(d_\mathrm{c}^{(\ell)}\right) + (1 - c_{i,j}(\ell)) \Big[1 - p_{\ell,\mathrm{noisy}}^{1|0}\left(d^{(\ell)}\right)\Big]\Big\} \\
+ (1-p_0) \sum_{i=0}^{k_\mathrm{f}-1} \sum_{j=1}^{n_\mathrm{c}} \prod_{\ell=1}^{n_\mathrm{c}} \Big\{ c_{i,j}(\ell) p_{\ell,}^{1|1}\left(d^{(\ell)}\right) \\
+ (1 - c_{i,j}(\ell)) \Big[1 - p_{\ell,}^{1|1}\left(d_\mathrm{c}^{(\ell)}\right)\Big]\Big\}. \qquad (7.1)
\end{aligned}
$$

At the left-hand side of (7.1), we have explicitly indicated that $P_\mathrm{e}$ depends on the observation quality (i.e., $\mathrm{SNR}_\mathrm{sensor}$) and the communication quality (i.e., $p$). Moreover, (7.1) reduces to (2.9) when $p = 0$:

$$
\begin{aligned}
P_\mathrm{e}(\mathrm{SNR}|\mathscr{D}) = p_0 \sum_{i=k}^{n_\mathrm{c}} \sum_{j=1}^{\binom{n_\mathrm{c}}{i}} \prod_{\ell=1}^{n} \Big\{ c_{i,j}(\ell) p_\ell^{1|0} + (1 - c_{i,j}(\ell))(1 - p_\ell^{1|0}) \Big\} \\
+ (1-p_0) \sum_{i=0}^{k_\mathrm{f}-1} \sum_{j=1}^{\binom{n_\mathrm{c}}{i}} \prod_{\ell=1}^{n} \Big\{ c_{i,j}(\ell) p_\ell^{1|1} + (1 - c_{i,j}(\ell))(1 - p_\ell^{1|1}) \Big\}. \qquad (7.2)
\end{aligned}
$$

### 7.1.2 The Access to the Channel

We consider a network composed of IEEE 802.15.4-compliant sensors, working in beacon-enabled mode [3]. Each FC coincides with a PAN coordinator, receiving data from sensors belonging to its PAN (i.e., its cluster). We assume that the different clusters use the same frequency channel, but different resources in terms of time. In other words, a time division between clusters is applied, so that sensors of different clusters do not interfere among them. We evaluate performance by fixing the total time made available to all sensors in the network (i.e., all the clusters) for transmitting data to the FCs. This means that performance is evaluated under a total achievable throughput constraint. We assume that no connectivity problems exist: each sensor can receive the query (i.e., the beacon packets) from the FC and can reach it. Nodes transmit packets with size $D \cdot 10$ bytes, being $D$ an integer parameter (see Chap. 5). As stated in Chap. 5, according to the IEEE 802.15.4 MAC protocol in beacon-enabled mode, the access to the channel is managed through a superframe, starting with the beacon transmitted by the PAN coordinator. The superframe may contain an inactive part, allowing sensors to enter in sleeping mode, whereas in the active part sensors use a slotted Carrier-Sense Multiple Access with Collision Avoidance (CSMA/CA) algorithm to transmit data. We recall that the duration of the active part, the superframe duration, and of the entire superframe, the beacon interval, depend on the value of two integer parameters ranging from 0 to 14, that are the superframe order, *SO*, and the beacon order, *BO*, respectively. We also recall that the superframe duration can be expressed as $960 \cdot 2^{SO} \cdot T_s$, where $T_s = 16\ \mu s$ is the symbol time, whereas the beacon interval is given by $960 \cdot 2^{BO} \cdot T_s$ (see Fig. 7.2).

Time division between clusters is performed as follows. The application sets the value of *BO*, that is the total time made available to the network for transmissions from sensors to FCs. If the AP does not know the clusters size, it allocates the same resource to all the clusters, that is the same value of *SO*. In particular, *SO* is set accordingly to the value of *BO* and the number of clusters, such that all clusters have a portion of the beacon interval allocated. If, instead, the AP is aware of the network topology, it may allocate resources according to the number of sensors in each cluster. In this case, the AP assigns different values of *SO* according to the clusters' sizes: the smaller is the cluster, the smaller is the value of *SO* assigned to it. Both the above mentioned resource allocation strategies will be considered in Sect. 7.3. The AP communicates to the FCs the values of *SO* and *BO* and the instant in which the superframe of each FC must start. In this way, the active parts of the superframes defined by the different FCs will not overlap and during transmissions within a given cluster, sensors belonging to the other clusters will be in sleeping mode, being in the inactive part of the superframe of their FCs (see Fig. 7.2). According to our application, each FC will send periodic queries,

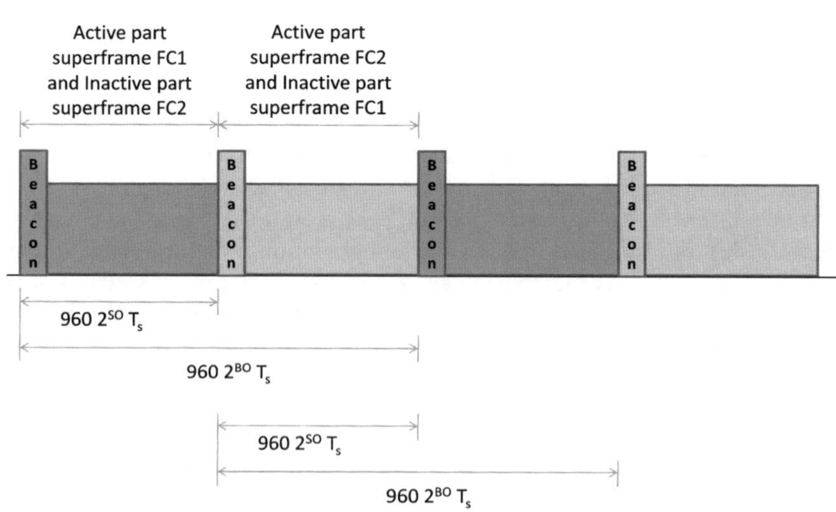


**Fig. 7.2** The time division between clusters, when two FCs are present

starting from the instant provided by the AP, and will wait for decisions coming from sensors. The application also requires that the data must be received by the FC by the end of the active part of the superframe defined by the FC. Therefore, each sensor has one packet to be transmitted per beacon received and has to transmit it by the end of the active part of the superframe defined by its FC.

The mathematical model proposed in Chap. 5 for 802.15.4 beacon-enabled mode is used in this chapter, in order to derive the success probability for a node using the slotted CSMA/CA algorithm to access the channel, denoted hereafter as $p_{\mathrm{MAC}}$. This probability represents the probability that a sensor transmits successfully the packet to its FC by the end of the active part of the superframe of its FC. $p_{\mathrm{MAC}}$ is the success probability, $p_{\mathrm{s}}$, derived in Chap. 5, given by Eq. 5.27, for the beacon-enabled case. We recall that a packet could be lost due to the following reasons: (i) a collision, (ii) the channel is sensed busy more than five consecutive times [3], or (iii) the available time ends before the channel is sensed idle. Note that retransmissions are not allowed in our scenario.

## 7.2 Impact of MAC on Decentralized Detection

In this section, we derive an analytical framework for the computation of the probability of decision error in the presence of the IEEE 802.15.4 MAC protocol. Each FC will receive a number of decisions smaller than the number of sensors in the cluster, owing to the contention-based nature of the protocol, that may cause collisions.

Equation 7.1 needs to be modified to take into account the presence of a non-ideal MAC protocol, characterized, concisely, by $p_{\mathrm{MAC}} \leq 1$.[1]

Being $p_{\mathrm{MAC}}(d_{\mathrm{c}})$ the success probability in a scenario with $d_{\mathrm{c}}$ competing sensors in a cluster and assuming that all transmissions are independent, it follows that the number of successful transmissions in the $j$th cluster can be modeled as a binomial random variable, denoted as $\mathscr{D}_{\mathrm{c}}^{(j)}$ $(j = 1, \ldots, n_{\mathrm{c}})$, with parameters $d_{\mathrm{c}}^{(j)}$ and $p_{\mathrm{MAC}}(d_{\mathrm{c}}^{(j)})$. Referring to the analysis in Sect. 7.1, the $n_{\mathrm{c}}$-dimensional vector, with the numbers of decisions received by the FCs, is a random vector[2] $\mathscr{D} \triangleq (\mathscr{D}_{\mathrm{c}}^{(1)}, \mathscr{D}_{\mathrm{c}}^{(2)}, \ldots, \mathscr{D}_{\mathrm{c}}^{(n_{\mathrm{c}})})$. Note that even through the clusters are uniform, the number of decisions received at the FCs may vary from cluster to cluster, being such number a random variable. Therefore, the true clustering configuration is non uniform.

---

[1] For the ease of clearness, in this section we refer to the generic scenario with noisy communication links. However, the same considerations apply in a scenario with ideal communication links, the only difference being that (7.1) has to be substituted by (7.2).

[2] The symbol $\mathscr{D}$ was used in Chap. 2 for a deterministic vector. With an abuse of notation, it now refers to a random vector. The context eliminates any ambiguity.

At this point, the probability of decision error depends on a realization of the random vector $\mathscr{D}$ which, in turns, depends on $\mathscr{P}^{1|1}$ and $\mathscr{P}^{1|0}$. The average probability of decision error, with respect to the clustering configuration, can then be computed as follows:

$$\overline{P}_{\mathrm{e}}(\mathrm{SNR}_{\mathrm{sensor}}, p) = \mathbb{E}_{\mathscr{D}}[P_{\mathrm{e}}(_{\mathrm{sensor}}, p|\mathscr{D})]. \tag{7.3}$$

After a few manipulations, one obtains

$$\begin{aligned}\overline{P}_{\mathrm{e}}(\mathrm{SNR}_{\mathrm{sensor}}, p) = &\sum_{i_1=0}^{d_{\mathrm{c}}^{(1)}}\sum_{i_2=0}^{d^{(2)}}\cdot\cdot\sum_{i_{n_{\mathrm{c}}}=0}^{d^{(n_{\mathrm{c}})}} \mathbb{P}\left\{\mathscr{D}^{(1)} = i_1\right\}\mathbb{P}\left\{\mathscr{D}^{(2)} = i_2\right\}\cdot\mathbb{P}\left\{\mathscr{D}^{(n_{\mathrm{c}})} = i_{n_{\mathrm{c}}}\right\}\\ &\times P\left(\mathrm{SNR}_{\mathrm{sensor}}, p|\mathscr{D}^{(1)} = i_1, \mathscr{D}^{(2)} = i_2, \ldots, \mathscr{D}_{\mathrm{c}}^{(n)} = i_{n_{\mathrm{c}}}\right)\end{aligned} \tag{7.4}$$

where the last probability at the right-hand side is given by (7.1) (with $d_{\mathrm{c}}^{(j)} = i_j$, $j = 1, \ldots, n_{\mathrm{c}}$) and

$$\mathbb{P}\left\{\mathscr{D}_{\mathrm{c}}^{(\ell)} = i_\ell\right\} = \binom{d_{\mathrm{c}}^{(\ell)}}{i_\ell}\left[p_{\mathrm{MAC}}\left(d_{\mathrm{c}}^{(\ell)}\right)\right]^{i_\ell}\left[1 - p_{\mathrm{MAC}}\left(d_{\mathrm{c}}^{(\ell)}\right)\right]^{d_{\mathrm{c}}^{(\ell)}-i_\ell}.$$

It would be interesting to preliminary evaluate a lower bound on the average probability of decision error, as the limiting average probability of decision error in an ideal scenario with no observation and communication noises, i.e., for $\mathrm{SNR}_{\mathrm{sensor}} \to \infty$ and $p = 0$. In this case, if at least one bit is delivered to the AP, then a correct decision will be made. At this point, there is a decision error *if and only if* no sensor decisions can be reliably sent to the AP. Therefore, an error happens only if $i_\ell = 0$, $\forall \ell \in \{1, \ldots, n_{\mathrm{c}}\}$. In this case, the AP decides randomly, thus obtaining:

$$\begin{aligned}\overline{P}_{\mathrm{e,lim}} &= \mathbb{P}\left\{\mathscr{D}^{(1)} = 0\right\}\mathbb{P}\left\{\mathscr{D}^{(2)} = 0\right\}\cdots\mathbb{P}\left\{\mathscr{D}^{(n_{\mathrm{c}})} = 0\right\}\\ &\quad\times \underbrace{P\left(\mathrm{SNR}_{\mathrm{sensor}}, p|\mathscr{D}^{(1)} = 0, \mathscr{D}^{(2)} = 0, \ldots, \mathscr{D}_{\mathrm{c}}^{(n)} = 0\right)}_{=1/2}\\ &= \frac{1}{2}\prod_{i=1}^{n}\left[1 - p_{\mathrm{MAC}}\left(d_{\mathrm{c}}^{(i)}\right)\right]^{d^{(i)}}.\end{aligned} \tag{7.5}$$

In the presence of uniform clustering, i.e., $d_{\mathrm{c}}^{(i)} = d_{\mathrm{c}}\ \forall i$, (7.5) reduces to

$$\overline{P}_{\mathrm{e,lim}} = \frac{1}{2}\left\{[1 - p_{\mathrm{MAC}}(d_{\mathrm{c}})]^{d_{\mathrm{c}}}\right\}^{n_{\mathrm{c}}} = \frac{1}{2}[1 - p_{\mathrm{MAC}}(d_{\mathrm{c}})]^{n} \tag{7.6}$$

where we have used the fact that $n_{\mathrm{c}} \cdot d_{\mathrm{c}} = n$, regardless of the (uniform) clustering configuration. It can be observed that expression (7.6) for $\overline{P}_{\mathrm{e,lim}}$ is a decreasing function of the number of clusters. On the opposite, in a scenario with an ideal

MAC protocol, this limiting probability *does not* depend on $d_c$ [1]. As an example, in the case $n = 64$, $D = 2$, $BO = 3$, when no clustering is applied, $\overline{P}_{e,lim}$ will be equal to $10^{-5}$ (see Fig. 7.4b). By increasing the number of clusters, $p_{MAC}$ gets larger and, therefore, the floor appears at very small (and not practical) values of the probability of decision error (e.g., in the case with eight uniform clusters we have $\overline{P}_{e,lim} = 6 \times 10^{-22}$). Finally, note that the limiting probability (7.5) equals to zero for any unbalanced configuration with at least one cluster with one sensor. In this case, in fact, the sensor directly connected to its FC always accesses the channel and, therefore, there is always at least one correct decision (sent by the corresponding FC to the AP) on the basis of which the AP can correctly estimate the phenomenon status.

## 7.3 Numerical Results

We now investigate performance of the proposed decentralized detection schemes. In particular, in the presence of IEEE 802.15.4 MAC protocol the value of $p_{MAC}$ is determined offline, for a given clustering configuration, by using the analytical framework presented in Sect. 5.4. The obtained value is then used in (7.4) and in our simulator. In particular, our C simulator is designed "ad hoc" as follows. The transmissions from the sensors to the FCs are represented as Bernoulli trials, each with parameter $p_{MAC}$. On the basis of the received packets in their cluster, the FCs perform a data fusion (with decision threshold set according to the number of received packets) and transmit their decisions to the AP.

Since each sensor must send only its decision (i.e., one bit) and since the model requires that nodes transmit packets of size multiple of 10 bytes [4], being the packet header equal to 19 bytes, we initially set $D = 2$, that is packets of 20 bytes are transmitted. In Fig. 7.11, the impact of different values of $D$ will be investigated.

In the following, we set $n = 64$ and the MAC parameters to the default values (i.e., $BE_{min} = 3$, $BE_{max} = 5$, $NB_{max} = 4$). We first consider uniform resources allocation among clusters. Then, in Fig. 7.6, we extend our approach to a scenario where resources are allocated accordingly to the cluster size. Note that in the first case, uniform clustering will be favored with respect to the non uniform case, since resources will be better used. By the way, in scenarios where the AP is not aware of the network topology, only the uniform resource allocation can be implemented.

In the case of uniform resource allocation among clusters (i.e., all clusters use the same value of $SO$), two values of $BO$, namely 3 and 4, are considered and the related values of $SO$, set according to the number of FCs, are reported in Table 7.1.[3] In particular, in the absence of clustering we set $SO = BO$

[3] Note that, for the ease of clearness, in this chapter a slightly different notation is used with respect to that used in Chap. 2 to denote uniformly clustered topologies.

**Table 7.1** Values of *SO* used in the different topologies for (i) $BO = 3$ and (ii) $BO = 4$

| Topology | $BO = 3$ | $BO = 4$ |
|---|---|---|
| No clustering | 3 | 4 |
| 2 × 32 (2 uniform clusters with 32 sensors each) | 2 | 3 |
| 56-4-4 | 1 | 2 |
| 40-8-8-8 | 1 | 2 |
| 4 × 16 (4 uniform clusters with 16 sensors each) | 1 | 2 |
| 32-8-8-8-8 | 0 | 1 |
| 8 × 8 (8 uniform clusters with 8 sensors each) | 0 | 1 |
| 16 × 4 (16 uniform clusters with 4 sensors each) | Not allowed | 0 |

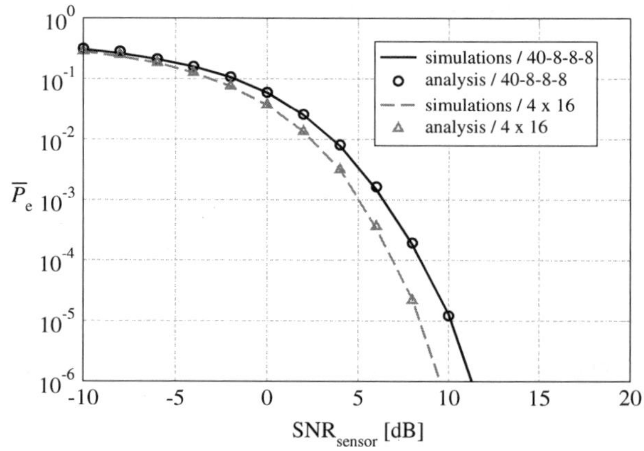


**Fig. 7.3** Comparison between analytical and simulation results in a scenario with ideal communication links and two possible clustering configurations

(i.e., we give to the single cluster all the available time). Moreover, when $n_c = 2$ we set *SO* so that two non overlapping active parts within the beacon interval are present), and so on. Note that in the cases $n_c = 3$ and $n_c = 5$ part of the beacon interval is not used by any cluster, and, therefore, some resources are wasted, due to the constraint that *SO* must be an integer. Note that the topology with 16 uniform clusters is not allowed when $BO = 3$, since there are not sufficient portions of the superframe to be allocated to the different clusters. In the case of non uniform resources allocation strategy only the case $BO = 3$ is considered and the related values of *SO* are reported below.

We first analyze a scenario with $BO = 3$. In Fig. 7.3, a comparison between analytical and simulation results in a scenario with ideal MAC and ideal communication links (i.e., no noisy communication links) and two possible clustering configurations, uniform (4 × 16) and non-uniform (40-8-8-8), is proposed. As expected, a good agreement between simulations and analytical results was found in both cases. In fact, the analysis carried out in Sect. 7.2 is exact and the simulator is implemented by exactly replicating the analysis conditions. In other

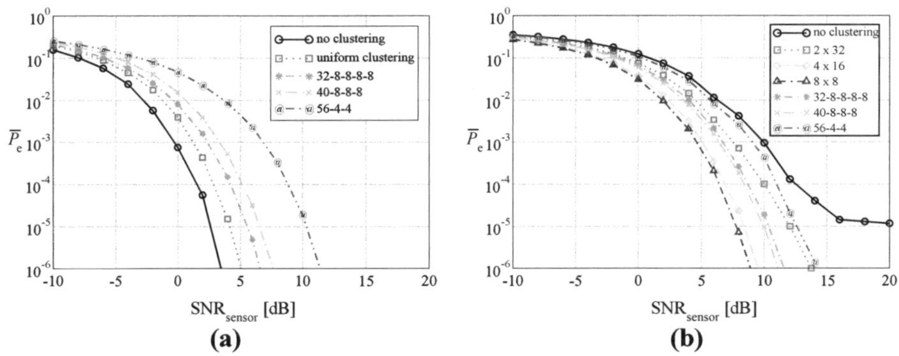


**Fig. 7.4** $\overline{P}_e$ as a function of the sensor SNR, for different clustering configurations when $BO = 3$ and ideal communication links. Case **a**: ideal MAC; case **b**: IEEE 802.15.4 MAC

words, this is a "sanity check," which allows us to use the simulator, especially to avoid numerical problems in the evaluation of the analytical formulas.

In Fig. 7.4, the probability of decision error is shown, as a function of the sensor SNR, for different clustering configurations. No noisy communication links are accounted for and ideal MAC is considered in case (a), whereas the IEEE 802.15.4 MAC protocol is accounted for in case (b). The use of the IEEE 802.15.4 MAC protocol leads to a performance degradation with respect to the case ideal MAC. The highest degradation is achieved with no clustering, since in this case a large number of sensors are competing for the radio resource. The best configuration, in the case with IEEE 802.15.4 MAC protocol is achieved for $n_c = 8$, where only eight sensors per cluster are competing for the channel, and even though $SO = 0$ (i.e., sensors have only approx. 15 ms to access the channel), the success probability is the largest. By comparing curves in (a) and (b) we can observe that, while distributed detection is mainly affected by the uniformity or non uniformity of clusters, rather than by the number of clusters itself, MAC losses strongly depend on the value of $n_c$. In fact, while in the ideal case the performance of uniform clustering does not depend on the specific configuration, this is no longer true in the presence of contention-based MAC protocols. Moreover, note that the case (40-8-8-8) outperforms the case (32-8-8-8-8), since even though more sensors are competing for the channel (in the largest cluster), sensors have more time to access the channel (i.e., $SO = 1$ instead of 0). In fact we have $p_{\mathrm{MAC}} = 0.23$ in the cluster with 40 sensors and $SO = 1$, and $p_{\mathrm{MAC}} = 0.13$ in the cluster with 32 sensors and $SO = 0$. This means that the best performance is achieved when a good balance between the number of sensors competing for the channel and the time made available to sensors for transmissions is reached.

The comparison made in Fig. 7.4 is done by assuming that all decisions coming from the FCs have the same reliability. This implies that the same weight is assigned to all FCs' decisions. However, in non uniform scenarios the decisions obtained by fusing a larger number of sensors' decisions are more reliable than those obtained by fusing a smaller number of sensors' decisions. Therefore, one

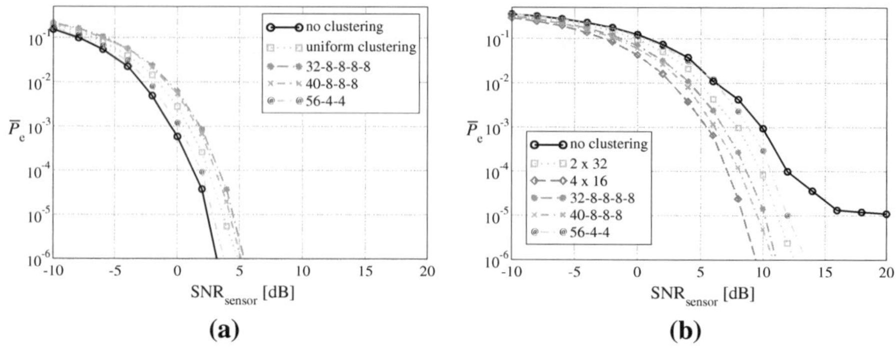


**Fig. 7.5** $\overline{P}_e$ as a function of the sensor SNR, for different clustering configurations when $BO = 3$, ideal communication links, and weighing strategy at the AP. Case **a**: ideal MAC; case **b**: IEEE 802.15.4 MAC

may resort to a weighing strategy, where the AP decides according to the following rule:

$$\Psi(y_1, \ldots, y_M) \triangleq \begin{cases} 0 & \text{if} \sum_{m=1}^{M} w_m y_m < 0 \\ 1 & \text{if} \sum_{m=1}^{M} w_m y_m \geq 0 \end{cases} \tag{7.7}$$

where $y_1, \ldots, y_M$ are the $M$ data ($y_m = 2x_m - 1$) to be fused together and $w_1, \ldots, w_M$ are the weights computed as the number of sensors in the cluster (which successfully access the channel) divided by the total number of sensors (which successfully access the channel). In Fig. 7.5, $\overline{P}_e$ is shown, as a function of the sensor SNR, for different clustering configurations, ideal communication links, and weighing strategy at the AP. Two scenarios for the MAC are considered: (a) ideal MAC protocol and (b) IEEE 802.15.4 MAC protocol. In the scenario with ideal MAC protocol, one can observe that the non uniform configurations experience the expected performance improvement. Moreover, the more higher is the non uniformity degree, the large is this improvement. On the other hand, when the IEEE 802.15.4 MAC protocol is considered, one can observe that the weighing strategy has no significant impact and the performance is the same predicted in Fig. 7.4b. This is probably due to the fact that, while in the case with no weighing strategy the performance is given by the average number of sensors accessing the channel, in the presence of weighing the performance is determined by the overall statistics of the number of sensors accessing the channel.

In Fig. 7.6, we show the performance achieved when resources (in terms of time) are allocated to clusters depending on their size. In particular, we set $BO = 3$ and we allocate $SO = 0$ to clusters with eight sensors, $SO = 1$ to clusters with 16 sensors, $SO = 2$ to clusters with 32 sensors and $SO = 3$ to the non clustering case. In this way, the resource available to each cluster is proportional to the cluster size and also no resources are wasted for the considered set of network topologies. As expected, the performance of the non uniform cases slightly improve with respect to those with uniform resource allocation (see, for example the case (32-8-8-8-8)

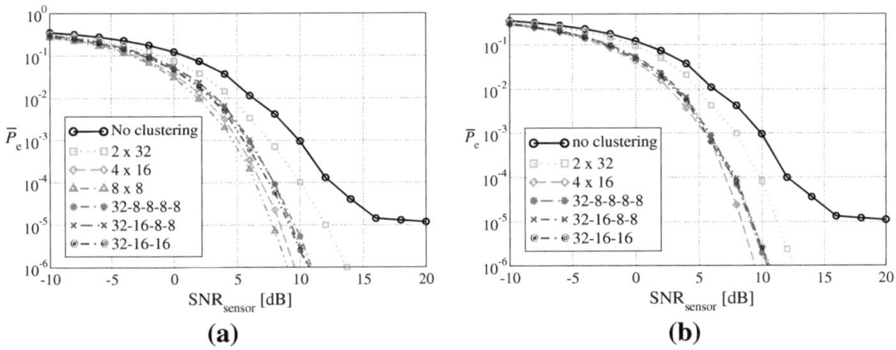


**Fig. 7.6** $\overline{P}_e$ as a function of the sensor SNR, for different clustering configurations, ideal communication links, and the IEEE 802.15.4 MAC protocol. Case **a**: absence of weighing; case **b**: presence of weighing

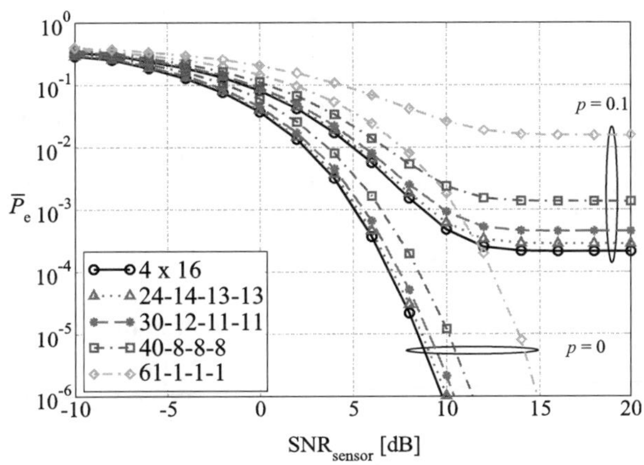


**Fig. 7.7** $\overline{P}_e$ as a function of the SNR, in a scenario with $n = 64$ and different clustering configurations with four FCs. Two scenarios are considered: $p = 0$ and $p = 0.1$

present in both the figures). In particular, when the weighing strategy is applied, the performance in the presence of non uniform scenarios provide approximatively the same results. However, uniform configurations are still to be preferred.

Since increasing the number of FCs will increase also the cost of the network, being FCs sensors with special functionalities, therefore high cost, it is of interest to investigate what is the best possible configuration for a fixed number of FCs. Only results in the presence of the IEEE 802.15.4 MAC protocol will be presented in the following figures. In Fig. 7.7, the probability of decision error is shown, as a function of the sensor SNR, in a scenario with $n = 64$ and four FCs are present. Two different values of $p$ are considered: 0 (ideal communication links) and 0.1 (high communication noise). In the ideal case, the uniform configuration is still to be preferred, thus confirming the results in [1] with an ideal MAC protocol. Moreover, the larger is the non-uniformity degree, the worse is performance. In fact, when clusters are balanced, the decisions coming from the FCs to the AP

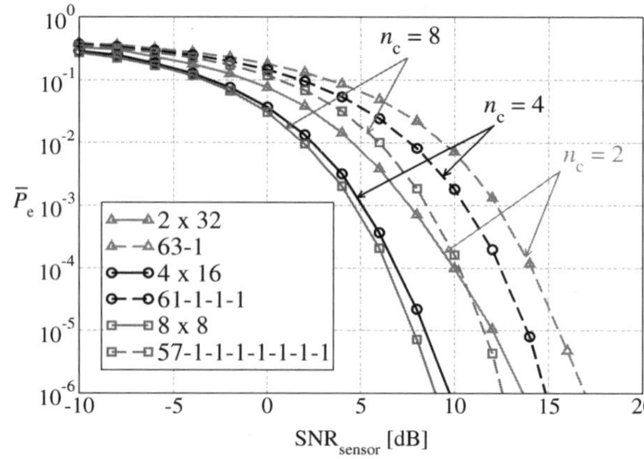


**Fig. 7.8** $\overline{P}_e$ as a function of the SNR, when applying 802.15.4 MAC and for ideal communication links

have approximately the same reliability, since the number of collisions is approximately the same in all clusters. On the other hand, with unbalanced clusters the decisions do not have the same reliability and, therefore, the quality of the AP decision worsens. In the scenario with $p = 0.1$, the impact of the communication noise on the probability of decision error is significant and the performance rapidly degrades. As predicted in Chap. 2, for large values of SNR curves present a floor, due to the communication noise, which is independent on the observation noise. Therefore, increasing more and more the observation quality does not lead to better performance, since this is also limited by the communication noise.

In Fig. 7.8, $\overline{P}_e$ is shown, as a function of the SNR, for different clustering configurations, considering ideal communication links and the IEEE 802.15.4 MAC protocol. For each value of $n_c$, the best and the worst configurations are shown. More precisely, the best possible configurations are uniform for all values of $n_c$: 32-32 for $n_c = 2$, $4 \times 16$ for $n_c = 4$, and $8 \times 8$ for $n_c = 8$. On the other hand, the worst possible configuration, for a given value of $n_c$, is that with one big cluster and the others with only one sensor, i.e., 63-1 for $n_c = 2$, 61-1-1-1 for $n_c = 4$, and 57-1-1-1-1-1-1-1 for $n_c = 8$. One should observe that the relative loss (in terms of sensor SNR) from the best to worst configuration is approximately constant, regardless of the value of $n_c$. For instance, at $\overline{P}_e = 10^{-3}$ this loss is around $4.5 \div 5$ dB. This implies that the gain brought by the use of uniform clustering is (more or less) the same, the only difference being the fact that the larger is the number of FCs (with a corresponding larger cost), the better is the performance.

Since other network topologies with larger values of $n_c$ cannot be analyzed considering $BO = 3$ (see Table 7.1), we now set $BO = 4$ such that we can evaluate performance also in the case of 16 clusters present. In Fig. 7.9, the probability of decision error is shown, as a function of the sensor SNR, for different clustering configurations. No noisy communication links are accounted for and ideal MAC is considered in case (a), whereas the IEEE 802.15.4 MAC protocol is accounted for in case (b). The considerations carried out for the same scenario and $BO = 3$ (see Fig. 7.4) still apply for $BO = 4$. However, one can observe that, in the case with

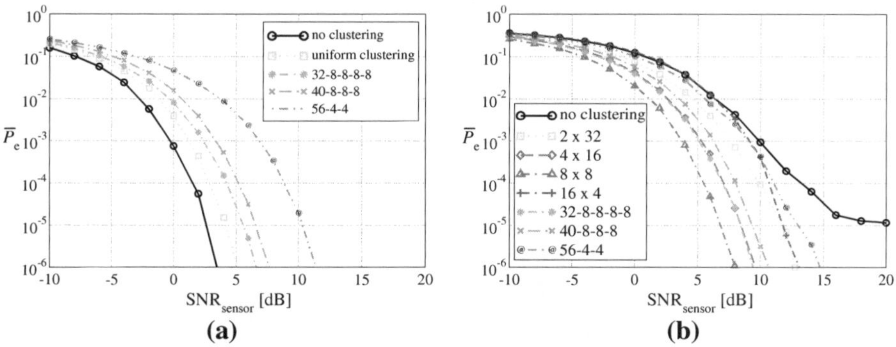


**Fig. 7.9** $\overline{P}_e$ as a function of the sensor SNR, for different clustering configurations when $BO = 4$, and ideal communication links. Case **a**: ideal MAC; case **b**: IEEE 802.15.4 MAC

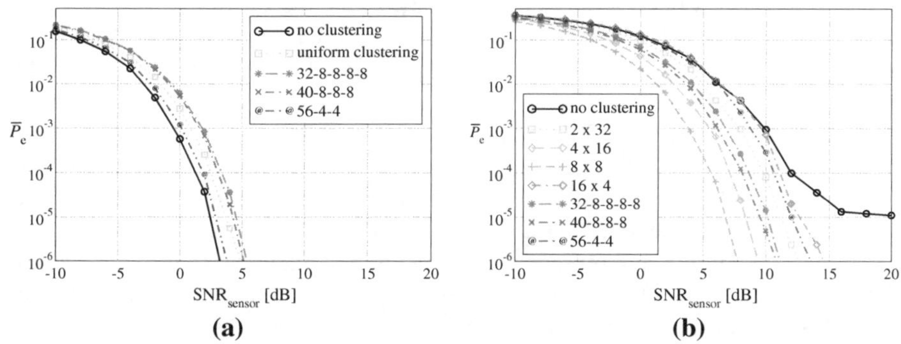


**Fig. 7.10** $\overline{P}_e$ as a function of the sensor SNR, for different clustering configurations when $BO = 4$, ideal communication links, and weighing strategy at the AP. Case **a**: ideal MAC; case **b**: IEEE 802.15.4 MAC

IEEE 802.15.4 MAC protocol, a best performance is achieved for $n_c = 8$ uniform clusters. In this case, in fact, only eight nodes per cluster are competing for the channel, and even though $SO = 1$ (i.e., nodes have only approx. 15 ms to access the channel), the success probability is large. Moreover, note that when $n_c = 16$ uniform clusters are considered, there is a performance degradation with respect to $n_c = 8$. The problem of an optimal network topology will be investigated in more detail in Fig. 7.11.

In Fig. 7.10, $\overline{P}_e$ is shown, as a function of the sensor SNR, for different clustering configurations, ideal communication links, and the weighing strategy (7.7) at the AP. Two scenarios for the MAC are considered: (a) ideal MAC protocol and (b) IEEE 802.15.4 MAC protocol. In this case as well, the impact of the weighing strategy is the same commented in Fig. 7.5 for the case with $BO = 3$. Moreover, it is still true that an optimal topology, in terms of best probability of decision error, arises. As in the case with no weighing strategy illustrated in Fig. 7.9, the best curve is obtained for the scenario with eight uniform clusters.

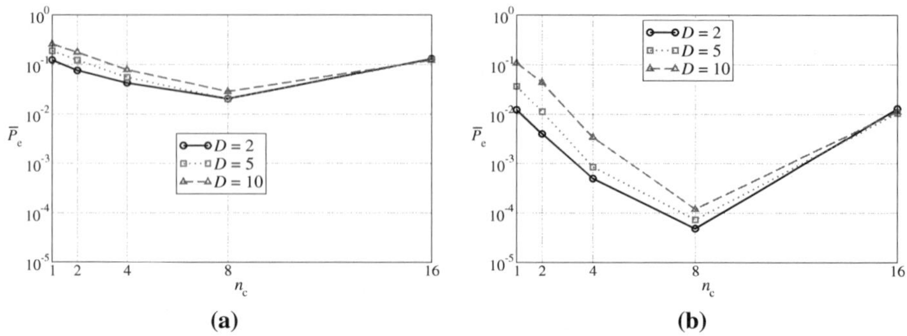


**Fig. 7.11** $\overline{P}_e$ for different uniformly clustered topologies and values of the parameter $D$: 2, 5, and 10. In case **a**, the sensor SNR is set to 0 dB, whereas in case **b** the sensor SNR is set to 6 dB

From the results in Figs. 7.9 and 7.10 it seems that the use of the IEEE 802.15.4 MAC protocol leads to the identification of an optimal distribution of the sensors among the clusters, which minimizes the probability of decision error. In Fig. 7.11, $\overline{P}_e$ is shown for different uniformly clustered topologies and values of the parameter $D$: 2, 5, and 10.[4] In case (a), the sensor SNR is set to 0 dB, whereas in case (b) the sensor SNR is set to 6 dB. Note that only uniform clustering configurations are considered, since the use of non uniform topologies leads to a waste of energy resources, as outlined in Table 7.1. Moreover, the same curve is associated to both absence and presence of the weighing strategy. One can observe that the best strategy is given by the topology composed by eight uniform clusters, regardless of the sensor SNR and the value of $D$. First, the fact that the best topology is uniform is expected, since in all cases, we have observed that uniform clustering leads to the best performance with respect to non-uniform clustering. Moreover, $n_c = 8$ is an optimum value, since it allows to obtain the best trade-off between the number of nodes competing in accessing the channel (given by the IEEE 802.15.4 MAC protocol) and the quality of the decisions taken by the FCs (given by the decentralized detection process).

## 7.4 Concluding Remarks

In this chapter, we have proposed a mathematical framework to study decentralized detection in IEEE 802.15.4 WSNs. In particular, on the basis of an analytical computation of the probability that a packet is correctly received at the destination when the IEEE 802.15.4 MAC protocol is used, we have evaluated the impact of the MAC protocol on a decentralized detection strategy. This analysis has been carried out considering different clustered topologies. Results show that the MAC

[4] Note that $n_c = 1$ denotes the topology with no clustering, that is only one cluster is present.

protocol has a relevant impact on performance. In particular, while the absence of clustering guarantees the best performance of a decentralized detection strategy in the presence of an ideal MAC, this leads to the worst performance with the 802.15.4 MAC protocol. In the latter case the best performance is achieved when the optimum balance between the number of clusters and the time allowed to cluster nodes for transmitting packets is reached. Finally, the presence of communication noise increases the probability of decision error floor induced by the MAC protocol, and this degradation is more pronounced, the higher is the non-uniformity degree among the clusters.

## References


1. G. Ferrari, M. Martalò, R. Pagliari, Decentralized detection in clustered sensor networks. IEEE Trans. Aerosp. Electron. Syst., **47**(2), April (2011)
2. G. Ferrari, R. Pagliari, Decentralized binary detection with noisy communication links. IEEE Trans. Aerosp. Electron. Syst. 42(4), 1554–1563 (2006)
3. IEEE 802.15.4 Std, *Wireless Medium Access Control (MAC) and Physical Layer (PHY) Specifications for Low-Rate Wireless Personal Area Networks (LR-WPANs)*. IEEE Computer Society Press, pp. 1–679, October 2003, ISBN: 0-7381-3677-5
4. C. Buratti, Performance analysis of IEEE 802.15.4 beacon-enabled mode. IEEE Trans. Veh. Technol. **54**(6), 2031–2045 (2010)

# Index